Analysis and Control of Nonlinear Systems with Stationary Sets
Time-Domain and Frequency-Domain Methods

Analysis and Control of Nonlinear Systems with Stationary Sets

Time-Domain and Frequency-Domain Methods

Jinzhi Wang • Zhisheng Duan
Ying Yang • Lin Huang

Peking University, China

World Scientific

NEW JERSEY · LONDON · SINGAPORE · BEIJING · SHANGHAI · HONG KONG · TAIPEI · CHENNAI

Published by
World Scientific Publishing Co. Pte. Ltd.
5 Toh Tuck Link, Singapore 596224
USA office: 27 Warren Street, Suite 401-402, Hackensack, NJ 07601
UK office: 57 Shelton Street, Covent Garden, London WC2H 9HE

British Library Cataloguing-in-Publication Data
A catalogue record for this book is available from the British Library.

ANALYSIS AND CONTROL OF NONLINEAR SYSTEMS WITH STATIONARY SETS
Time-Domain and Frequency-Domain Methods

Copyright © 2009 by World Scientific Publishing Co. Pte. Ltd.

All rights reserved. This book, or parts thereof, may not be reproduced in any form or by any means, electronic or mechanical, including photocopying, recording or any information storage and retrieval system now known or to be invented, without written permission from the Publisher.

For photocopying of material in this volume, please pay a copying fee through the Copyright Clearance Center, Inc., 222 Rosewood Drive, Danvers, MA 01923, USA. In this case permission to photocopy is not required from the publisher.

ISBN-13 978-981-281-469-2
ISBN-10 981-281-469-8

Printed in Singapore.

Preface

Control theory has been well developed in the last 80 years ever since the establishment of the Nyquist criterion in the 1930s. A central issue in control science is to analyze the stability of specific motions and to design controllers to realize those motions even when the systems are subject to uncertainties or perturbations. Using a simple algebraic technique, the aforementioned problem can be transformed into the stability problem of a system at its equilibrium, and the design of a controller is to make the corresponding close-loop system stable. This approach has come to be well-known today as the stabilization problem, and therefore it is believed that the most fundamental problem in control science is the stability analysis of the system equilibrium and how to design a controller to stabilize it.

The concept of motion or equilibrium stability was formulated and carefully studied by Lyapunov more than a century ago. It originated from the continuous dependence of the solutions of an ordinary differential equation (ODE) on its initial values, and was then developed by extending the time-domain to the infinite interval. This concept and various corresponding analytic methods stimulated the rapid and fruitful developments of stability theory to the benefit of control science. As a result, much of modern control theory focuses on stability and stabilization issues.

There are two methods to describe a control system. One is formulated in the time-domain, often referred to as the state-space method, which considers time as an independent variable and uses differential or other appropriate equations to describe the system dynamics. In the state-space setting, both the input and the output signals are expressed as time-dependent functions. The relationship between the input and the output is expressed via system state equations indirectly, and the characteristics of the system are usually not represented explicitly by the equations. Even so, this way of description is widely adopted by control scientists and engineers, since there are a lot of mature mathematical tools that can be utilized, especially for LTI systems where many powerful mathematical methods, such as linear algebra and numerical mathematics, can be applied. The other one is formulated in the frequency-domain, which is very popular in the study of LTI systems. Theoretically, it can be obtained from integral transformation (Fourier transform) of time-domain equations because of its time-invariance. Essentially, it gives the input/output relationship in the frequency-domain based on the fact that for LTI systems a harmonic input will generate a harmonic steady output of the same frequency. This relationship can also be interpreted as that the system steady output is the product of the system frequency characteristic and the system input. This model uses the frequency of harmonic wave as the independent variable and has clear physical meaning. Therefore, it has been widely used in control engineering. For a practical system, usually only the input and the output information can be obtained. The system states and the state equations are determined only in the sense of mathematical equivalence. For instance, when there is a revertible linear transformation on the system states in an LTI system, although the equations that describe the system have been changed, the basic input-output relationship through the equations remain unchanged. Obviously, the approach based on frequency characteristics or transformation functions has the superiority for describing the input-output relationship with more explicit physical interpretation. Moreover, frequency-domain descriptions can be obtained via experiments, and there are also a handful of approximate but effective engineering methods for systems analysis and design. Therefore, frequency methods are welcomed by researchers and practitioners in control system applications. With the rapid development of computer science, many effective computing methods have been applied to solving complicated and large-scale problems in science and engineering, e.g., using algorithms related to numerical linear algebra, which makes the state-space approach even more efficient and effective. Using these methods, it is possible to design controllers without considering their physical meanings, then verify and improve the original ideas through simulations. In this way, both the frequency-domain methods and state-space methods can be further developed swiftly.

In the history of control science development, there were several successful encounters between frequency-domain methods and state-space methods. These encounters have led to a spurt in the evolvement of control science, and have established a strong link between frequency-domain methods, which possess physical and practical characteristics, and the state-space methods, that have the powerful support of mathematical tools, thus being able to handle time-varying and even nonlinear systems. This important link has actually become a new growing point of control science.

As early as the time when Wiener proposed the filtering and prediction problems of stationary stochastic processes, the solutions of the integral equations describing the process could be archived via Fourier transform in the frequency-domain framework. At that time, this method was just an interesting experiment. The real encounter between time-domain and frequency-domain methods took place through the important research on absolute stability of control systems.

The notion of absolute stability was first proposed by Lur'e and Postnikov in the 1940s. Since then, a large number of papers and monographs have appeared which investigate the problem of absolute stability. The basic system considered is composed of a linear time-invariant feed-forward part and a nonlinear memoryless feedback part, subject to a sector-bounded constraint. The system is said to be absolute stable if for any nonlinear function satisfying the sector-bounded constraint the system is globally asymptotically stable in the sense of Lyapunov. The fundamental problem in the study of absolute stability is to establish conditions of absolute stability for the system. The conditions should consist of parameters of the sector-bounded constraint and the information provided by the linear part of the given system.

From the very beginning, Lur'e studied this problem in the state-space framework. Therefore, a natural idea is to construct a quadratic Lyapunov function containing the states and the nonlinear characteristics of the system to determine the asymptotic stability of the system. However, Lyapunov equations and inequalities had not been fully studied at that time. Most of the research was carried out based on the Jordan canonical form through linear transformations, and the conditions on absolute stability was reduced to the existence of solutions to some algebraic equations. For more than ten years after the problem was proposed, it had been widely believed that Lyapunov method is the most appropriate and perhaps the only effective way to solve the problem of absolute stability.

In 1960, the First International Federation Automatic Control (IFAC) World Congress was held in Moscow. It symbolized the globalization of control science in the scientific world. In that conference, V. M. Popov presented an amazing result on the frequency-domain criterion of absolute stability derived by using only Fourier transformation. Thereafter, this result was developed and finally formulated as the Popov criterion and Circle criterion. This is a "strong stimulus" which motivated many people to try to find out the basic relationship between the frequency-domain criteria and the time-domain Lyapunov methods. A naturally occurring problem is, which of these two methods is more effective? If one method can be used to test the system stability, can the other one do the same? With a long-time effort made by many scientists from different countries worldwide, this intrinsic relationship was gradually revealed. It is now known that the frequency-domain criterion for absolute stability is formulated in terms of some frequency-domain inequalities about the frequency response of the linear part of the system and the parameters of the nonlinear sector-bounded constraints; while in the Lyapunov function method, the problem of absolute stability is reduced to the existence of a positive-definite matrix P to a matrix inequality with system matrices and parameters of the sector-bounded constraints as coefficients. The feasibility of the LMI (Linear Matrix Inequalities) condition ensures the negative definiteness of the total derivative of the Lyapunov function $x^T P x$. Now, it is also known that the Kalman-Yakubovic-Popov (KYP) Lemma bridges the gap between these two methods and establishes the equivalence relationship between the frequency-domain and time-domain inequalities. The well-known positive real lemma and the bounded real lemma can be regarded as special forms of the KYP lemma.

The dynamics of a system should be reflected by the global nature of the direction of the trajectory flow in the system state space. To describe the global nature, a Lyapunov function in terms of the system states is usually considered. The value of the function determines a hypersurface in the space. If the time derivative of the function along all the trajectories has a fixed sign (for instance, a negative sign), then one knows the global flow direction of all points on the hypersurface along the moving direction of the system trajectories. Due to this geometric view, the Lyapunov method has become an effective tool for determining the asymptotic stability or the instability of a system. For linear systems, when the quadratic Lyapunov function is positive definite, the corresponding hypersurface is an ellipsoid and the system is asymptotically stable. If the function has a negative value, then the system is unstable. In fact, whether the total derivative (along all of the system trajectories) has a fixed sign or not is the essence of the Lyapunov method, which can be used to discuss not only the asymptotic stability or instability of a single equilibrium, but also other system properties such as boundedness of trajectories.

Linear system models are comparatively simple, since the principle of linear superposition holds for linear models. The dynamic characteristics of a linear system in the whole state space can be obtained by investigating that in the neighborhood of the equilibrium, the origin of coordinates. For time-invariant linear systems, when a quadratic Lyapunov function is adopted, an important result on the system dynamics is described in the following theorem.

Theorem. For an n-dimensional time-invariant linear system

$$\dot{x} = Ax, \quad x \in \mathbb{R}^n, \quad A \in \mathbb{R}^{n \times n}$$

where A is a constant real matrix, suppose that A has k eigenvalues with negative real parts and $n - k$ eigenvalues with positive real parts. If there exists a symmetric matrix P such that

$$PA + A^T P = -Q, \quad Q > 0,$$

then P has $n - k$ negative eigenvalues and k positive eigenvalues.

This theorem is an extension of a corresponding classical Lyapunov result. It can also be used to discuss some other global properties of nonlinear systems.

When a global property of a system is considered, it should describe the global nature of the system rather than just some peculiar properties of a specific solution of the system; for instance, the existence of multiple equilibrium states in the system, if the system is just with a single equilibrium state and it is asymptotically stable, the boundedness of all solutions, the existence of auto-oscillations, and the nonexistence of chaos, etc. Meanwhile, such a property should be operational, or in other words mathematically provable or computationally tractable. In this book, we mainly study global properties by means of the Lyapunov function method, dynamic system analysis and ordinary differential equations theory, where the time-domain results will be interpreted in the frequency-domain framework via the KYP Lemma.

The main difficulties of nonlinear problems come from two aspects. One is the dimensional difficulty. For a one-dimensional system, $\dot{\xi} = \xi^3$, for instance, although the equation is nonlinear, it can be easily checked as if its solution is asymptotically stable without using advanced mathematics. For a one-dimensional time-variant system, $\dot{\xi} = a(t)\xi$, for instance with $a(t) \leq -\beta < 0$, the asymptotic stability of the solution can be obtained immediately. It also indicates that the frozen-coefficient method is applicable in this case. When the dimension is greater than one, the solution of the system may generate "rotation", and the above discussion may not be valid. In the case when the dimension is only two, there still exist some theoretical results based on the qualitative theory of planar dynamical systems. When the dimension is greater than two, however, difficulties encountered in analysis are far more than one can imagine. The other aspect is the essential nonlinearity in the system, which is the main difficulty in nonlinear analysis which typically cause many complicated dynamic behaviors that would never happen in linear systems. A convenient method for a nonlinear analysis is to use the same framework for linear systems, but it cannot solve the essential nonlinearity problem in nonlinear systems. Similar to the most fundamental non-convexity difficulty in nonlinear programming, there are no available mathematical tools for effectively handling the essential nonlinearity in the study of general nonlinear control systems.

Nonlinear systems theory originated from research on nonlinear oscillations, since auto-oscillation is the most common nonlinear phenomenon found in nonlinear mechanical systems, such as the escapements frequently used in clocks and watches, or such as found in the van der Pol circuit. This phenomenon corresponds to an isolated periodic solution of the system which cannot exist in linear systems. The research on auto-oscillations was the first hot topic in the field of nonlinear systems research, from the 1940s to the 1960s. But the work was restrained to second-order systems due to the lack of powerful computational tools and mathematical analysis methods.

The second phase in the development of nonlinear science starts from the finding of chaos, which is far more complicated than auto-oscillations. The occurrence of auto-oscillations did not go beyond common imaginations, because this phenomenon frequently emerges, in both natural and artificial systems such as the beating of the heart and the swinging of pendulums in mechanical clocks. In geometry, it is just a circle repeating itself constantly, therefore a kind of regular motion as compared to chaotic dynamics. What is needed to study is why this periodic dynamic behavior can be produced and how to produce it in a specific nonlinear system without external periodic excitement. However, this is not the case for chaos, which demonstrates some so-called fantastic nonlinear properties and completely altered people's conventional view from several aspects. First of all, the existence of chaos indicates that a deterministic system can produce stochastic-like dynamical behaviors with the ergodicity property. Secondly, it shows high sensitivity to initial conditions that linear and general nonlinear systems do not have. Thirdly, it typically demonstrates a strange attractor of the system, which is not the usual fixed point (zero-dimensional) and limit cycle (one-dimensional) but a set of points with a very complicated structure having a fractional dimension. Such a special dynamic process has attracted a lot of interest from almost all scientific communities in the past half a century. However, for the same reason as of lacking powerful mathematical tools, although there are many qualitative results for low-dimensional systems, quantitative analysis of higher-dimensional chaotic systems basically relies on numerical computations and approximate analytic approaches today.

From the control theoretic point of view, the main research interest on auto-oscillations or chaos is focused on how to design a controller to affect the dynamic behavior; that is, to produce or eliminate such a behavior in the given system by means of control. Both auto-oscillations and chaos are non-convergent but bounded evolutionary processes. Auto-oscillations also have the property of isolation. Neither of these two phenomena can exist in linear systems, but exist only in two kinds of bounded evolutionary processes: one is the trivial equilibrium point, which is either a unique point or a subspace; the other is a compound oscillation composing of one or several simple harmonic oscillations. A constant multiple of a same class of compound oscillations is still a possible oscillation and all of them together still compose a subspace. Apart from these two kinds of bounded dynamical behaviors, the solutions of a linear system can be divided into two parts, one is the convergent solutions and the other is unbounded solutions. When the system does not have eigenvalues located on the imaginary axis, all its bounded solutions will converge. This property is known as the property of dichotomy. For a linear system, dichotomy is not a crucial property, but for a nonlinear system this property can be used to exclude all processes that are bounded but not convergent, such as auto-oscillations and chaos. When a system does not have eigenvalues located on the imaginary axis, from the premier theorem the distribution of the eigenvalues of the solution matrix P to the Lyapunov equation is converted to that of the system relative to the imaginary axis. This fact provides a fundamental principle for, and indeed facilitates, the study of dichotomy and the construction of the existing area of limit cycles.

The demands on control systems are numerous and in various forms, many of their problems cannot be reduced to the stability of a single equilibrium. Sometimes, demanding stability is hard to satisfy, at other times, it may be unnecessary. Since some nonlinear characteristics of the system are uncertain, the equilibrium position may move constantly. In this case, it is unreasonable to design a controller to stabilize such an equilibrium state, and the primary demand would be that all the solutions of the system are bounded, i.e., there are no divergent solutions in the system. Furthermore, one may require every solution be convergent to a certain equilibrium. This property is known as a gradient-like behavior in systems with multiple isolated equilibria. An interesting fact for a time-invariant nonlinear system is that when the system is gradient-like, there must exist at least one equilibrium that is not asymptotically stable in the sense of Lyapunov. This conclusion is drawn from the following contradiction: if all the equilibria are asymptotically stable, then each equilibrium has an open set of domain of attraction, while denumerable open sets cannot cover the whole space. It is also known that even though a system is gradient-like, the characteristics of the trajectories and the equilibria of the system in the phase space are considerably complicated. Such a complex situation is very common in electric power systems, for instance, and is a key to further understand the modern power systems.

With so many and so complex non-conventional dynamic characteristics, it is hard to make progress if one addresses the problems based on a very general model of nonlinear systems, since the available mathematical theory can only provide general conclusions rather than concrete details. Therefore, the theme of this book will be focused on a commonly used model in control science and engineering — the Lur'e-type systems, which is composed of a linear feed-forward part and a nonlinear feedback part, as depicted by the following block diagram:

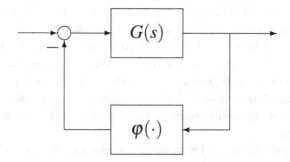

This model was first proposed by Lur'e, with the background that a hydraulic valve was used as the actuator in a driving system at that time. Since the characteristic curve of the hydraulic valve is nonlinear with uncertainty, $\varphi(\sigma)$ is an uncertain function with constraints. The feature that the nonlinearity can be separated from the linear part brought convenience to the understanding and study of many physical systems. In fact, many typical nonlinear systems, such as Chua's circuit and the Lorenz system, can be classified into this type. As a matter of fact, Lur'e system has become a preferable model for investigating nonlinear dynamics such as chaos today. During the first two decades since the proposal of the Lur'e system framework, the research focus was on the absolute stability. Afterwards, Leonov and some other researchers further extended the research scope to other properties of the Lur'e systems, including dichotomy, gradient-like behaviors and auto-oscillations. These efforts made it possible to design controllers guaranteeing or eliminating some dynamical properties of the given system. The main controller design methods adopted in this book is to embed the controller parameters into some matrix inequalities by means of the KYP lemma and to reduce the problem to the feasibility of solving such matrix inequalities. By taking advantage of the peculiar merits of matrix inequality methods, this book will further extend the frequency-domain results on global properties, developed by Leonov *et al.*, to controllers design and robustness analysis, which will provide a basic theoretical and methodological framework for future investigations and applications.

This book is organized as follows: Chapters 1–2 are two introductory chapters, presenting basic definitions and the major analytical tools that will be used to study systems with stationary equilibrium sets. More specifically, Chapter 1 reviews some basic system concepts and formulas related to linear matrix inequalities; Chapter 2 introduces some useful tools in control systems synthesis and the linear matrix inequality approach to the standard optimal and suboptimal H_∞ control theory. Chapter 3 discusses analysis and control problems for positive realness. In Chapter 4, a unified framework is proposed for analyzing the absolute stability and dichotomy of Lur'e systems. Chapter 5 introduces two kinds of special forms of pendulum-like feedback systems and gives both time-domain and frequency-domain conditions on global properties of such systems. Chapter 6 is devoted to controllers design for a class of pendulum-like systems, which can ensure some global properties and preserve physical and dynamical phenomena of the pendulum-like systems simultaneously. Chapter 7 studies control problems for a class of systems with input nonlinearities. In Chapter 8, a time-domain approach to robust analysis and control for uncertain feedback nonlinear systems is presented and discussed. Chapter 9 is devoted to robust analysis and synthesis on the nonexistence of periodic oscillations in nonlinear Lur'e systems. Chapter 10 considers interconnected systems and discusses the effects of interconnections on system stability and performances. Chapter 11 demonstrates some applications of the theories established in the previous chapters using Chua's circuit as the main example. Finally, a bibliography and index are provided to complete the presentation of the entire book.

Some acknowledgements are in order. First, we would like to thank the sustained financial supports from the National Natural Science Foundation of China under grants 60334030, 60774005, 60674093 and 60874011, which enabled us to carry on related research and to develop a new research area in control science and engineering. We would also like to thank Professor Zhongqin Xu for his help and support during this book's publication. We are especially indebted to Professor Iven Mareels at the University of Melbourne for his careful review of the English writing of a large part of the manuscript and for his helpful comments and suggestions. We are grateful to Professor Guanrong Chen at the City University of Hong Kong for revising the Preface. In addition, we thank Professor Zhiyong Geng, Dr. Xinbin Li, Dr. Pingli Lu, Dr. Xian Liu, Dr. Chao Liu, Dr. Dun Ao and Mr. Xiaoshan Yang from Peking University, for their valuable feedback throughout many seminars in the past. The academic atmosphere and harmony around us was of great importance to complete this joint work, which we particularly enjoy and appreciate.

Contents

Preface   v

Notation and Symbols   xxi

1. Linear Systems and Linear Matrix Inequalities   1
    1.1 Controllability and observability of linear systems   1
        1.1.1 Controllability and observability   2
        1.1.2 Stabilizability and detectability   6
    1.2 Algebraic Lyapunov equations and Lyapunov inequalities   7
        1.2.1 Continuous-time algebraic Lyapunov equations . .   7
        1.2.2 Continuous-time Lyapunov inequalities   10
        1.2.3 Discrete-time algebraic Lyapunov equations and inequalities   11
    1.3 Formulation related to linear matrix inequalities   12
        1.3.1 Schur complements   12
        1.3.2 Projection lemma   13
    1.4 The S-procedure   15
        1.4.1 The S-procedure for nonstrict inequalities   15
        1.4.2 The S-procedure for strict inequalities   15
    1.5 Kalman-Yakubovič-Popov (KYP) lemma and its generalized forms   16
    1.6 Notes and references   21

2. LMI Approach to H_∞ Control   23
    2.1 $\mathcal{L}_\infty$ norm and $\mathcal{H}_\infty$ norm of the systems   23
        2.1.1 $\mathcal{L}_\infty$ and $\mathcal{H}_\infty$ spaces   24
        2.1.2 Computing $\mathcal{L}_\infty$ and $\mathcal{H}_\infty$ norms   25

	2.2	Linear fractional transformations	27
	2.3	Redheffer star product	29
	2.4	Algebraic Riccati equations	30
		2.4.1 Solvability conditions for Riccati equations	31
		2.4.2 Discrete Riccati equations	33
	2.5	Bounded real lemma	34
	2.6	Small gain theorem	36
	2.7	LMI approach to H_∞ control	37
		2.7.1 Continuous-time $\mathcal{H}_\infty$ control	37
		2.7.2 Discrete-time $\mathcal{H}_\infty$ control	42
	2.8	Notes and references	43
3.	Analysis and Control of Positive Real Systems		45
	3.1	Positive real systems	45
	3.2	Positive real lemma	52
	3.3	LMI approach to control of SPR	63
	3.4	Relationship between SPR control and SBR control	66
	3.5	Multiplier design for SPR	69
	3.6	Notes and references	73
4.	Absolute Stability and Dichotomy of Lur'e Systems		75
	4.1	Circle criterion of SISO Lur'e systems	75
	4.2	Popov criterion of SISO Lur'e systems	80
	4.3	Aizerman and Kalman conjectures	82
	4.4	MIMO Lur'e systems	84
	4.5	Dichotomy of Lur'e systems	89
	4.6	Bounded derivative conditions	93
	4.7	Notes and references	97
5.	Pendulum-like Feedback Systems		99
	5.1	Several examples	99
	5.2	Pendulum-like feedback systems	102
		5.2.1 The first canonical form of pendulum-like feedback system	103
		5.2.2 The second canonical form of pendulum-like feedback system	105
		5.2.3 The relationship between the first and the second forms of pendulum-like feedback systems	106

5.3	Dichotomy of pendulum-like feedback systems		107
	5.3.1	Dichotomy of the second form of autonomous pendulum-like feedback systems	107
	5.3.2	Dichotomy of the first form of pendulum-like feedback systems	112
5.4	Gradient-like property of pendulum-like feedback systems		114
	5.4.1	Gradient-like property of the second form of pendulum-like feedback systems	114
	5.4.2	Gradient-like property of the first form of pendulum-like feedback systems	117
5.5	Lagrange stability of pendulum-like feedback systems		118
5.6	Bakaev stability of pendulum-like feedback systems		124
5.7	Notes and references		129

6. **Controller Design for a Class of Pendulum-like Systems** — 131

   6.1 Controller design with dichotomy or gradient-like property — 131
       6.1.1 Controller design with dichotomy — 131
       6.1.2 Controller design with gradient-like property — 137
   6.2 Controller design with Lagrange stability — 139
   6.3 Notes and references — 147

7. **Controller Designs for Systems with Input Nonlinearities** — 149

   7.1 Lagrange stabilizing for systems with input nonlinearities — 149
   7.2 Bakaev stabilizing for systems with input nonlinearities — 155
   7.3 Control for systems with input nonlinearities guaranteeing dichotomy — 159
   7.4 Notes and references — 162

8. **Analysis and Control for Uncertain Feedback Nonlinear Systems** — 163

   8.1 Dichotomy of systems with norm bounded uncertainties — 163
       8.1.1 Robust analysis for dichotomy — 164
       8.1.2 Robust control for systems with dichotomy — 168
   8.2 Dichotomy of pendulum-like systems with uncertainties — 174
   8.3 Controller design with dichotomy for uncertain pendulum-like systems — 179
   8.4 Lagrange stability for uncertain pendulum-like systems — 184
   8.5 Gradient-like property for pendulum-like systems with uncertainties — 187

8.6	Control of uncertain systems guaranteeing gradient-like property	191
8.7	Gradient-like property of systems with norm bounded uncertainties	199
8.8	Notes and references	205

9. Control of Periodic Oscillations in Nonlinear Systems 207

9.1	Periodic solutions in systems with cylindrical phase space	207
9.2	Nonexistence of periodic solutions in Lur'e systems	211
	9.2.1 LMI-based conditions for nonexistence of periodic solutions	211
	9.2.2 Robustness analysis	213
	9.2.3 Robust synthesis	214
9.3	Nonexistence of cycles of the second kind in interconnected systems	218
	9.3.1 Nonexistence of cycles of the second kind in interconnected systems	220
	9.3.2 Nonlinear interconnection design	224
9.4	Cycle slipping in phase synchronization systems	228
9.5	Notes and references	236

10. Interconnected Systems 237

10.1	Linearly interconnected systems	238
	10.1.1 The effect of the unstable subsystem	238
	10.1.2 Interconnected feedbacks	241
	10.1.3 Decentralized controller design	244
	10.1.4 The effect of small gain theorem	246
10.2	Interconnected Lur'e systems	250
10.3	Lagrange stability of a generalized smooth Chua circuit	252
10.4	Input and output coupled nonlinear systems	257
10.5	Notes and references	264

11. Chua's Circuit 267

11.1	Chua's circuit	267
11.2	Dichotomy: application to chaos control for Chua's circuit system	270
11.3	Kalman conjecture: application to the stabilization of Chua's circuit	280

11.4	An extended Chua circuit	286
11.5	Coupled Chua circuit	288
11.6	Notes and references	293

Bibliography   295

Index   309

Notation and Symbols

$\mathbb{R}$ and $\mathbb{C}$	field of real and complex numbers
$\mathbb{R}_+$	the set of nonnegative real numbers
$\mathbb{Z}$	the set of integer numbers
$\mathbb{R}^{n \times m}$ and $\mathbb{C}^{n \times m}$	$n \times m$ real and complex matrices
$j\mathbb{R}$	imaginary axis
$:=$	defined as
$\overline{\alpha}$	complex conjugate of $\alpha \in \mathbb{C}$
$|\alpha|$	absolute value of $\alpha \in \mathbb{C}$
I	identical matrix
$\mathrm{diag}(a_1, \cdots, a_n)$	an $n \times n$ diagonal matrix with a_i as its ith diagonal element
A^T and A^*	transpose and complex conjugate transpose of A
$\mathbf{Re}\{\alpha\}$	real part of $\alpha \in \mathbb{C}$
$\mathbf{Im}\{\alpha\}$	imaginary part of $\alpha \in \mathbb{C}$
$\mathbf{Re}\{G\}$	$G + G^*$ for $G \in \mathbb{C}^{n \times n}$
A^{-1} and A^+	inverse and pseudo inverse of A
$\det(A)$	determinant of A
$\mathrm{trace}(A)$	trace of A
$\lambda_i(A)$	eigenvalue of A
$\rho(A)$	spectral radius of A
$\overline{\sigma}(A)$ and $\underline{\sigma}(A)$	the largest and the smallest singular values of A
$\|x\|$	vector norm of $x \in \mathbb{C}^n$
$\|A\|$	spectral norm of A, i.e., $\|A\| = \overline{\sigma}(A)$
$\Lambda(A)$	the set of eigenvalues of A
$\mathrm{Im}(A)$, $\mathrm{R}(A)$	image or range space of A
$\mathrm{N}(A)$, $\mathrm{Ker}(A)$	the null space or kernel space of A
$\mathcal{X}_-(A)$	stable invariant subspace of A

$A^{\frac{1}{2}}$	the unique nonnegative definite square root of A
$\text{Ric}(H)$	the stabilizing solution of an ARE
$A^{\perp}$	a matrix with properties: $N(A^{\perp}) = R(A)$ and $A^{\perp} A^{\perp T} > 0$
$A > 0$	the matrix A is positive definite
$A \geq 0$	the matrix A is positive semi-definite
$A < 0$	the matrix A is negative definite
$A \leq 0$	the matrix A is negative semi-definite
$\otimes$	Kronecker product
$\cup$	set union
$\mathcal{L}_\infty(j\mathbb{R})$	functions bounded on imaginary axis including at ∞
$\mathcal{RL}_\infty$	the set of all proper and real rational transfer matrices with no poles on the imaginary axis
$\mathcal{H}_\infty$	the set of $\mathcal{L}_\infty(j\mathbb{R})$ functions analytic in $\mathbf{Re}(s) > 0$
$\mathcal{RH}_\infty$	the set of all proper and real rational stable transfer matrices
$\mathcal{H}_\infty^-$	the set of $\mathcal{L}_\infty(j\mathbb{R})$ functions analytic in $\mathbf{Re}(s) < 0$
$\left(\begin{array}{c\|c} A & B \\ \hline C & D \end{array}\right)$	shorthand for state space realization $C(sI - A)^{-1} B + D$
$\mathcal{F}_\ell(M, Q)$	lower linear fractional transformation
$\mathcal{F}_u(M, Q)$	upper linear fractional transformation
$S(P, Q)$	the star product of P and Q
ARE	algebraic Riccati equation
SISO	single-input single-output
MIMO	multi-input multi-output
LMI	linear matrix inequality
PR	positive real
SPR	strictly positive real
BR	bounded real
SBR	strictly bounded real
KYP	Kalman-Yakubovič-Popov

Chapter 1

Linear Systems and Linear Matrix Inequalities

This chapter reviews some basic system concepts and formulas related to linear matrix inequalities (LMIs) which are widely used in system analysis and design. The notions of controllability, observability, stabilizability and detectability are defined and conditions of characterizing those notions are summarized. The theory of Lyapunov equation and Lyapunov inequality are then introduced. Schur complements and conditions for solvability of some kinds of LMIs are provided. Finally, the results of S-procedure and Kalman-Yakubovič-Popov (KYP) lemma as well as the generalized KYP lemma are stated.

1.1 Controllability and observability of linear systems

We first give the descriptions of linear systems then introduce some important concepts in linear system theory and related criteria for the given notions.

A finite-dimensional linear time invariant dynamical system is described as:

$$\dot{x} = Ax + Bu, \qquad (1.1)$$
$$y = Cx + Du, \qquad (1.2)$$

where $x(t) \in \mathbb{R}^n$ is the system state, $u(t) \in \mathbb{R}^m$ is the system input, and $y(t) \in \mathbb{R}^p$ is the system output. A, B, C and D are real matrices of appropriate dimensions. A dynamical system with single-input ($m = 1$) and single-output ($p = 1$) is called a SISO (single-input and single-output) system. The transfer matrix $G(s)$ from u to y is defined by

$$Y(s) = G(s)U(s)$$

where $U(s)$ and $Y(s)$ are the Laplace transforms of $u(t)$ and $y(t)$ with zero initial condition $x(0) = 0$. From (1.1) and (1.2), $G(s)$ can be expressed as

$$G(s) = C(sI - A)^{-1}B + D.$$

1.1.1 Controllability and observability

The controllability of state characterizes the dominating capability of input for state variables. It gives an answer for the problem whether the state vector can be transferred arbitrarily by means of input. The observability of state reflects the estimated capacity of the output for state variables. It gives an answer for the problem whether the state vector can be determined by measurements of the output.

Definition 1.1. The linear system (1.1) or the pair (A, B) is said to be controllable if, for any initial state $x(0) = x_0$, and any instant of time $t_1 > 0$ and final state x_1, there exists an input $u(\cdot)$ such that the solution of equation (1.1) satisfies $x(t_1) = x_1$. Otherwise, the system or the pair (A, B) is said to be uncontrollable.

In other words the linear system (1.1) is controllable if it may be transferred from any given state into any other state at a given period of time in virtue of appropriate input.

Definition 1.2. The linear system (1.1) and (1.2) or the pair (A, C) is said to be observable if, for any $t_1 > 0$, the initial state $x(0) = x_0$ can be determined uniquely from the input $u(t)$ and output $y(t)$ on the interval $[0, t_1]$. Otherwise, the system or the pair (A, C) is said to be unobservable.

The observability of system indicates for any given period of time $[0, t_1]$, the initial state $x(0) = x_0$ can be determined uniquely by input and output on the interval $[0, t_1]$.

The controllability and observability are structural properties of system. Four kinds of states including controllable and observable parts, controllable but unobservable parts, uncontrollable but observable parts and uncontrollable and unobservable parts are all reflected in the form of state space representation (1.1) and (1.2). Compared with the state space description of a system the transfer matrix is a kind of incomplete description that only characterizes the property of the parts of states with controllability and observability. In other words, if a system is not controllable and observable, then the order of denominator polynomial in the transfer matrix

1.1. Controllability and observability of linear systems

is less than n which is the dimension of the state vector in (1.1) and (1.2).

The Kalman duality principle states the relationship between controllability and observability.

Theorem 1.1. *(Kalman duality principle) The pair (A, C) is observable if and only if the pair (A^*, C^*) is controllable.*

Some algebraic and geometric criteria for controllability of a system are summarized as follows.

Theorem 1.2. *The following statements are equivalent:*

(i) (A, B) is controllable;
(ii) (A^*, B^*) is observable;
(iii) The matrix
$$W_c(t) = \int_0^t e^{A\tau} BB^* e^{A^*\tau} d\tau$$
is positive definite for any $t > 0$;
(iv) The controllability matrix
$$G_c = (B \quad AB \quad A^2 B \quad \cdots \quad A^{n-1} B)$$
has full-row rank or $\langle A|Im(B)\rangle = \sum_{i=1}^n Im(A^{i-1}B) = \mathbb{R}^n$;
(v) The matrix $(A - \lambda I, B)$ has full-row rank for all $\lambda \in \mathbb{C}$;
(vi) Let λ and x be any eigenvalue and any corresponding left eigenvector of A, i.e., $x^* A = x^* \lambda$. Then $x^* B \neq 0$.
(vii) The eigenvalues of $A + BF$ can be freely assigned by choosing F;
(viii) The matrix $(Q_1 \quad Q_2 \quad \cdots \quad Q_n)$ has full-row rank, where Q_k ($k = 1, \cdots, n$) are matrices with size $n \times m$ which are defined by the coefficient of $Q(s) = det(sI - A)(sI - A)^{-1}B$, i.e.,
$$Q(s) = det(sI - A)(sI - A)^{-1}B = Q_1 s^{n-1} + \cdots + Q_{n-1} s + Q_n.$$

Combining Theorem 1.2 with the Kalman duality principle equivalent conditions for observability of a system can be stated as follows.

Theorem 1.3. *The following statements are equivalent:*

(i) (A, C) is observable;
(ii) (A^*, C^*) is controllable;
(iii) The matrix
$$W_o(t) = \int_0^t e^{A^*\tau} C^* C e^{A\tau} d\tau$$
is positive definite for any $t > 0$;

(iv) The observability matrix

$$G_0 = \begin{pmatrix} C \\ CA \\ \vdots \\ CA^{n-1} \end{pmatrix}$$

has full-column rank or $\cap_{i=1}^{n} Ker(CA^{i-1}) = \{0\}$;

(v) The matrix $\begin{pmatrix} A - \lambda I \\ C \end{pmatrix}$ has full-column rank for all $\lambda \in \mathbb{C}$;

(vi) Let λ and y be any eigenvalue and any corresponding right eigenvector of A, i.e., $Ay = \lambda y$. Then $Cy \neq 0$;

(vii) The eigenvalues of $A + LC$ can be freely assigned by choosing L;

(viii) The matrix $(Q_1' \quad Q_2' \quad \cdots \quad Q_n')$ has full-row rank, where Q_k' ($k = 1, \cdots, n$) are matrices with size $n \times p$ which are defined by the coefficient of $Q'(s) = det(sI - A^*)(sI - A^*)^{-1}C^*$, i.e.,

$$Q(s) = det(sI - A^*)(sI - A^*)^{-1}C^* = Q_1' s^{n-1} + \cdots + Q_{n-1}' s + Q_n'.$$

Definition 1.3. Let λ be an eigenvalue of A or, equivalently, a mode of the system. Then the mode λ is said to be controllable (observable) if $x^* B \neq 0$ ($Cx \neq 0$) for all left (right) eigenvectors of A associated with λ; that is, $x^* A = \lambda x^*$ ($Ax = \lambda x$) and $0 \neq x \in \mathbb{C}^n$. Otherwise, the mode is said to be uncontrollable (unobservable).

It follows that a system is controllable (observable) if and only if every mode is controllable (observable).

For single-input single-output (SISO) system the transfer function from the input u to the output y has the form

$$G(s) = C(sI - A)^{-1}B + D = \frac{\alpha(s)}{\delta(s)} \tag{1.3}$$

where $\delta(s) = det(sI - A)$ and the degree of $\alpha(s)$ is not more than n.

Definition 1.4. The transfer function defined by (1.3) is said to be non-degenerate if $\alpha(s)$ and $\delta(s)$ are co-prime polynomials.

Theorem 1.4. *The transfer function of SISO system is non-degenerate if and only if (A, B) is controllable and (A, C) is observable.*

In studying global properties of uncertain systems with some nonlinearities the assumptions of controllability and observability of corresponding

1.1. Controllability and observability of linear systems

linear systems with uncertainties are often required. But there are few results on it. The sufficient conditions of controllability and observability for SISO uncertain linear systems are given in [Yang (2005)].

Consider uncertain linear systems

$$\begin{cases} \dot{x} = (A + \Delta A)x + (b + \Delta b)u \\ y = (c + \Delta c)x + du \end{cases} \quad (1.4)$$

where $x \in \mathbb{R}^n, u \in \mathbb{R}$ and $y \in \mathbb{R}$. The matrices $\Delta A, \Delta b$ and Δc denote plant uncertainties for A, b and c respectively.

Theorem 1.5. *For $A \in \mathbb{R}^{n \times n}, \Delta A \in \mathbb{R}^{n \times n}, b \in \mathbb{R}^{n \times 1}, \Delta b \in \mathbb{R}^{n \times 1}$, where $n \geq 2$, assume that the system (1.4) satisfies the following assumptions:*

(i) (A, b) is controllable, i.e., rank $Q(A, b) = n^2$;

(ii) $max\{||\Delta A||_1, ||\Delta b||_1\} < \dfrac{1}{||Q^{-1}(A,b)||_1}.$

Then $(A + \Delta A, b + \Delta b)$ is controllable, where

$$Q(A,b) = \begin{bmatrix} I_n & 0 & \cdots & 0 & 0 & 0 & \cdots & 0 & b \\ -A & I_n & \ddots & 0 & 0 & 0 & \cdots & b & 0 \\ 0 & \ddots & \ddots & 0 & \vdots & \vdots & \ddots & \vdots & \vdots \\ \vdots & \ddots & \ddots & I_n & 0 & b & \vdots & 0 & 0 \\ 0 & \cdots & 0 & -A & b & 0 & \cdots & 0 & 0 \end{bmatrix}_{n^2 \times n^2} \quad (1.5)$$

and norm $||F||_1$ of the matrix $F = (f_{ij}) \in \mathbb{R}^{n \times m}$ is defined by $||F||_1 = \max\limits_{1 \leq j \leq m} \sum_{i=1}^{n} |f_{ij}|.$

Theorem 1.6. *For $A \in \mathbb{R}^{n \times n}, \Delta A \in \mathbb{R}^{n \times n}, c \in \mathbb{R}^{1 \times n}, \Delta c \in \mathbb{R}^{1 \times n}$, where $n \geq 2$, assume that the system (1.4) satisfies the following assumptions:*

(i) (A, c) is observable, i.e., rank $H(A, c) = n^2$;

(ii) $max\{||\Delta A||_\infty, ||\Delta c||_\infty\} < \dfrac{1}{||H^{-1}(A,c)||_1}.$

Then $(A + \Delta A, c + \Delta c)$ is observable, where

$$H(A,c) = \begin{bmatrix} I_n & 0 & \cdots & 0 & 0 & 0 & \cdots & 0 & c^T \\ -A^T & I_n & \ddots & 0 & 0 & 0 & \cdots & c^T & 0 \\ 0 & \ddots & \ddots & 0 & \vdots & \vdots & \ddots & \vdots & \vdots \\ \vdots & \ddots & \ddots & I_n & 0 & c^T & \vdots & 0 & 0 \\ 0 & \cdots & 0 & -A^T & c^T & 0 & \cdots & 0 & 0 \end{bmatrix}_{n^2 \times n^2}, \quad (1.6)$$

and norm $||F||_\infty$ of the matrix $F = (f_{ij}) \in \mathbb{R}^{n \times m}$ is defined by $||F||_\infty = \max\limits_{1 \le i \le n} \sum_{j=1}^{m} |f_{ij}|$.

1.1.2 Stabilizability and detectability

From Theorem 1.2, if (A, B) is controllable, then there exists a matrix F such that all the eigenvalues of $A + BF$ lie in the open left-half plane, that is, there exists a state feedback $u = Fx$ such that the system is stable. Compared with the controllability, a weak condition can also leads to the same result, that is, the stabilizability which we introduce in the following.

Definition 1.5. A linear dynamical system $\dot{x} = Ax$ is said to be stable if all the eigenvalues of A have negative real parts. A matrix with such a property is said to be stable (or Hurwitz stable).

Definition 1.6. The dynamical system of equation (1.1) or the pair (A, B) is said to be stabilizable if there exists a state feedback $u = Fx$ such that the system is stable, i.e., $A + BF$ is stable.

We also consider the dual notion: detectability of the system (1.1) and (1.2).

Definition 1.7. The system or the pair (A, C) is said to be detectable if there exists a matrix L such that $A + LC$ is stable.

The following theorem is a consequence of Theorem 1.2.

Theorem 1.7. *The following are equivalent:*

(i) (A, B) *is stabilizable;*
(ii) (A^*, B^*) *is detectable;*
(iii) The matrix $(A - \lambda I, B)$ *has full-row rank for all* $\mathbf{Re}\{\lambda\} \ge 0$;
(iv) For all λ *and* x *such that* $x^*A = x^*\lambda$ *and* $\mathbf{Re}\{\lambda\} \ge 0$, $x^*B \ne 0$;
(v) Linear matrix inequalities

$$P > 0, \quad AP + PA^T < BB^T$$

are feasible for matrix variable P.

Theorem 1.8. *The following are equivalent:*

(i) (A, C) *is detectable;*
(ii) The matrix $\begin{pmatrix} A - \lambda I \\ C \end{pmatrix}$ *has full-column rank for all* $\mathbf{Re}\{\lambda\} \ge 0$;

(iii) For all λ and x such that $Ax = \lambda x$ and $\mathbf{Re}\{\lambda\} \geq 0$, $Cx \neq 0$;
(iv) (A^*, C^*) is stabilizable;
(v) Linear matrix inequalities

$$Q > 0, \quad QA + A^T Q < C^T C$$

are feasible for matrix variable Q.

From Definition 1.3 and above theorems we know that a system is stabilizable (detectable) if and only if every unstable mode is controllable (observable).

1.2 Algebraic Lyapunov equations and Lyapunov inequalities

This section introduces the main results of algebraic Lyapunov equations and Lyapunov inequalities.

1.2.1 *Continuous-time algebraic Lyapunov equations*

For a linear time invariant system $\dot{x} = Ax$, the quadratic form $V(x) = x^T P x$ is often chosen as Lyapunov function in using Lyapunov method to study the stability of zero solution. With use of $\dot{x} = Ax$ leads to $\dot{V}(x) = x^T (A^T P + PA)x = -x^T Q x$ which is also quadratic form. The equation

$$A^T P + PA = -Q \tag{1.7}$$

is called algebraic Lyapunov equation defined by the matrix A.

Solving P for a given matrix Q is equivalent to solving a set of linear equations. In order to obtain the existence conditions of solution to Lyapunov equation (1.7) the Kronecker product was introduced. It can be found in many texts dealing with linear algebra, e.g., [Lancaster and Tismenetsky (1985); Huang (2003, 1990)].

Definition 1.8. Let $A = (a_{ij}) \in \mathbb{R}^{m \times n}$ and $B = (b_{ij}) \in \mathbb{R}^{k \times l}$. The symbol $\otimes$ denotes the Kronecker product, defined as

$$A \otimes B = \begin{pmatrix} a_{11}B & a_{12}B & \cdots & a_{1n}B \\ a_{21}B & a_{22}B & \cdots & a_{2n}B \\ \vdots & \vdots & \ddots & \vdots \\ a_{n1}B & a_{n2}B & \cdots & a_{nn}B \end{pmatrix}. \tag{1.8}$$

The Kronecker product has the following property

$$(A \otimes B)(C \otimes D) = (AC) \otimes (BD).$$

The general method for solving (1.7) based on the Kronecker product is described in [Barnett and Man (1970)]. Let $P = (p_{ij}), Q = (q_{ij})$. Then (1.7) is converted into an equivalent system which can be described as the following form of linear equations

$$\mathcal{A}x = b \tag{1.9}$$

where

$$\mathcal{A} = I \otimes A^T + A^T \otimes I \tag{1.10}$$

and

$$b^T = [q_{11}, q_{12}, \cdots, q_{1n}, q_{21}, \cdots, q_{2n}, \cdots, q_{n1}, \cdots, q_{nn}],$$
$$x^T = [p_{11}, p_{12}, \cdots, p_{1n}, p_{21}, \cdots, p_{2n}, \cdots, p_{n1}, \cdots, p_{nn}].$$

From the property of the matrix $\mathcal{A}$ defined in (1.10), we have

Theorem 1.9. *Let Q be a given matrix. Then the Lyapunov equation (1.7) has a unique solution if and only if*

$$\lambda_i + \lambda_j \neq 0, \quad \forall \lambda_i, \lambda_j \in \mathbf{\Lambda}(A). \tag{1.11}$$

If (1.11) holds, then $P = P^T$ when $Q = Q^T$.

Corollary 1.1. *Suppose that all the eigenvalues of the matrix A have negative real parts. Then Lyapunov equation (1.7) has a unique solution*

$$P = \int_0^\infty e^{A^T t} Q e^{At} dt.$$

The condition (1.11) in Theorem 1.9 is necessary and sufficient for a unique solution to exist. It does not mean that the Lyapunov equation has no solutions when (1.11) is not satisfied. The following example explains this situation.

Example 1.1. Consider Lyapunov equation $A^T P + PA + I = 0$, where $A = \begin{pmatrix} -1 & 0 \\ 0 & 1 \end{pmatrix}$. It is obvious that the condition (1.11) of Theorem 1.9 is not satisfied. But it is easy to test that any matrix in the form $P = \begin{pmatrix} \frac{1}{2} & \alpha \\ \alpha & -\frac{1}{2} \end{pmatrix}$ is the solution to above Lyapunov equation for any $\alpha \in \mathbb{R}$.

1.2. Algebraic Lyapunov equations and Lyapunov inequalities

In the case that $Q \geq 0$ and (A, Q) is observable the solution of the Lyapunov equation has the following properties [Huang (2003); Yakubovich (1973b)].

Theorem 1.10. *Assume $Q \geq 0$, (A, Q) is observable and the Lyapunov equation (1.7) has a solution P, then*

(a) A has no eigenvalues with zero real part,
(b) $\det(P) \neq 0$, and
(c) The number of negative eigenvalues for P is equal to the number of eigenvalues for A with positive real parts.

Corollary 1.2. *All the eigenvalues of the matrix A have negative real parts if and only if the solution of the Lyapunov equation (1.7) is positive definite for any given matrix $Q \geq 0$ and (A, Q) being observable.*

Corollary 1.3. *All the eigenvalues of the matrix A have negative real parts if and only if the Lyapunov equation (1.7) has a unique positive definite solution $P > 0$ for any given matrix $Q > 0$.*

Remark 1.1. As discussed above, the Kronecker product can be used to solve the Lyapunov equation. In addition, if all the eigenvalues of the matrix A have negative real parts then Corollary 1.1 can also be used to derive the solution to (1.7). In this case the Lyapunov equation (1.7) has a unique solution in the form

$$P = \int_0^\infty e^{A^T t} Q e^{At} dt$$

and thus solving the Lyapunov equation (1.7) is realized by calculating the matrix exponential $e^{A^T t}$ [Mori et al. (1986)].

Using the Cayley-Hamilton theorem $e^{A^T t}$ can be expressed as

$$e^{A^T t} = a_1(t) I_n + a_2(t) A^T + \cdots + a_n(t) (A^T)^{n-1}. \tag{1.12}$$

Let $Q = \Gamma\Gamma^T, \Gamma \in \mathbb{R}^{n \times r}$ and

$$M = \begin{pmatrix} \Gamma & A^T \Gamma & (A^T)^2 \Gamma & \cdots & (A^T)^{n-1} \Gamma \end{pmatrix}. \tag{1.13}$$

Then $P = M(G \otimes I_r) M^T$, where $G = G^T = \{g_{ij}\} \in \mathbb{R}^{n \times n}$, $g_{ij} = \int_0^\infty a_i(t) a_j(t) dt$.

1.2.2 Continuous-time Lyapunov inequalities

Note that if $Q > 0$ or $Q \geq 0$ in (1.7), then the Lyapunov equation (1.7) can be equivalently written as Lyapunov inequality:

$$A^T P + PA < 0 \tag{1.14}$$

or

$$A^T P + PA \leq 0. \tag{1.15}$$

The method of standard convex programming can be used to solve the Lyapunov inequalities efficiently by computer.

In terms of the properties of Lyapunov equation the conditions of solvability for (1.14) and (1.15) can be stated as follows [Boyd et al. (1994)].

Theorem 1.11. *The Lyapunov inequality (1.14) is feasible for $P > 0$ if and only if all the eigenvalues of A have negative real parts.*

Theorem 1.12. *The Lyapunov inequality (1.15) is feasible for $P > 0$ if and only if the eigenvalues of A have nonpositive real part, and the size of Jordan blocks for each eigenvalue with zero real part is one.*

Remark 1.2. Suppose the Lyapunov inequality (1.15) is feasible for $P > 0$. Then solving (1.15) can be converted into solving a strict Lyapunov inequality with less variables.

In fact, there exists a nonsingular matrix T such that

$$T^{-1}AT = diag\left(\begin{pmatrix} 0 & \omega_1 I_{k_1} \\ -\omega_1 I_{k_1} & 0 \end{pmatrix}, \cdots, \begin{pmatrix} 0 & \omega_r I_{k_r} \\ -\omega_r I_{k_r} & 0 \end{pmatrix}, 0_{k_{r+1}}, A_{stab}\right)$$

where $0 < \omega_1 < \cdots < \omega_r, 0_{k_{r+1}}$ denotes the zero matrix with size $\mathbb{R}^{k_{r+1} \times k_{r+1}}$, and all the eigenvalues of $A_{stab} \in \mathbb{R}^{s \times s}$ have negative real parts. Theorem 1.11 implies that there exists a matrix $P_{stab} > 0$ satisfying $A_{stab}^T P_{stab} + P_{stab} A_{stab} < 0$. Let

$$P = T^{-T} daig\left(\begin{pmatrix} I_{k_1} & 0 \\ 0 & I_{k_1} \end{pmatrix}, \cdots, \begin{pmatrix} I_{k_r} & 0 \\ 0 & I_{k_r} \end{pmatrix}, I_{k_{r+1}}, P_{stab}\right) T^{-1}. \tag{1.16}$$

Then $P > 0$ satisfies

$$A^T P + PA = T^{-T} diag\left(0, A_{stab}^T P_{stab} + P_{stab} A_{stab}\right) T^{-1} \leq 0,$$

where the zero matrix has size $2k_1 + \cdots + 2k_r + k_{r+1}$. (1.16) gives the solutions to Lyapunov inequality (1.15).

1.2.3 Discrete-time algebraic Lyapunov equations and inequalities

In this subsection we consider the discrete-time algebraic Lyapunov equation for discrete-time systems given by

$$A^T P A + Q = P, \qquad (1.17)$$

where $P \in \mathbb{R}^{n \times n}$ is solution matrix. $A \in \mathbb{R}^{n \times n}$ and $Q \in \mathbb{R}^{n \times n}$ are given matrices. Two kinds of bilinear transformation are introduced here to convert the discrete-time Lyapunov equation into the continuous-time form.

(I) **The first form of bilinear transformation.**

Let $B = (A - I)^{-1}(A + I)$. Then the discrete-time Lyapunov equation (1.17) is converted into an equivalent form

$$B^T P_b + P_b B + Q = 0, \qquad (1.18)$$

where P_b and P are related by

$$P = \frac{1}{2}(B - I)^T P_b (B - I).$$

The matrix P is a solution of the discrete-time Lyapunov equation (1.17) if and only if the matrix P_b is a solution of the continuous Lyapunov equation (1.18).

(II) **The second form of bilinear transformation.**

Another form of the bilinear transformation was derived in [Popov (1964)] which can be described as

$$B = (A - I)(A + I)^{-1},$$
$$C = 2(A^T + I)^{-1} Q (A + I)^{-1}.$$

Then (1.17) is converted into the equivalent form

$$B^T P + PB + C = 0. \qquad (1.19)$$

Note that in this case the solution of the continuous-time Lyapunov equation (1.19) is the same as the solution P of the original discrete-time Lyapunov equation (1.17).

The properties of the solutions for the discrete-time Lyapunov equations can be derived in terms of above bilinear transformations and properties of continuous-time Lyapunov equations. The corresponding results are not stated in here. The discrete-time Lyapunov equations can be obtained by sampling the continuous-time systems and relation between them are studied in [Troch (1988)].

Remark 1.3. By using the same bilinear transformations as given above the discrete-time Lyapunov inequalities $A^T P A - P + Q < 0 \,(\leq 0)$ can also be converted into the continuous-time Lyapunov inequalities.

1.3 Formulation related to linear matrix inequalities

This section lists some common formulae related to LMIs including Schur complement and project lemma. These basic results are widely used in the problems of analysis and synthesis from system and control theory.

1.3.1 Schur complements

Schur complements are often used to convert nonlinear (convex) inequalities to LMIs. The case of strict inequalities for Schur complements is stated as follows which is easily shown by congruence transformation.

Theorem 1.13 (Schur complement). *Suppose R and S are Hermitian. Then the following conditions are equivalent:*

(i) $\begin{pmatrix} S & G^T \\ G & R \end{pmatrix} < 0;$

(ii) $R < 0, \quad S - G^T R^{-1} G < 0;$

(iii) $S < 0, \quad R - G S^{-1} G^T < 0.$

Schur complement for the non-strict inequalities is described as follows [Boyd et al. (1994)].

Theorem 1.14. *Suppose R and S are Hermitian. Then the following conditions are equivalent:*

(i) $R \leq 0, \quad S - G^T R^+ G \leq 0, \quad (I - R R^+) G = 0,$
where R^+ denotes the pseudo-inverse of R;

(ii) $\begin{pmatrix} S & G^T \\ G & R \end{pmatrix} \leq 0.$

1.3.2 Projection lemma

Consider a LMI
$$Q + UGV^T + VG^TU^T < 0, \qquad (1.20)$$
where $G \in \mathbb{R}^{m \times k}$ is a matrix variable and $Q = Q^T \in \mathbb{R}^{n \times n}, U \in \mathbb{R}^{n \times m}$ and $V \in \mathbb{R}^{n \times k}$ do not depend on G.

The inequality (1.20) is often encountered in studying the problems of controller synthesis using LMI. The solvability of this inequality for variable G can be characterized by equivalent inequalities without variable G. The results can be found in e.g. [Gahinet (1992); Iwasaki and Skelton (1994); Boyd et al. (1994)].

Lemma 1.1 (Projection lemma). *Let matrices U, V and Q be given. Suppose that $\operatorname{rank}(U) < n$ and $\operatorname{rank}(V) < n$. Then (1.20) holds if and only if*
$$U^\perp Q U^{\perp T} < 0, \quad V^\perp Q V^{\perp T} < 0 \qquad (1.21)$$
holds.

Furthermore, if (1.21) holds and $V^T V > 0$, then
$$G = -\rho U^T \Phi V \Upsilon + \Omega^{\frac{1}{2}} F \Upsilon^{\frac{1}{2}}, \quad \|F\| < \rho,$$
where scalar ρ and the matrix F are free parameters, and
$$\Phi := \left(UU^T - \frac{1}{\rho}Q\right)^{-1} > 0,$$
$$\Omega := I - U^T(\Phi - \Phi V \Upsilon V^T \Phi)U,$$
$$\Upsilon := (V^T \Phi V)^{-1}.$$

Remark 1.4. There exists another parameterized form of solutions for (1.20), the details can be found in [Skelton and Iwasaki (1995)].

Remark 1.5. If $\operatorname{rank}(U) = n$ or $\operatorname{rank}(V) = n$, then the results of Lemma 1.1 still hold but the first or second inequality in (1.21) disappears [Boyd et al. (1994)].

Next, we consider non-strict LMI
$$Q + UGV^T + VG^TU^T \leq 0. \qquad (1.22)$$
For this non-strict case, if (1.22) holds, then
$$U^\perp Q U^{\perp T} \leq 0, \quad V^\perp Q V^{\perp T} \leq 0 \qquad (1.23)$$
holds. But the converse is not true generally. When the subspaces spanned by U and V are linearly independent the equivalence between (1.22) and (1.23) is true. The corresponding results can be stated as the following lemma [Helmersson (1995)].

Lemma 1.2 (Non-strict LMI solvability). *Let matrices $U \in \mathbb{R}^{n \times m}$, $V \in \mathbb{R}^{n \times k}$ and $Q = Q^T \in \mathbb{R}^{n \times n}$ be given such that range U and range V are linearly independent. Then (1.22) is solvable for $G \in \mathbb{C}^{m \times k}$, if and only if*

$$U^{\perp} Q U^{\perp T} \leq 0, \quad V^{\perp} Q V^{\perp T} \leq 0. \tag{1.24}$$

Furthermore, if (1.24) holds, then all solutions that solve (1.22) are parameterized by

$$G = G_1 + G_2 L G_3$$

with $\sigma_{max}(L) \leq 1$, where

$$G_1 = Q_{23} Q_{33}^+ Q_{13}^T - Q_{12}^T,$$
$$G_2 = \left((Q_{22} - Q_{23} Q_{33}^+ Q_{23}^T)^+\right)^{\frac{1}{2}},$$
$$G_3 = \left((Q_{11} - Q_{13} Q_{33}^+ Q_{13}^T)^+\right)^{\frac{1}{2}}$$

in which

$$\begin{pmatrix} Q_{11} & Q_{12} & Q_{13} \\ Q_{12}^T & Q_{22} & Q_{23} \\ Q_{13}^T & Q_{23}^T & Q_{33} \end{pmatrix} = \begin{pmatrix} U_X \\ V_X \\ X^{\perp} \end{pmatrix} Q \begin{pmatrix} U_X \\ V_X \\ X^{\perp} \end{pmatrix}^T,$$

and

$$X^{\perp} = \begin{pmatrix} U & V \end{pmatrix}^{\perp}, \quad U_X = \begin{pmatrix} U & X^{\perp T} \end{pmatrix}^{\perp}, \quad V_X = \begin{pmatrix} V & X^{\perp T} \end{pmatrix}^{\perp}.$$

If Q_{33} is nonexistent then $G = -Q_{12}^T + (Q_{22}^+)^{\frac{1}{2}} L (Q_{11}^+)^{\frac{1}{2}}$.

In the following we present another preliminary lemma which gives the existence condition of solution for inequality $Q - \mu N N^T < 0$ with variable μ [Iwasaki and Skelton (1994)].

Lemma 1.3. *Let matrices $N \in \mathbb{R}^{n \times m}$ and $Q \in \mathbb{R}^{n \times n}$ be given. Suppose rank$(N) < n$ and $Q = Q^T$. Let (N_R, N_L) be any full rank factor of N, i.e., $N = N_L N_R$, and define $D := (N_R N_R^T)^{-\frac{1}{2}} N_L^+$. Then*

$$Q - \mu N N^T < 0 \tag{1.25}$$

holds for some $\mu \in \mathbb{R}$ if and only if

$$P := N^{\perp} Q N^{\perp T} < 0 \tag{1.26}$$

holds, in which case, all such μ are given by

$$\mu > \mu_{min} := \lambda_{max} \left[D(Q - Q N^{\perp T} P^{-1} N^{\perp} Q) D^T \right]. \tag{1.27}$$

The equivalence between the conditions (1.25) and (1.26) is first through Finsler's lemma [Finsler (1937); Schweppe (1973)] which is often used to eliminate variables in some matrix inequalities as in the references [Peterson and Hollot (1986); Khargonekar and Rotea (1988)]. It is closely related to the S-procedure which will be introduced in next section.

1.4 The S-procedure

Constraint problems related to quadratic forms are often encountered in control theory. Such problems usually require some quadratic form to be negative whenever other quadratic forms are negative. In some cases, the S-procedure can be used to deal with this kind of constraint problem.

1.4.1 *The S-procedure for nonstrict inequalities*

Let $F_0, F_1 \cdots, F_n$ be quadratic functions of the variable $x \in \mathbb{R}^n$ defined as
$$F_i(x) = x^T T_i x + 2u_i^T x + v_i, \quad i = 0, 1, \cdots, p.$$
without loss of generality $T_i = T_i^T$. Consider the following two conditions (I) and (II).

(I) $F_0(x) \geq 0$ for all x such that $F_i(x) \geq 0$, $i = 1, \cdots, n$.
(II) There exist $\tau_1 \geq 0, \cdots, \tau_p \geq 0$ such that for all x,
$$F_0(x) - \sum_{i=1}^{p} \tau_i F_i(x) \geq 0.$$

It is obvious that if the condition (II) holds, then the condition (I) holds. But in general, the converse is not true except for $p = 1$. The case for $p = 1$ can be described as the following theorem. The proof can be found in [Yakubovich (1971, 1973a, 1977); Fradkov and Yakubovich (1979); Huang (2003)].

Proposition 1.1. *Assume that there exists some x_0 such that $F_1(x_0) > 0$. Then $F_0(x) \geq 0$ for all x such that $F_1(x) \geq 0$ if and only if there exists $\tau \geq 0$ such that for all x, $F_0(x) - \tau F_1(x) \geq 0$.*

1.4.2 *The S-procedure for strict inequalities*

Consider another form of the S-procedure which involves quadratic forms and strict inequalities. Let $F_0, F_1, \cdots, F_p \in \mathbb{R}^{n \times n}$ be symmetric matrices. We still consider the following conditions.

(I) $x^T F_0 x > 0$ for all $x \in \mathbb{R}^n$, $x \neq 0$ such that $x^T F_i x \geq 0$, $i = 1, \cdots, p$.
(II) There exists $\tau_1 \geq 0, \cdots, \tau_p \geq 0$ such that for all $x \neq 0$,
$$x^T F_0 x - \sum_{i=1}^{p} \tau_i x^T F_i x > 0.$$

Obviously, if the condition (II) holds, then the condition (I) holds. But the converse is not true.

Similar to the case of non-strict inequalities, the equivalence between the condition (I) and (II) for $p = 1$ can also be found in [Yakubovich (1971, 1973a, 1977); Fradkov and Yakubovich (1979); Huang (2003)]. The corresponding result can be summarized as follows.

Proposition 1.2. *Assume that there exists some x_0 such that $x_0^T F_1 x_0 > 0$. Then $x^T F_0 x > 0$ for all $x \neq 0$ such that $x^T F_1 x \geq 0$ if and only if there exists $\tau \geq 0$ such that for all $x \neq 0$, $x^T F_0 x - \tau x^T F_1 x > 0$.*

1.5 Kalman-Yakubovič-Popov (KYP) lemma and its generalized forms

The celebrated Kalman-Yakubovič-Popov (KYP) lemma [Willems (1971); Rantzer (1996)] originates from Popov's criterion [Popov (1962)] and the positive real lemma [Yakubovich (1962); Kalman (1963); Anderson (1967)]. It has been recognized as one of the most basic tools of system theory that establishes the equivalence between a frequency-domain inequality and existence of a Lyapunov function of certain form. The latter can be expressed as a linear matrix inequality. The conversion between frequency-domain inequalities and real-domain conditions for absolute stability of Lur'e systems and the bounded real lemma can be realized by KYP Lemma. Some generalized forms of KYP lemma presented in [Iwasaki and Hara (2005a); Iwasaki et al. (2005b, 2003)] give a unified form for continuous and discrete time systems. The equivalence between the frequency domain inequalities restricted on a certain frequency range and LMIs are also covered. In the following we first introduce the generalized forms of KYP lemma [Iwasaki and Hara (2005a)].

Lemma 1.4. *Let matrices $A, E \in \mathbb{C}^{n \times n}$, $B, N \in \mathbb{C}^{n \times m}$, $\Pi = \Pi^* \in \mathbb{C}^{(n+m) \times (n+m)}$, and $\Phi = \Phi^* \in \mathbb{C}^{2 \times 2}, \Psi = \Psi^* \in \mathbb{C}^{2 \times 2}$ be given, and a set of complex numbers $\Lambda(\Phi, \Psi)$ and $\bar{\Lambda}(\Phi, \Psi)$ are defined as*

$$\Lambda(\Phi, \Psi) := \{\lambda \in \mathbb{C} | \sigma(\lambda, \Phi) = 0,\ \sigma(\lambda, \Psi) \geq 0\} \qquad (1.28)$$

and

$$\bar{\Lambda}(\Phi, \Psi) := \begin{cases} \Lambda(\Phi, \Psi), & (if\ \Lambda(\Phi, \Psi)\ is\ bounded) \\ \Lambda(\Phi, \Psi) \cup \infty. & (otherwise) \end{cases} \qquad (1.29)$$

Suppose that:

1.5. Kalman-Yakubovič-Popov (KYP) lemma and its generalized forms

(a) $\Lambda(\Phi, \Psi)$ represents curves on the complex plane,
(b) $\det(\lambda E - A) \neq 0$ for all $\lambda \in \Lambda(\Phi, \Psi)$, and
(c) either E is nonsingular or $\Lambda(\Phi, \Psi)$ is bounded.

Then the following statements are equivalent.

(i) For $G(\lambda) := (\lambda E - A)^{-1}(B - \lambda N)$, we have

$$\sigma(G, \Pi) := \begin{pmatrix} G(\lambda) \\ I \end{pmatrix}^* \Pi \begin{pmatrix} G(\lambda) \\ I \end{pmatrix} < 0$$

for all $\lambda \in \bar{\Lambda}(\Phi, \Psi)$;

(ii) There exist $P = P^* \in \mathbb{C}^{n \times n}, Q = Q^* \in \mathbb{C}^{n \times n}$ such that $Q > 0$ and

$$\begin{pmatrix} A & B \\ E & 0 \end{pmatrix}^* (\Phi \otimes P + \Psi \otimes Q) \begin{pmatrix} A & B \\ E & 0 \end{pmatrix} + \Pi < 0.$$

The case of nonstrict inequality is described as follows [Iwasaki and Hara (2005a)].

Lemma 1.5. *Let matrices* $A \in \mathbb{C}^{n \times n}, B \in \mathbb{C}^{n \times m}$, $\Pi = \Pi^* \in \mathbb{C}^{(n+m) \times (n+m)}$, and $\Phi = \Phi^* \in \mathbb{C}^{2 \times 2}, \Psi = \Psi^* \in \mathbb{C}^{2 \times 2}$ be given, $\Lambda(\Phi, \Psi)$ and $\bar{\Lambda}(\Phi, \Psi)$ are defined by (1.28) and (1.29), respectively. Suppose that:

(a) $\Lambda(\Phi, \Psi)$ represents curves on the complex plane, and
(b) the pair (A, B) is controllable.

Let Ω be the set of eigenvalues of A in $\Lambda(\Phi, \Psi)$. Then the following are equivalent.

(i) For each $\lambda \in \bar{\Lambda}(\Phi, \Psi) \backslash \Omega$, we have

$$\begin{pmatrix} (\lambda I - A)^{-1} B \\ I \end{pmatrix}^* \Pi \begin{pmatrix} (\lambda I - A)^{-1} B \\ I \end{pmatrix} \leq 0;$$

(ii) There exist $P = P^* \in \mathbb{C}^{n \times n}, Q = Q^* \in \mathbb{C}^{n \times n}$ such that $Q \geq 0$ and

$$\begin{pmatrix} A & B \\ I & 0 \end{pmatrix}^* (\Phi \otimes P + \Psi \otimes Q) \begin{pmatrix} A & B \\ I & 0 \end{pmatrix} + \Pi \leq 0.$$

Let

$$N = 0, \quad E = I, \quad \Phi = \begin{pmatrix} 0 & 1 \\ 1 & 0 \end{pmatrix}, \quad \Psi = 0_{2 \times 2}$$

in Lemma 1.4 and Lemma 1.5, the corresponding results are the standard version of the KYP lemma which is expressed as follows. The results can also be found in [Rantzer (1996)].

Corollary 1.4 (KYP lemma for continuous-time systems). *Given $A \in \mathbb{C}^{n \times n}, B \in \mathbb{C}^{n \times m}, \Pi = \Pi^* \in \mathbb{C}^{(n+m) \times (n+m)}$, with $det(j\omega I - A) \neq 0$ for all $\omega \in \mathbb{R}$. Assume that (A,B) is controllable. Then the following statements are equivalent.*

(i) $\left(\begin{array}{c} (j\omega I - A)^{-1} B \\ I \end{array} \right)^* \Pi \left(\begin{array}{c} (j\omega I - A)^{-1} B \\ I \end{array} \right) \leq 0$
for all $\omega \in \mathbb{R} \cup \{\infty\}$;

(ii) *There exists a matrix $P = P^* \in \mathbb{C}^{n \times n}$ such that*

$$\left(\begin{array}{cc} A^*P + PA & PB \\ B^*P & 0 \end{array} \right) + \Pi \leq 0.$$

P is real matrix when A, B and Π are real matrices. The corresponding equivalence for strict inequalities holds even if (A, B) is not controllable. Furthermore, if A is Hurwitz stability and the upper left corner of Π is positive semidefinite, then $P \geq 0$.

Similarly, for the discrete-time systems, choosing

$$N = 0, \quad E = I, \quad \Phi = \left(\begin{array}{cc} -1 & 0 \\ 0 & 1 \end{array} \right), \quad \Psi = 0_{2 \times 2}$$

in Lemma 1.4 and Lemma 1.5, the corresponding results are the KYP lemma of the discrete-time systems which is described by following corollary.

Corollary 1.5 (KYP lemma for discrete-time systems). *Given $A \in \mathbb{C}^{n \times n}, B \in \mathbb{C}^{n \times m}, \Pi = \Pi^* \in \mathbb{C}^{(n+m) \times (n+m)}$, with $det(e^{j\omega} I - A) \neq 0$ for all $\omega \in \mathbb{R}$. Assume that (A, B) is controllable. Then the following statements are equivalent.*

(i) $\left(\begin{array}{c} (e^{j\omega} I - A)^{-1} B \\ I \end{array} \right)^* \Pi \left(\begin{array}{c} (e^{j\omega} I - A)^{-1} B \\ I \end{array} \right) \leq 0$
for all $\omega \in \mathbb{R}$;

(ii) *There exists a matrix $P = P^* \in \mathbb{C}^{n \times n}$ such that*

$$\left(\begin{array}{cc} A^*PA - P & A^*PB \\ B^*PA & B^*PB \end{array} \right) + \Pi \leq 0.$$

P is real matrix when A, B and Π are real matrices. The corresponding equivalence for strict inequalities holds even if (A, B) is not controllable.

1.5. Kalman-Yakubovič-Popov (KYP) lemma and its generalized forms

Note that the frequency-domain inequalities are required to be true for all $\omega \in \mathbb{R}$ in KYP lemma. But for some problems in systems and control it is often required to consider the frequency restriction in frequency-domain inequalities, for instance, some property only holds for all ω within a high or low frequency range. In this case it corresponds to a restriction on the class of input signals for time-domain inequalities. The equivalence between the frequency-domain inequalities with high or low frequency limits and LMIs can also be obtained by Lemmas 1.4 and 1.5.

Let

$$N = 0, \quad E = I, \quad \Phi = \begin{pmatrix} 0 & 1 \\ 1 & 0 \end{pmatrix}, \quad \Psi = \begin{pmatrix} \pm 1 & j\omega_o \\ -j\omega_o & \pm\omega_1\omega_2 \end{pmatrix}$$

in Lemma 1.4 and Lemma 1.5, where $\omega_0 := (\omega_1 + \omega_2)/2$. Then

$$\Lambda(\Phi, \Psi) = \{j\omega | \tau(\omega - \omega_1)(\omega - \omega_2) \leq 0\}, \quad \tau = 1 \text{ or } -1$$

and thus the corresponding results can be described as follows.

Corollary 1.6. *Let complex matrices A, B, Π, and real scalars ω_1, ω_2 be given. Let τ be $+1$ or -1, and define*

$$\Omega := \{\omega \in \mathbb{R} | \tau(\omega - \omega_1)(\omega - \omega_2) \leq 0\}.$$

Suppose that $\Pi = \Pi^*$, (A, B) is controllable, and Ω has a nonempty interior. Then the following statements are equivalent.

(i) The frequency-domain inequality

$$\begin{pmatrix} (j\omega I - A)^{-1}B \\ I \end{pmatrix}^* \Pi \begin{pmatrix} (j\omega I - A)^{-1}B \\ I \end{pmatrix} \leq 0 \quad (1.30)$$

holds for all $\omega \in \Omega$ such that $\det(j\omega I - A) \neq 0$;
(ii) There exist matrices $P = P^*$ and $Q = Q^*$ such that

$$\tau Q \geq 0 \quad (1.31)$$

and

$$\begin{pmatrix} A & B \\ I & 0 \end{pmatrix}^* \begin{pmatrix} -Q & P + j\omega_0 Q \\ P - j\omega_0 Q & -\omega_1\omega_2 Q \end{pmatrix} \begin{pmatrix} A & B \\ I & 0 \end{pmatrix} + \Pi \leq 0 \quad (1.32)$$

hold.

The corresponding equivalence for strict inequalities of (1.30), (1.31) and (1.32) holds even if (A, B) is not controllable.

Similarly, for the discrete-time setting, choosing

$$N = 0, \quad E = I, \quad \Phi = \begin{pmatrix} -1 & 0 \\ 0 & 1 \end{pmatrix}, \quad \Psi = \begin{pmatrix} 0 & e^{j\theta_o} \\ e^{-j\theta_o} & -\gamma \end{pmatrix}$$

in Lemma 1.4 and Lemma 1.5, where $\theta_0 := (\theta_1 + \theta_2)/2$, $\gamma := 2\cos\theta_c$ and $\theta_c := (\theta_2 - \theta_1)/2$. Then

$$\Lambda(\Phi, \Psi) = \{e^{j\theta} | \theta_1 \leq \theta \leq \theta_2\},$$

and thus we have

Corollary 1.7. *Let complex matrices A, B, Π, and real scalars θ_1, θ_2 be given. Suppose that $\Pi = \Pi^*$, (A, B) is controllable, and $0 < \theta_2 - \theta_1 \leq 2\pi$. Define*

$$\Theta := \{\theta \in \mathbb{R} | \theta_1 \leq \theta \leq \theta_2\}.$$

The following are equivalent.

(i) The frequency-domain inequality

$$\begin{pmatrix} (e^{j\theta}I - A)^{-1}B \\ I \end{pmatrix}^* \Pi \begin{pmatrix} (e^{j\theta}I - A)^{-1}B \\ I \end{pmatrix} \leq 0 \quad (1.33)$$

holds for all $\theta \in \Theta$ such that $\det(e^{j\theta}I - A) \neq 0$;
(ii) There exist matrices $P = P^*$ and $Q = Q^*$ such that

$$\tau Q \geq 0 \quad (1.34)$$

and

$$\begin{pmatrix} A & B \\ I & 0 \end{pmatrix}^* \begin{pmatrix} -P & e^{j\theta_o}Q \\ e^{-j\theta_o}Q & P - \gamma Q \end{pmatrix} \begin{pmatrix} A & B \\ I & 0 \end{pmatrix} + \Pi \leq 0 \quad (1.35)$$

holds.

The corresponding equivalence for strict inequalities of (1.33), (1.34) and (1.35) holds even if (A, B) is not controllable.

In terms of Lemma 1.5 and Schur complement the corresponding result follows for the case that $G(\lambda)$ is a proper transfer matrix.

Corollary 1.8. *Let $G(\lambda) = C(sI - A)^{-1}B + D$ and the matrix $\Pi = \Pi^* \in \mathbb{C}^{(m+p)\times(m+p)}$ be such that*

$$\Pi = \begin{pmatrix} \Pi_{11} & \Pi_{12} \\ \Pi_{12}^* & \Pi_{22} \end{pmatrix}, \quad \Pi_{11} = \Pi_{11}^* \in \mathbb{C}^{p \times p}, \quad \Pi_{11} \geq 0.$$

The condition

$$\begin{pmatrix} G(\lambda) \\ I \end{pmatrix}^* \Pi \begin{pmatrix} G(\lambda) \\ I \end{pmatrix} \leq 0$$

holds for all $\lambda \in \Lambda(\Phi, \Psi)$ if and only if there exist matrices $P = P^*$ and $Q = Q^* \geq 0$ such that

$$\begin{pmatrix} \Gamma(P,Q,C,D) & [C \quad D]^* \Pi_{11} \\ \Pi_{11}[C \quad D] & -\Pi_{11} \end{pmatrix} \leq 0$$

holds, where

$$\Gamma(P,Q,C,D) := \begin{pmatrix} A & B \\ I & 0 \end{pmatrix}^* (\Phi \otimes P + \Psi \otimes Q) \begin{pmatrix} A & B \\ I & 0 \end{pmatrix}$$
$$+ \begin{pmatrix} 0 & C^* \Pi_{12} \\ \Pi_{12}^* C & D^* \Pi_{12} + \Pi_{12}^* D + \Pi_{22} \end{pmatrix}.$$

Remark 1.6. For the strict inequalities, the case where $G(\lambda)$ is a nonproper transfer matrix can also be treated similarly by using Lemma 1.4.

1.6 Notes and references

The equivalent conditions for the controllability, observability, stabilizability and detectability are summarized in many literature, see [Zhou et al. (1996); Leonov et al. (1996); Huang (2003)]. The results of the controllability and observability for SISO uncertain linear systems are given in [Yang (2005)]. The concept of Kronecker product can be found in large texts dealing with linear algebra e.g. [Lancaster and Tismenetsky (1985); Huang (1990, 2003)]. The methods for solving the Lyapunov equations based on the Kronecker product is from [Barnett and Man (1970)] and by calculating the matrix exponential is from [Mori et al. (1986)]. The various properties of the Lyapunov equation are closely related to the eigenvalue bounds of solution matrix. Readers can refer to the literature [Barnett and Man (1970); Peng (1972); Shapiro (1974); Montemayor and Womack (1975); Kwon and Pearson (1977); Fahmy and Hanafy (1981); Karanam (1982); Wimmer (1975)] if interested in. The conditions of solvability for Lyapunov inequalities and the structure of solution set for nonstrict Lyapunov inequalities are discussed in [Boyd et al. (1994)]. The transformation between continuous-time and discrete-time Lyapunov equations can be found in [Popov (1964); Gajić (1995)] and references therein. Schur complement for strict inequalities are from [Helmersson (1995)] and the case with nonstrict inequalities is from [Boyd et al. (1994)]. The projection lemma for strict and nonstrict inequalities [Boyd et al. (1994); Gahinet (1992); Iwasaki and Skelton (1994); Helmersson (1995); Skelton and Iwasaki (1995)] are used to eliminate variables in certain matrix inequalities. It dates back to the Finsler's Lemma [Finsler (1937)]. It is also related to the S-procedure. A survey article on S-procedure is by [Uhlig (1979)], see also [Horn and Johnson (1991); Yakubovich (1971, 1973a); Fradkov and Yakubovich (1979); Huang (2003)] for proofs of various S-procedure results. The celebrated Kalman-Yakubovič-Popov (KYP) lemma [Willems (1971); Rantzer (1996)] originates from Popov's criterion [Popov (1962)] and the positive real lemma [Yakubovich (1962); Kalman (1963); Anderson (1967)]. It has been recognized as one of the most basic tools of system theory that establishes the equivalence between a frequency-domain inequality and a linear matrix inequality. Some generalized forms of KYP lemma presented in [Iwasaki and Hara (2005a); Iwasaki et al. (2005b, 2003)] give a unified form for continuous and discrete time systems. The equivalence between the frequency domain inequalities restricted on a certain frequency range and LMIs are also covered.

Chapter 2

LMI Approach to H_∞ Control

In this chapter, first we introduce two classes of important Banach spaces: $\mathcal{L}_\infty$ and $\mathcal{H}_\infty$ spaces and give a state space method of computing $\mathcal{L}_\infty$ norm and $\mathcal{H}_\infty$ norm of real rational transfer matrix. Then we review matrix linear fractional transformation (LFT). Many analysis and synthesis problems discussed in this book can be solved on a unified LFT framework. As a byproduct, the notion of redheffer star product is introduced. Then we give solvability conditions of continuous-time and discrete-time algebric Ricatti equations (AREs) respectively, which are most useful in control system synthesis. Bounded real lemma and small gain theorem are also introduced in this chapter. The last section addresses linear matrix inequality approach to standard optimal $\mathcal{H}_\infty$ control and suboptimal $\mathcal{H}_\infty$ control.

2.1 $\mathcal{L}_\infty$ norm and $\mathcal{H}_\infty$ norm of the systems

Consider a linear finite dimensional system as shown in the following diagram with input u, output z, and transfer matrix G, where G is often assumed to be proper and real rational stable transfer matrix.

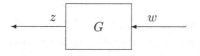

Fig. 2.1 A linear system.

If G is considered as an operator from the input space to the output space, then a norm can be induced on G which measures the size of the output for a given input u. From the view of control engineering practice, these norms can indicate to what extent the output signals can be affected by different classes of input disturbances and/or system parameter variations. This section is concerned about $\mathcal{L}_\infty$ and $\mathcal{H}_\infty$ norms of G.

2.1.1 $\mathcal{L}_\infty$ and $\mathcal{H}_\infty$ spaces

Let X be a vector space over $\mathbb{C}$. A norm on X is a function $x \to \|x\|$ from X to $\mathbb{R}$ having the following four properties:

(i) $\|x\| \geq 0$;
(ii) $\|x\| = 0$ if and only if $x = 0$;
(iii) $\|cx\| = |c| \|x\|$, $c \in \mathbb{C}$;
(iv) $\|x + y\| \leq \|x\| + \|y\|$.

Then X is a normed space. A sequence $\{x_k\}$ in X converges to x in X, and x is the limit of the sequence, if the sequence of real numbers $\{\|x_k - x\|\}$ converges to zero; if such x exists, then the sequence is convergent. A sequence $\{x_k\}$ is a *Cauchy sequence* if

$$(\forall \epsilon)(\exists \text{ integer } n) : i, k > n \Rightarrow \|x_i - x_k\| < \epsilon.$$

A normed space X is said to be *complete* if every Cauchy sequence in X converges in X. A complete normed space is called a *Banach space*. For example, $\mathbb{R}^n$ and $\mathbb{C}^n$ with the usual spatial p-norm, $\|\cdot\|_p$ for $1 \leq p \leq \infty$, are Banach spaces.

$\mathcal{L}_\infty$ and $\mathcal{H}_\infty$ spaces are two frequently used Banach spaces related to complex matrix functions bounded on imaginary axis.

$\mathcal{L}_\infty(j\mathbb{R})$ **Space**
$\mathcal{L}_\infty(j\mathbb{R})$ or simply $\mathcal{L}_\infty$ is a Banach space of matrix-valued (or scalar-valued) functions that are (essentially) bounded on $j\mathbb{R}$, with norm

$$\|F\|_\infty := \operatorname*{ess\,sup}_{\omega \in \mathbb{R}} \bar{\sigma}[F(j\omega)].$$

The rational subspace of $\mathcal{L}_\infty$, denoted by $\mathcal{RL}_\infty(j\mathbb{R})$ or simply $\mathcal{RL}_\infty$, consists of all proper and real rational transfer matrices with no poles on the imaginary axis.

$\mathcal{H}_\infty(j\mathbb{R})$ **Space**
$\mathcal{H}_\infty$ is a (closed) subspace of $\mathcal{L}_\infty$ with functions that are analytic and bounded in the open right-half plane. The $\mathcal{H}_\infty$ norm is defined as

$$\|F\|_\infty := \sup_{\operatorname{Re}(s)>0} \bar{\sigma}[F(s)] = \sup_{\omega \in \mathbb{R}} \bar{\sigma}[F(j\omega)].$$

2.1. $\mathcal{L}_\infty$ norm and $\mathcal{H}_\infty$ norm of the systems

A proof for the second equality can be seen in [Boyd and Desoer (1985)].The real rational subspace of $\mathcal{H}_\infty$ is denoted by $\mathcal{RH}_\infty$, which consists of all proper and real rational stable transfer matrices.

$\mathcal{H}_\infty^-(j\mathbb{R})$ Space

$\mathcal{H}_\infty^-$ is a (closed) subspace of $\mathcal{L}_\infty$ with functions that are analytic and bounded in the open left-half plane. The $\mathcal{H}_\infty^-$ norm is defined as

$$\|F\|_\infty := \sup_{\text{Re}(s)<0} \bar{\sigma}[F(s)] = \sup_{\omega\in\mathbb{R}} \bar{\sigma}[F(j\omega)].$$

The real rational subspace of $\mathcal{H}_\infty^-$ is denoted by $\mathcal{RH}_\infty^-$ which consists of all proper and real rational transfer matrices with all poles in the open right half plane.

Let G and H be two complex matrix functions. Note that the following submultiplicative property

$$\|GH\|_\infty \leq \|G\|_\infty \|H\|_\infty$$

is satisfied. If G is a complex scalar function, then

$$\|GH\|_\infty < 1$$

if and only if

$$\bar{\sigma}[H(j\omega)] < \frac{1}{|G(j\omega)|}, \forall \omega \in [0,\infty).$$

When G is rational, $G \in \mathcal{L}_\infty$ if and only if G has no poles on the imaginary axis. In this case $\bar{\sigma}(G(j\omega))$ is a continuous function of ω and

$$\|G\|_\infty < \gamma \Leftrightarrow \bar{\sigma}(G(j\omega)) < \gamma, \forall \omega \in \mathbb{R} \cup \infty.$$

Thus bounds on $\|G\|_\infty$ are equivalent to uniform bounds on $\bar{\sigma}(G(j\omega))$, thereby allowing us to write objectives of the form $\bar{\sigma}(G(j\omega)) < \gamma$ for all ω using the more compact notation $\|G\|_\infty < \gamma$. The design problems discussed in this Chapter can be expressed in terms of bounds on the infinity norm of various closed-loop transfer function matrices.

When G is rational, $G \in \mathcal{H}_\infty$ if and only if G has no poles in the closed right half plane.

2.1.2 Computing $\mathcal{L}_\infty$ and $\mathcal{H}_\infty$ norms

Let $G \in \mathcal{L}_\infty$. A control engineering interpretation of the infinity norm of a scalar transfer function G is the distance in the complex plane from the origin to the farthest point on the Nyquist plot of G, and it also appears as the peak value on the Bode magnitude plot of $|G(j\omega)|$. Hence the infinity norm of a transfer function can in principle be obtained graphically.

To get an estimate, set up a fine grid of frequency points $\{\omega_1, \cdots, \omega_N\}$, then an estimate for $\|G\|_\infty$ is

$$\max_{1 \leq k \leq N} \bar{\sigma}\{G(j\omega_k)\}.$$

This value is usually read directly from a Bode singular value plot. The $\mathcal{L}_\infty$ norm can also be computed from the state space realization if G is rational.

Lemma 2.1. *Let $\gamma > 0$ and*

$$G = \left(\begin{array}{c|c} A & B \\ \hline C & D \end{array}\right) \in \mathcal{RL}_\infty.$$

Then $\|G\|_\infty < \gamma$ if only if $\bar{\sigma}(D) < \gamma$ and the Hamiltonian matrix H has no eigenvalues on the imaginary axis, where

$$H := \begin{pmatrix} A + BR^{-1}D^T C & BR^{-1}B^T \\ -C^T(I + DR^{-1}D^T)C & -(A + BR^{-1}D^T C)^T \end{pmatrix}$$

and $R = \gamma^2 I - D^T D$.

A bisection algorithm is presented by Boyd and Balakrishnan [Boyd et al. (1989)] and Robel [Robel (1989)], to compute the $\mathcal{RL}_\infty$ norm with grarranteed accuracy, using the relation between the singular values of the transfer function matrix and the eigenvalues of a related Hamiltonian matrix as follows

(a) select an upper bound γ_u and a lower bound γ_l such that $\gamma_u \leq \|G\|_\infty \leq \gamma_l$;
(b) if $(\gamma_u - \gamma_l)/\gamma_l \leq$ specified level, stop; $\|G\|_\infty \approx (\gamma_u + \gamma_l)/2$, otherwise go to the next step;
(c) set $\gamma = (\gamma_u + \gamma_l)/2$;
(d) test if $\|G\|_\infty < \gamma$ by calculating the eigenvalues of H for the given γ: if H has an eigenvalue on $j\mathbb{R}$, set $\gamma_l = \gamma$; otherwise set $\gamma_u = \gamma$; go back to step (b).

This bisection algorithm is simple and much more efficient than a search over frequencies, but for repeated use as well as for very large systems, it is still not very fast. A two-step algorithms is presented in [Bruinsma and Steinbuch (1990)], which is guaranteed to converge and considerably faster than the bisection algorithm.

2.2 Linear fractional transformations

A mapping $F : \mathbb{C} \mapsto \mathbb{C}$ of the form
$$F(s) = \frac{as+b}{cs+d}$$
with $a, b, c,$ and $d \in \mathbb{C}$ is called a linear fractional transformation. In the case that $d \neq 0$, $F(s)$ can be written as
$$F(s) = bd^{-1} + (a - bd^{-1}c)s(1 + d^{-1}cs)^{-1}d^{-1}$$
This is the form we will usually use for LFTs, because it corresponds naturally to input-output block diagram representations of control systems, which we will see in the later part of this section. The linear fractional transformation described above for scalars can be generalized to the matrix case.

Definition 2.1. Let M be a complex matrix partitioned as
$$M = \begin{pmatrix} M_{11} & M_{12} \\ M_{21} & M_{22} \end{pmatrix} \in \mathbb{C}^{(p_1+p_2) \times (q_1+q_2)}$$
and let $\Delta_\ell \in \mathbb{C}^{q_2 \times p_2}$ and $\Delta_u \in \mathbb{C}^{q_1 \times p_1}$ be two other complex matrices. Then we can formally define a lower LFT with respect to Δ_ℓ as the map
$$\mathcal{F}_\ell(M, \bullet) : \mathbb{C}^{q_2 \times p_2} \mapsto \mathbb{C}^{p_1 \times q_1}$$
with
$$\mathcal{F}_\ell(M, \Delta_l) := M_{11} + M_{12}\Delta_\ell(I - M_{22}\Delta_\ell)^{-1}M_{21}$$
provided that the inverse $(I - M_{22}\Delta_\ell)^{-1}$ exists. We can also define an upper LFT with respect to Δ_u as
$$\mathcal{F}_u(M, \bullet) : \mathbb{C}^{q_1 \times p_1} \mapsto \mathbb{C}^{p_2 \times q_2}$$
with
$$\mathcal{F}_u(M, \Delta_u) := M_{22} + M_{21}\Delta_u(I - M_{11}\Delta_u)^{-1}M_{12}$$
provided that the inverse $(I - M_{11}\Delta_u)^{-1}$ exists.

The matrix M in the above LFTs is called the coefficient matrix. Figure 2.2 shows the structures of lower and upper LFTs. The diagram on the left represents the following set of equations:
$$\begin{pmatrix} z_1 \\ y_1 \end{pmatrix} = M \begin{pmatrix} w_1 \\ u_1 \end{pmatrix} = \begin{pmatrix} M_{11} & M_{12} \\ M_{21} & M_{22} \end{pmatrix} \begin{pmatrix} w_1 \\ u_1 \end{pmatrix}$$
$$u_1 = \Delta_\ell y_1$$

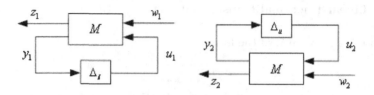

Fig. 2.2 Linear fractional transformations.

while the diagram on the right represents

$$\begin{pmatrix} y_2 \\ z_2 \end{pmatrix} = M \begin{pmatrix} u_2 \\ w_2 \end{pmatrix} = \begin{pmatrix} M_{11} & M_{12} \\ M_{21} & M_{22} \end{pmatrix} \begin{pmatrix} u_2 \\ w_2 \end{pmatrix}$$

$$u_2 = \Delta_u y_2$$

It is easy to verify that the mapping defined on the left diagram is equal to $\mathcal{F}_\ell(M, \Delta_l)$ and the mapping defined on the right diagram is equal to $\mathcal{F}_u(M, \Delta_u)$. So from the above diagrams, $\mathcal{F}_\ell(M, \Delta_\ell)$ is a transformation obtained from closing the lower loop on the left diagram; similarly $\mathcal{F}_u(M, \Delta_u)$ is a transformation obtained from closing the upper loop on the right diagram.

The physical meaning of an LFT in control science is obvious if we take M as a proper transfer matrix. In that case, the LFTs defined above are simply the closed-loop transfer matrices from $w_1 \mapsto z_1$ and $w_2 \mapsto z_2$, respectively, i.e.,

$$T_{z_1 w_1} = \mathcal{F}_\ell(M, \Delta_\ell), \qquad T_{z_2 w_2} = \mathcal{F}_u(M, \Delta_u)$$

where M may be the controlled plant and Δ may be either the system model uncertainties or the controllers. In robust control theory, LFT is a useful way to standardize block diagrams for robust control analysis and design.

Lemma 2.2. *Consider a feedback system shown in Figure 2.3 where N is a suitably partitioned transfer matrix*

$$N(s) = \begin{pmatrix} N_{11} & N_{12} \\ N_{21} & N_{22} \end{pmatrix}$$

Then the closed-loop transfer matrix from w to z is given by

$$T_{zw} = \mathcal{F}_\ell(N, Q) = N_{11} + N_{12} Q (I - N_{22} Q)^{-1} N_{21}$$

Assume that the feedback loop is well-posed, i.e., $\det(I - N_{22}(\infty) Q(\infty)) \neq 0$, and either $N_{21}(j\omega)$ has full row rank for all $\omega \in \mathbb{R} \cup \infty$ or $N_{12}(j\omega)$ has full column rank for all $\omega \in \mathbb{R} \cup \infty$ and $\|N\|_\infty \leq 1$; then $\|\mathcal{F}_\ell(N, Q)\|_\infty < 1$ if $\|Q\|_\infty < 1$.

2.3. Redheffer star product

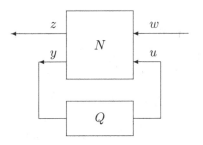

Fig. 2.3 A feedback system.

2.3 Redheffer star product

Since any interconnection of LFTs is also an LFT, most of the interconnection structures, e.g., feedback and cascade, can be viewed as special cases of the so-called *star product*.

Suppose that P and K are compatibly partitioned matrices

$$P = \begin{pmatrix} P_{11} & P_{12} \\ P_{21} & P_{22} \end{pmatrix}, K = \begin{pmatrix} K_{11} & K_{12} \\ K_{21} & K_{22} \end{pmatrix}$$

such that the matrix product $P_{22}K_{11}$ is well defined and square, and assume further that $I - P_{22}K_{11}$ is invertible. Then the star product of P and K with respect to this partition is defined as

$$\mathcal{S}(P, K) := \begin{pmatrix} \mathcal{F}_l(P, K_{11}) & P_{12}(I - K_{11}P_{22})^{-1}K_{12} \\ K_{21}(I - P_{22}K_{11})^{-1}P_{21} & \mathcal{F}_u(K, P_{22}) \end{pmatrix}.$$

This definition is dependent on the partitioning of the matrices P and K above, as shown in Figure 2.4.

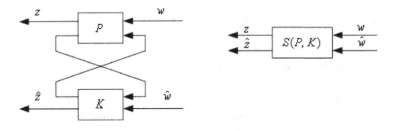

Fig. 2.4 Interconnection of LFTs.

Now suppose that P and K are transfer matrices with state space representations:

$$P = \begin{pmatrix} A & B_1 & B_2 \\ \hline C_1 & D_{11} & D_{12} \\ C_2 & D_{21} & D_{22} \end{pmatrix}, K = \begin{pmatrix} A_K & B_{K1} & B_{K2} \\ \hline C_{K1} & D_{K11} & D_{K12} \\ C_{K2} & D_{K21} & D_{K22} \end{pmatrix}.$$

Then the transfer matrix

$$\mathcal{S}(P,K) : \begin{pmatrix} w \\ \hat{w} \end{pmatrix} \mapsto \begin{pmatrix} z \\ \hat{z} \end{pmatrix}$$

has a state space representation

$$\mathcal{S}(P,K) = \begin{pmatrix} \bar{A} & \bar{B}_1 & \bar{B}_2 \\ \hline \bar{C}_1 & \bar{D}_{11} & \bar{D}_{12} \\ \bar{C}_2 & \bar{D}_{21} & \bar{D}_{22} \end{pmatrix} = \begin{pmatrix} \bar{A} & \bar{B} \\ \hline \bar{C} & \bar{D} \end{pmatrix},$$

where

$$\bar{A} = \begin{pmatrix} A + B_2 \tilde{R}^{-1} D_{K11} C_2 & B_2 \tilde{R}^{-1} C_{K1} \\ B_{K1} R^{-1} C_2 & A_K + B_{K1} R^{-1} D_{22} C_{K1} \end{pmatrix},$$

$$\bar{B} = \begin{pmatrix} B_1 + B_2 \tilde{R}^{-1} D_{K11} D_{21} & B_2 \tilde{R}^{-1} D_{K12} \\ B_{K1} R^{-1} D_{21} & B_{K2} + B_{K1} R^{-1} D_{22} D_{K12} \end{pmatrix},$$

$$\bar{C} = \begin{pmatrix} C_1 + D_{12} D_{K11} R^{-1} C_2 & D_{12} \tilde{R}^{-1} C_{K1} \\ D_{K21} R^{-1} C_2 & C_{K2} + D_{K21} R^{-1} D_{22} C_{K1} \end{pmatrix},$$

$$\bar{D} = \begin{pmatrix} D_{11} + D_{12} D_{K11} R^{-1} D_{21} & D_{12} \tilde{R}^{-1} D_{K12} \\ D_{K21} R^{-1} D_{21} & D_{K22} + D_{K21} R^{-1} D_{22} D_{K12} \end{pmatrix},$$

$$R = I - D_{22} D_{K11}, \quad \tilde{R} = I - D_{K11} D_{22}.$$

The following proposition can be easily verified by the definition of redheffer star product.

Proposition 2.1. $\mathcal{F}_\ell(P, \mathcal{F}_\ell(K,Q)) = \mathcal{F}_\ell(\mathcal{S}(P,K), Q).$

2.4 Algebraic Riccati equations

Let A, Q, and R be real $n \times n$ matrices with Q and R symmetric. Then an algebraic Riccati equation is the following quadratic matrix equation:

$$A^T X + XA + XRX + Q = 0. \tag{2.1}$$

Algebraic Riccati equation is important for control design as Lyapunov equation for control analysis. The quadratic optimal control for a linear

2.4. Algebraic Riccati equations

time-invariant system is to find a control law u such that the following quadratic cost function

$$J = \int_0^\infty (x^T Q x + u^T R u) dt$$

is minimum, where $Q = Q^T \geq 0$ and $R = R^T > 0$. The existence of the optimal control law can be reduce to the existence of the solution to (2.1). Algebric Riccati equation also play an important role in $\mathcal{H}_\infty$ control theory. A state space method to standard $\mathcal{H}_2$ and $\mathcal{H}_\infty$ control was presented in [Doyle et al. (1989)], where an $\mathcal{H}_\infty$ (sub)optimal controller can be obtained through solving two algebric Riccati equations. Since $\mathcal{H}_\infty$ criterion is established in frequency domain, this method builds the connection between the frequency domain conditions and the time domain conditions for $\mathcal{H}_\infty$ control.

Associated with this Riccati equation is a $2n \times 2n$ matrix:

$$H := \begin{pmatrix} A & R \\ -Q & -A^T \end{pmatrix}. \tag{2.2}$$

A matrix of this form is called a Hamiltonian matrix. Note that $\Lambda(H)$ (the spectrum of H) is symmetric about the imaginary axis. The matrix H in (2.2) will be used to obtain the solutions to the equation (2.1).

2.4.1 Solvability conditions for Riccati equations

The following theorem gives a way of constructing solutions to (2.1) in terms of invariant subspaces of H.

Theorem 2.1. *Let $\mathcal{V} \subset \mathbb{C}^{2n}$ be an n-dimensional invariant subspace of H, and let $X_1, X_2 \in \mathbb{C}^{n \times n}$ be two complex matrices such that*

$$\mathcal{V} = \mathrm{Im}\begin{pmatrix} X_1 \\ X_2 \end{pmatrix}.$$

If X_1 is invertible, then $X := X_2 X_1^{-1}$ is a solution to the Riccati equation (2.1) and $\Lambda(A + RX) = \Lambda(H|_\mathcal{V})$. Furthermore, the solution X is independent of a specific choice of bases of $\mathcal{V}$.

The converse of the theorem also holds.

Theorem 2.2. *If $X \in \mathbb{C}^{n \times n}$ is a solution to the Riccati equation (2.1), then there exist matrices $X_1, X_2 \in \mathbb{C}^{n \times n}$, with X_1 invertible, such that $X = X_2 X_1^{-1}$ and the columns of $\begin{pmatrix} X_1 \\ X_2 \end{pmatrix}$ form a basis of an n-dimensional invariant subspace of H.*

Assume H has no eigenvalues on the imaginary axis. Then it must have n eigenvalues in $\operatorname{Re} s < 0$ and n in $\operatorname{Re} s > 0$. Consider the two n-dimensional spectral subspaces $\mathcal{X}_-(H)$ and $\mathcal{X}_+(H)$, the former is the invariant subspace corresponding to eigenvalues in $\operatorname{Re} s < 0$, and the latter corresponds to eigenvalues in $\operatorname{Re} s > 0$. By finding a basis for $\mathcal{X}_-(H)$, stacking the basis vectors up to form a matrix, and partitioning the matrix, we get

$$\mathcal{X}_-(H) = \operatorname{Im}\begin{pmatrix} X_1 \\ X_2 \end{pmatrix},$$

where $X_1, X_2 \in \mathbb{C}^{n \times n}$. If X_1 is nonsingular or, equivalently, if the two subspaces

$$\mathcal{X}_-(H), \operatorname{Im}\begin{pmatrix} 0 \\ I \end{pmatrix} \tag{2.3}$$

are complementary, we can set $X := X_2 X_1^{-1}$. Then X is uniquely determined by H, i.e. $H \mapsto X$ is a function, which will be denoted by Ric. We will take the domain of Ric, denoted $dom(Ric)$, to consist of Hamiltonian matrices H with two properties: H has no eigenvalues on the imaginary axis and the two subspaces in (2.3) are complementary. This solution will be called the stabilizing solution. Thus, $X = Ric(H)$ and

$$Ric : dom(Ric) \subset \mathbb{R}^{2n \times 2n} \mapsto \mathbb{R}^{n \times n}.$$

Theorem 2.3. *Suppose $H \in dom(Ric)$ and $X = Ric(H)$. Then*

(i) X is real symmetric;

(ii) X satisfies the algebraic Riccati equation

$$A^T X + XA + XRX + Q = 0;$$

(iii) $A + RX$ is stable.

The following theorem gives the necessary and sufficient conditions for the existence of a unique stabilizing solution of (2.1) under certain restrictions on the matrix R.

Theorem 2.4. *Suppose H has no imaginary eigenvalues and R is either positive semi-definite or negative semi-definite. Then $H \in dom(Ric)$ if and only if (A, R) is stabilizable.*

Theorem 2.5. *Suppose H has the form*

$$H := \begin{pmatrix} A & -BB^T \\ -C^T C & -A^T \end{pmatrix}$$

2.4. Algebraic Riccati equations

then $H \in \text{dom(Ric)}$ if and only if (A, B) is stabilizable and (A, C) has no unobservable modes on the imaginary axis. Furthermore, $X \in \text{Ric}(H) \geq 0$ if $H \in \text{dom(Ric)}$, and $\text{Ker}(X) = \{0\}$ if and only if (A, C) has no stable unobservable modes.

Corollary 2.1. *Suppose that (A, B) is stabilizable and (A, C) is detectable. Then the Riccati equation*

$$A^T X + XA - XBB^T X + C^T C = 0$$

has a unique positive semidefinite solution. Moreover, the solution is stabilizing.

2.4.2 Discrete Riccati equations

Let a matrix $S \in \mathbb{R}^{2n \times 2n}$ be partitioned into four $n \times n$ blocks as

$$S := \begin{pmatrix} S_{11} & S_{12} \\ S_{21} & S_{22} \end{pmatrix} \quad (2.4)$$

and let $J = \begin{pmatrix} 0 & -I \\ I & 0 \end{pmatrix} \in \mathbb{R}^{2n \times 2n}$, then S is called *simplectic* if $J^{-1} S^T J = S^{-1}$. A simplectic matrix has no eigenvalues at the origin, and, furthermore, it is easy to see that if λ is an eigenvalue of a simplectic matrix S, then $\bar{\lambda}, 1/\lambda$, and $1/\bar{\lambda}$ are also eigenvalues of S.

Let A, Q, and G be real $n \times n$ matrices with Q and G symmetric and A nonsingular. A discrete Riccati equation has the following form of

$$A^T X A - X - A^T X G (I + XG)^{-1} X A + Q = 0.$$

Define a $2n \times 2n$ matrix

$$S := \begin{pmatrix} A + G(A^T)^{-1} Q & -G(A^T)^{-1} \\ -(A^T)^{-1} Q & (A^T)^{-1} \end{pmatrix}. \quad (2.5)$$

Then S is a simplectic matrix. Assume that S has no eigenvalues on the unit circle. Then it must have n eigenvalues in $|z| < 1$ and n in $|z| > 1$. Consider the two n-dimensional spectral subspaces $\mathcal{X}_-(S)$ and $\mathcal{X}_+(S)$: the former is the invariant subspace corresponding to eigenvalues in $|z| < 1$, and the latter corresponds to eigenvalues in $|z| > 1$. After finding a basis for $\mathcal{X}_-(S)$, stacking the basis vectors up to form a matrix, and partitioning the matrix, we get

$$\mathcal{X}_-(S) = \text{Im} \begin{pmatrix} T_1 \\ T_2 \end{pmatrix},$$

where $T_1, T_2 \in \mathbb{R}^{n \times n}$. If T_1 is nonsingular or, equivalently, if the two subspaces

$$\mathcal{X}_-(S), \operatorname{Im} \begin{pmatrix} 0 \\ I \end{pmatrix}$$

are complementary, we can set $X := T_2 T_1^{-1}$. Then X is uniquely determined by S.

Definition 2.2. The domain of Ric, denoted by $dom(Ric)$, consists of all $2n \times 2n$ simplectic matrices S such that S has no eigenvalues on the unit circle and the two subspaces $\mathcal{X}_-(S)$ and $\operatorname{Im} \begin{pmatrix} 0 \\ I \end{pmatrix}$ are complementary.

Theorem 2.6. Let S be defined in (2.5) and suppose $S \in dom(Ric)$ and $X = Ric(S)$. Then

(i) X is unique and symmetric;
(ii) $I + XG$ is invertible and X satisfies the algebraic Riccati equation

$$A^T X A - X - A^T X G (I + XG)^{-1} X A + Q = 0;$$

(iii) $A - G(I + XG)^{-1} X A = (I + GX)^{-1} A$ is stable.

2.5 Bounded real lemma

For a system G with the following realization

$$G = \left(\begin{array}{c|c} A & B \\ \hline C & D \end{array} \right)$$

the bounded real lemma gives the equivalent matrix characterization of the stability of A and the $\mathcal{H}_\infty$ norm constraint $\|G\|_\infty < \gamma$.

Theorem 2.7 (Bounded real lemma). Let $\gamma > 0, G \in RH_\infty$, and

$$H := \begin{pmatrix} A + BR^{-1}D^T C & BR^{-1}B^T \\ -C^T(I + DR^{-1}D^T)C & -(A + BR^{-1}D^T C)^T \end{pmatrix}$$

where $R = \gamma^2 I - D^T D$. Then the following conditions are equivalent:

(i) $\|G\|_\infty < \gamma$;
(ii) there exists an $X > 0$ such that

$$\begin{pmatrix} XA + A^T X & XB & C^T \\ B^T X & -\gamma I & D^T \\ C & D & -\gamma I \end{pmatrix} < 0; \tag{2.6}$$

2.5. Bounded real lemma

(iii) $\bar{\sigma}(D) < \gamma$ and H has no eigenvalues on the imaginary axis;
(iv) $\bar{\sigma}(D) < \gamma$ and $H \in \text{dom}(\text{Ric})$;
(v) $\bar{\sigma}(D) < \gamma$ and $H \in \text{dom}(\text{Ric})$ and $\text{Ric}(H) \geq 0$ ($\text{Ric}(H) > 0$ if (A, C) is observable);
(vi) $\bar{\sigma}(D) < \gamma$ and there exists an $X \geq 0$ such that

$$X(A + BR^{-1}D^TC) + (A + BR^{-1}D^TC)^TX + XBR^{-1}B^TX$$
$$+C^T(I + DR^{-1}D^T)C = 0$$

and $A + BR^{-1}D^TC + BR^{-1}B^TX$ has no eigenvalues on the imaginary axis;
(vii) $\bar{\sigma}(D) < \gamma$ and there exists an $X > 0$ such that

$$X(A + BR^{-1}D^TC) + (A + BR^{-1}D^TC)^TX + XBR^{-1}B^TX$$
$$+C^T(I + DR^{-1}D^T)C < 0.$$

The following theorem gives the nonstrict version of the bounded real lemma.

Theorem 2.8 (Nonstric bounded real lemma). *(i) If there exists $X \geq 0$ such that*

$$\begin{pmatrix} XA + A^TX & XB & C^T \\ B^TX & -\gamma I & D^T \\ C & D & -\gamma I \end{pmatrix} \leq 0,$$

then $G(s)$ has no poles on the imaginary axis and $\|G\|_\infty \leq \gamma$;
(ii) If A is stable, $\|G\|_\infty \leq \gamma$ and $\|D\| < \gamma$, then there exists $X \geq 0$ such that

$$\begin{pmatrix} XA + A^TX & XB & C^T \\ B^TX & -\gamma I & D^T \\ C & D & -\gamma I \end{pmatrix} \leq 0.$$

Explicit formulae for discrete-time H_∞ controllers are obtained by similar manipulations on the discrete-time version of the bounded real lemma.

Theorem 2.9 (Discrete bounded real lemma). *The following two statements are equivalent:*

(i) A is stable and $\|C(sI - A)^{-1}B + D\|_\infty < \gamma$;
(ii) there exists a symmetric matrix $P > 0$ such that

$$\begin{pmatrix} A^TPA - P & A^TPB & C^T \\ B^TPA & B^TPB - \gamma^2 I & D^T \\ C & D & -I \end{pmatrix} < 0.$$

2.6 Small gain theorem

The small gain theorem is an important result for stability test of a nominally stable system under unstructured uncertainties. Essentially, it states that if a feedback loop consists of stable systems and the loop gain product is less than unity, then the feedback loop is internally stable. There are various versions of the small gain theorem in the literature. For more general versions, readers are referred to [Desoer and Vidyasagar (1975)], [Mareels and Hill (1992)],[Jiang et al. (1994)].

Consider the feedback system shown in Figure 2.5 with $M(s)$ a stable $p \times q$ transfer matrix.

Definition 2.3. A feedback system is said to be well-posed if all closed-loop transfer matrices are well-defined and proper.

Theorem 2.10 (Small gain theorem). *Suppose $M \in \mathcal{RH}_\infty$ and let $\gamma > 0$, then the interconnected system shown in Figure 2.5 is well-posed and internally stable for all $\Delta \in \mathcal{RH}_\infty$ with*

(i) $\|\Delta\|_\infty \leq \dfrac{1}{\gamma}$ if and only if $\|M\|_\infty < \gamma$;

(ii) $\|\Delta\|_\infty < \dfrac{1}{\gamma}$ if and only if $\|M\|_\infty \leq \gamma$.

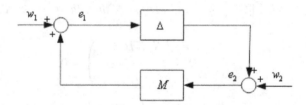

Fig. 2.5 Feedback loop for the small gain theorem.

The above theorem still holds when Δ and M are infinite dimensional.

Corollary 2.2. *The following statements are equivalent:*

(i) the system is well-posed and internally stable for all $\Delta \in \mathcal{H}_\infty$ with $\|\Delta\|_\infty < 1/\gamma$;

(ii) the system is well-posed and internally stable for all $\Delta \in \mathcal{RH}_\infty$ with $\|\Delta\|_\infty < 1/\gamma$;

(iii) the system is well-posed and internally stable for all $\Delta \in \mathbb{C}^{q \times p}$ with $\|\Delta\| < 1/\gamma$;

(iv) $\|M\|_\infty \leq \gamma$.

In [Desoer and Vidyasagar (1975)], it is shown that the small gain condition is sufficient to guarantee internal stability even if Δ is a nonlinear and time-varying stable operator with an appropriately defined stability notion.

2.7 LMI approach to H_∞ control

2.7.1 *Continuous-time $\mathcal{H}_\infty$ control*

Consider the system described by the block diagram shown in Figure 2.6 where the plant P and controller K are assumed to be real rational and

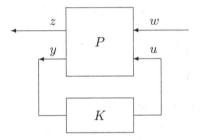

Fig. 2.6 The closed-loop system.

proper. It will be assumed that state-space models of P and K are available and that their realizations are assumed to be stabilizable and detectable. A controller is said to be admissible if it internally stabilizes the system. A standard optimal $\mathcal{H}_\infty$ control problem is: Find all admissible controllers K such that $\|T_{zw}\|_\infty$ is minimized, where T_{zw} is the closed loop transfer function from w to z. Alternatively, we can specify a certain value γ and find a stabilizing controller K such that $\|T_{zw}\|_\infty < \gamma$. This is known as the suboptimal $\mathcal{H}_\infty$ control problem, and γ is called the prescribed performance.

Suppose the plant P has the following state-space form by

$$\begin{aligned}
\dot{x} &= Ax + B_1 w + B_2 u \\
z &= C_1 x + D_{11} w + D_{12} u \\
y &= C_2 x + D_{21} w
\end{aligned} \qquad (2.7)$$

where x is the state of the plant, w is the exogenous input, u is the control input, z is the controlled output, y is the measurement output, and $A \in \mathbb{R}^{n \times n}, B_1 \in \mathbb{R}^{n \times m_1}, B_2 \in \mathbb{R}^{n \times m_2}, C_1 \in \mathbb{R}^{p_1 \times n}, D_{11} \in \mathbb{R}^{p_1 \times m_1}, D_{12} \in \mathbb{R}^{p_1 \times m_2}, C_2 \in \mathbb{R}^{p_2 \times n}, D_{21} \in \mathbb{R}^{p_2 \times m_1}$. Suppose the controller K has the following realization

$$K(s) := C_k(sI - A_k)^{-1} B_k + D_k, A_k \in \mathbb{R}^{n \times n}$$

A realization of the closed-loop transfer function from w to z is given by

$$T_{zw}(s) := C_{cl}(sI - A_{cl})^{-1} B_{cl} + D_{cl},$$

where

$$A_{cl} = \begin{pmatrix} A + B_2 D_k C_2 & B_2 C_k \\ B_k C_2 & A_k \end{pmatrix}, \quad B_{cl} = \begin{pmatrix} B_1 + B_2 D_k D_{21} \\ B_k D_{21} \end{pmatrix}, \quad (2.8)$$

$$C_{cl} = \begin{pmatrix} C_1 + D_{12} D_k C_2 & D_{12} C_k \end{pmatrix}, \quad D_{cl} = D_{11} + D_{12} D_k D_{21}.$$

According to the bounded real lemma, internal stability and the $\|T_{zw}\|_\infty < \gamma$ are jointly equivalent to the existence of $X_{cl} > 0$ of dimensions $2n \times 2n$ such that

$$\begin{pmatrix} X_{cl} A_{cl} + A_{cl}^T X_{cl} & X_{cl} B_{cl} & C_{cl}^T \\ B_{cl}^T X_{cl} & -\gamma I & D_{cl}^T \\ C_{cl} & D_{cl} & -\gamma I \end{pmatrix} < 0. \quad (2.9)$$

The above inequality can be written as

$$\hat{B} K \hat{C} + (\hat{B} K \hat{C})^T + Q < 0, \quad (2.10)$$

where

$$\left(\begin{array}{c|c} \hat{B} & Q \\ \hline * & \hat{C} \end{array}\right) = \left(\begin{array}{c|ccc} X_{cl} \hat{B}_2 & X_{cl} \hat{A} + \hat{A}^T X_{cl} & X_{cl} \hat{B}_1 & \hat{C}_1^T \\ 0 & \hat{B}_1^T X_{cl} & -\gamma I & \hat{D}_{11}^T \\ \hat{D}_{12} & \hat{C}_1 & \hat{D}_{11} & -\gamma I \\ \hline * & \hat{C}_2 & \hat{D}_{21} & 0 \end{array}\right),$$

$$\left(\begin{array}{c|c|c} \hat{A} & \hat{B}_1 & \hat{B}_2 \\ \hline \hat{C}_1 & \hat{D}_{11} & \hat{D}_{12} \\ \hline \hat{C}_2 & \hat{D}_{21} & K^T \end{array}\right) = \left(\begin{array}{cc|c|cc} A & 0 & B_1 & B_2 & 0 \\ 0 & 0 & 0 & 0 & I \\ \hline C_1 & 0 & D_{11} & D_{12} & 0 \\ C_2 & 0 & D_{21} & D_k^T & B_k^T \\ 0 & I & 0 & C_k^T & A_k^T \end{array}\right).$$

Partitioning X_{cl} and X_{cl}^{-1} as

$$X_{cl} \triangleq \begin{pmatrix} S & N \\ N^T & * \end{pmatrix}, X_{cl}^{-1} \triangleq \begin{pmatrix} R & M \\ M^T & * \end{pmatrix}, R, S, M, N \in \mathbb{R}^{n \times n}. \quad (2.11)$$

2.7. LMI approach to H_∞ control

Using the projection lemma of Chapter 1, (2.10) can be reduced to an LMI problem by elimination of the controller matrices. The following theorem establishes LMI-based existence conditions for a full order suboptimal $\mathcal{H}_\infty$ controller.

Theorem 2.11. *Consider a proper continuous-time plant P of order n and realization (2.7) and let $\mathcal{N}_{12}$ and $\mathcal{N}_{21}$ denote orthonormal bases of the null spaces of (B_2^T, D_{12}^T) and (C_2, D_{21}) respectively. Then there exists an admissible controller K such that $\|T_{zw}\|_\infty < \gamma$ if and only if there exist two symmetric matrices $R, S \in \mathbb{R}^{n \times n}$ satisfying the following system of LMIs:*

$$\begin{pmatrix} \mathcal{N}_{12} & 0 \\ 0 & I \end{pmatrix}^T \begin{pmatrix} AR + RA^T & RC_1^T & B_1 \\ C_1 R & -\gamma I & D_{11} \\ B_1^T & D_{11}^T & -\gamma I \end{pmatrix} \begin{pmatrix} \mathcal{N}_{12} & 0 \\ 0 & I \end{pmatrix} < 0, \quad (2.12)$$

$$\begin{pmatrix} \mathcal{N}_{21} & 0 \\ 0 & I \end{pmatrix}^T \begin{pmatrix} A^T S + SA & SB_1 & C_1^T \\ B_1^T S & -\gamma I & D_{11}^T \\ C_1 & D_{11} & -\gamma I \end{pmatrix} \begin{pmatrix} \mathcal{N}_{21} & 0 \\ 0 & I \end{pmatrix} < 0, \quad (2.13)$$

$$\begin{pmatrix} R & I \\ I & S \end{pmatrix} > 0. \quad (2.14)$$

Computing solutions (R, S) of the LMI system (2.12)-(2.14) is a convex optimization problem, which can be implemented by efficient polynomial-time algorithms. For any feasible solutions R and S, a set of full-order suboptimal controllers as can be obtained as follows. First compute two invertible matrices $M, N \in \mathbb{R}^{n \times n}$ such that

$$MN^T = I - RS.$$

The closed-loop matrix variable X_{cl} is then uniquely determined by (2.11). Specifically, X_{cl} is the unique solution of the linear equation

$$X_{cl} \begin{pmatrix} R & I \\ M^T & 0 \end{pmatrix} = \begin{pmatrix} I & S \\ 0 & N^T \end{pmatrix}$$

Once X_{cl} is determined, an adequate full-order controller is any solution of the controller LMI (2.10). For the parameterized form of the set of the controller one can refer to [Iwasaki and Skelton (1994)].

Remark 2.1. Theorem 2.11 gives LMI-based existence conditions for a full-order suboptimal $\mathcal{H}_\infty$ controller. It should be noted that for any $\mathcal{H}_\infty$ controller of order $n_k > n$, one can always construct a controller of order

$n_k = n$ which achieves the same $\mathcal{H}_\infty$ norm bound. When the controller order n_k is not fixed *a priori*, inequality (2.14) in the theorem should be written in the nonstrict form $\begin{pmatrix} R & I \\ I & S \end{pmatrix} \geq 0$. Adding another condition $\mathrm{rank}(R - S^{-1}) = n_k$, one can obtain the existence condition for a fixed order $\mathcal{H}_\infty$ controller.

The following corollary gives LMI formulations for standard optimal $\mathcal{H}_\infty$ control problem with the $\mathcal{H}_\infty$ performance that is directly optimized by solving a set of LMIs.

Corollary 2.3. *There exists an admissible controller K such that $\|T_{zw}\|_\infty$ is minimum if and only if there exist two symmetric matrices $R, S \in \mathbb{R}^{n \times n}$ satisfying the following optimization problem of LMI:*

$$\min \gamma.$$
$$s.t. (2.12), (2.13)(2.14)$$

Reference [Masubuchi *et al.* (1998)] presents another LMI-based approach for $\mathcal{H}_\infty$ controller synthesis problem.

Suppose the order of the controller $n_k \geq n$. Denote by $\mathcal{P}(n_k)$ the set of parameters $\mathbf{p} = \{P_f, P_g, P_h, W_f, W_g, W_h, L\}$, $P_f, P_g \in \mathbb{R}^{n \times n}$, $P_h \in \mathbb{R}^{(n_k-n) \times (n_k-n)}$, $W_f \in \mathbb{R}^{m_2 \times n_k}$, $W_g \in \mathbb{R}^{n_k \times p_2}$, $W_h \in \mathbb{R}^{m_2 \times p_2}$, $L \in \mathbb{R}^{n_k \times n_c}$ with P_f and P_g satisfying

$$\begin{pmatrix} P_f & I \\ I & P_g \end{pmatrix} > 0, \quad P_h > 0.$$

We also denote

$$L = \begin{pmatrix} L_{11} & L_{12} \\ L_{21} & L_{22} \end{pmatrix}, \quad W_f = \begin{pmatrix} W_{f1} & W_{f2} \end{pmatrix}, \quad W_g = \begin{pmatrix} W_{g1} \\ W_{g2} \end{pmatrix},$$

where $L_{11} \in \mathbb{R}^{n \times n}, W_{f1} \in \mathbb{R}^{m_2 \times n}, W_{g1} \in \mathbb{R}^{n \times p_2}$.

The parameter set $\mathcal{P}(n_c)$ is an open convex subset of $\mathbb{R}^{N(n_k)}$, where $N(n_k) := n(n+1) + (n_c - n)(n_k - n + 1)/2 + (n_k + m_2)(n_k + p_2)$. Define the following matrix-valued affine functions on $\mathcal{P}(n_k)$:

2.7. LMI approach to H_∞ control

$$M_p(\mathbf{p}) \triangleq \begin{pmatrix} P_f & I & 0 \\ I & P_g & 0 \\ 0 & 0 & P_h \end{pmatrix},$$

$$M_A(\mathbf{p}) \triangleq \begin{pmatrix} AP_f + B_2 W_{f1} & A + B_2 W_h C_1 & B_2 W_{f2} \\ L_{11} & P_g A + W_{g1} C_1 & L_{12} \\ L_{21} & W_{g2} C_1 & L_{22} \end{pmatrix},$$

$$M_C(\mathbf{p}) \triangleq \begin{pmatrix} C_2 P_f & C_2 \end{pmatrix},$$

$$M_B(\mathbf{p}) \triangleq \begin{pmatrix} B_1 + B_2 W_h D_{11} \\ P_g B_1 + W_{g1} D_{11} \\ W_{g2} D_{11} \end{pmatrix},$$

$$M_D(\mathbf{p}) \triangleq D_{21},$$

$$\Phi^*(\mathbf{p}) \triangleq M_p(\mathbf{p}) \oplus \begin{pmatrix} -M_A(\mathbf{p}) - M_A^T(\mathbf{p}) & M_B(\mathbf{p}) & M_C^T(\mathbf{p}) \\ M_B^T(\mathbf{p}) & \gamma I & -M_D^T(\mathbf{p}) \\ M_C(\mathbf{p}) & -M_D(\mathbf{p}) & \gamma I \end{pmatrix}.$$

(2.15)

where $\oplus$ means the direct sum of matrices, i.e,

$$A_1 \oplus \cdots \oplus A_n = \begin{pmatrix} A_1 & & \\ & \ddots & \\ & & A_n \end{pmatrix}.$$

The following theorem shows a convex parametrization of admissible controllers to the suboptimal $\mathcal{H}_\infty$ synthesis problem.

Theorem 2.12. *For any $n_k \geq n$, if there exists $\mathbf{p} \in \mathcal{P}_{n_k}$ such that*

$$\Phi^*(\mathbf{p}) > 0 \qquad (2.16)$$

then there exists an admissible controller K such that $\|T_{zw}\|_\infty < \gamma$. Further, the state space realization of the controller can be derived by

$$\left(\begin{array}{c|c} D_k & C_k \\ \hline B_k & A_k \end{array}\right) = \begin{pmatrix} I & 0 & 0 \\ \hline B_2 & -P_g^{-1} & 0 \\ \mathbf{0} & \mathbf{0} & I \end{pmatrix} \begin{pmatrix} W_h & W_{f1} & W_{f2} \\ \hline W_{g1} & L_{11} - P_g A_t P_f & L_{12} \\ W_{g2} & L_{21} & L_{22} \end{pmatrix}$$

$$\times \begin{pmatrix} I & -C_1 P_f S_f^{-1} & 0 \\ \hline 0 & S_f^{-1} & 0 \\ 0 & 0 & P_h^{-1} \end{pmatrix};$$

(2.17)

where $S_f := P_f - P_g^{-1} > 0$.

In reference [Masubuchi et al. (1998)], the equivalence of (2.9) and (2.16) is proved. One may observe that the above method is a unified approach to linear controller synthesis that employs various LMI conditions to represent control specifications. By defining a comprehensive class of LMIs, which contain the LMIs that have been conventionally used in linear control design, a convex parametrization of all controllers as well as the existence conditions can be obtained.

2.7.2 Discrete-time $\mathcal{H}_\infty$ control

The counterpart of the solvability conditions of Theorem 2.11 for discrete-time systems are given by the following theorem.

Theorem 2.13. *Consider a proper discrete-time plant $P(z)$ of order n with state-space equations:*

$$\begin{aligned} x_{k+1} &= Ax_k + B_1 w_k + B_2 u_k, \\ z_k &= C_1 x_k + D_{11} w_k + D_{12} u_k, \\ y_k &= C_2 x_k + D_{21} w_k. \end{aligned} \quad (2.18)$$

Let $\mathcal{N}_{12}$ and $\mathcal{N}_{21}$ denote orthonormal bases of the null spaces of (B_2^T, D_{12}^T) and $(C_2, D_{21},)$, respectively. The discrete-time suboptimal H_∞ problem of performance γ is solvable if and only if there exist two symmetric matrices $R, S \in \mathbb{R}^{n \times n}$ satisfying the following system of LMIs:

$$\begin{pmatrix} \mathcal{N}_{12} & 0 \\ 0 & I \end{pmatrix}^T \begin{pmatrix} ARA^T - R & ARC_1^T & B_1 \\ C_1 RA^T & -\gamma I + C_1 RC_1^T & D_{11} \\ B_1^T & D_{11}^T & -\gamma I \end{pmatrix} \begin{pmatrix} \mathcal{N}_{12} & 0 \\ 0 & I \end{pmatrix} < 0, \quad (2.19)$$

$$\begin{pmatrix} \mathcal{N}_{21} & 0 \\ 0 & I \end{pmatrix}^T \begin{pmatrix} A^T SA - S & A^T SB_1 & C_1^T \\ B_1^T SA & -\gamma I + B_1^T SB_1 & D_{11}^T \\ C_1 & D_{11} & -\gamma I \end{pmatrix} \begin{pmatrix} \mathcal{N}_{21} & 0, \\ 0 & I \end{pmatrix} < 0, \quad (2.20)$$

$$\begin{pmatrix} R & I \\ I & S \end{pmatrix} > 0. \quad (2.21)$$

Let (R, S) be any solution of this system of LMIs and define the notation

$$\Delta = \Delta^T := \begin{pmatrix} -R & 0 & A + B_2 D_k C_2 & B_1 + B_2 D_k D_{21} \\ * & -\gamma I & C_1 + D_{12} D_k C_2 & D_{11} + D_{12} D_k D_{21} \\ * & * & -S & 0 \\ * & * & 0 & -\gamma I \end{pmatrix}. \quad (2.22)$$

Then, a full-order γ-suboptimal controller $K(z) = D_k + C_k(zI - A_k)^{-1} B_k$ can be obtained as follows:

(a) compute any solution D_k of $\Delta > 0$ with Δ given by (2.22);
(b) compute the least-norm solutions $\begin{pmatrix} \Theta_B \\ * \end{pmatrix}$ and $\begin{pmatrix} \Theta_C \\ * \end{pmatrix}$ of the linear equations

$$\begin{pmatrix} 0 & 0\ 0\ C_2\ D_{21} \\ \hline 0 & \\ 0 & \\ C_2^T & -\Delta \\ D_{21}^T & \end{pmatrix} \begin{pmatrix} \Theta_B \\ \hline * \end{pmatrix} = - \begin{pmatrix} 0 \\ -I \\ 0 \\ A^T S \\ B_1^T S \end{pmatrix}, \quad (2.23)$$

$$\begin{pmatrix} 0 & B_2^T\ D_{12}^T\ 0\ 0 \\ \hline B_2 & \\ D_{12} & \\ 0 & -\Delta \\ 0 & \end{pmatrix} \begin{pmatrix} \Theta_C \\ \hline * \end{pmatrix} = - \begin{pmatrix} 0 \\ AR \\ C_1 R \\ -I \\ 0 \end{pmatrix}; \quad (2.24)$$

(c) factorize $I - RS$ as MN^T, with M and N invertible, and compute A_k, B_k and C_k by solving

$$NB_k = -SB_2 D_k + \Theta_B^T,$$
$$C_k M^T = -D_K C_2 R + \Theta_C,$$
$$-NA_K M^T = SB_2 \Theta_C + \Theta_B^T C_2 R + S(A - B_2 D_k C_2)R$$
$$+ \begin{pmatrix} -I \\ 0 \\ A^T S + C_2^T \Theta_B \\ B_1^T S + D_{21}^T \Theta_B \end{pmatrix}^T \Delta^{-1} \begin{pmatrix} AR + B_2 \Theta_C \\ C_1 R + D_{12} \Theta_C \\ -I \\ 0 \end{pmatrix}.$$

2.8 Notes and references

The basic concept of function spaces can be found in functional analysis textbooks or books on $\mathcal{H}_\infty$ control theory, such as [Zhou et al. (1996)]. The Bisection $\mathcal{H}_\infty$ algorithm is first developed in [Boyd et al. (1988)] and a more efficient norm computational algorithm is presented in [Bruinsma and Steinbuch (1990)]. The notion of LFT is first introduced in [Doyle et al. (1991)]. Basic principles and some of its properties are presented in [Zhou et al. (1996); Zhou and Doyle (1998)]. Solvability conditions for algebraic Riccati equations are given by [Huang (1964); Martensson (1971); Scherer (1990)]. Bounded real lemma for continuous-time linear time-invariant systems can be found in [Zhou and Khargonekar (1988); Sampei et al. (1990);

Scherer (1990)]. The discrete version of bounded real lemma can be seen in [de Souza and Xie (1992)]. [Zhang et al. (2008)] builds new bounded real lemma for discrete-time singular systems. Small gain theorem was first proposed in [Zames (1966)] and developed in [Anderson (1972); Desoer and Vidyasagar (1975); Dragan (1993); Green and Limebeer (1995)]. LMI approach to standard optimal $\mathcal{H}_\infty$ control and suboptimal $\mathcal{H}_\infty$ control are presented in [Iwasaki and Skelton (1994); Gahinet and Apkarian (1994); Gahinet (1996); Jia (2007); Feng et al. (1995); Shen (1996); Zhou et al. (1996); Zhou and Doyle (1998); Masubuchi et al. (1998); de Oliveira et al. (2002c)] and the references therein. Nonlinear H_∞ control theory can be found in [Hong and Cheng (2005)] and the references therein.

Chapter 3

Analysis and Control of Positive Real Systems

The notion of positive realness plays a very important role in systems and control theory. The concept of positive real functions originally arose in the context of circuit theory, where the transfer function of a passive network (passive in the sense that no energy is generated in the network, e.g., a network consisting of only inductors, resistors, and capacitors) is rational and positive real. This chapter discusses analysis and control problems for positive realness. Section 3.1 introduces the concepts of positive real systems and the relationships between different definitions of positive realness. Section 3.2 discusses the positive real lemma. Sections 3.3 discusses LMI method for strictly positive real (SPR) controller design. Section 3.4 gives the relationship between SPR control and strictly bounded real (SBR) control. Section 3.5 discusses multiplier design problems, and Section 3.6 concludes the chapter.

3.1 Positive real systems

Consider a finite dimensional linear system described by

$$\begin{cases} \dot{x} = Ax + Bu, \\ y = Cx + Du, \end{cases} \quad (3.1)$$

where $A \in \mathbb{R}^{n \times n}$, $B \in \mathbb{R}^{n \times m}$, $C \in \mathbb{R}^{m \times n}$ and $D \in \mathbb{R}^{m \times m}$. The transfer function from u to y is $G(s) = C(sI - A)^{-1}B + D$.

In the literature, one can find several different definitions of a positive real system. The reader may consult [Khalil (2002); Lozano et al. (2000); Sun et al. (1994); Huang (2003)] and references therein for more details. Throughout this book, we use the following definition [Khalil (2002); Huang (2003)].

Definition 3.1. System (3.1) or $G(s)$ is called positive real (PR) if $G(s)$ satisfies:

(i) poles of all elements of $G(s)$ are in $\mathbf{Re}\{s\} \leq 0$,
(ii) for all real w for which jw is not a pole of any element of $G(s)$, $G(jw) + G^*(jw) \geq 0$, and
(iii) any imaginary pole jw of any element of $G(s)$ is a simple pole and the residue matrix $\lim_{s \to jw}(s - jw)G(s)$ is positive semidefinite.

We can also give another definition for positive realness (PR).

Definition 3.2. System (3.1) or $G(s)$ is called positive real (PR) if $G(s)$ satisfies:

(i) poles of all elements of $G(s)$ are in $\mathbf{Re}\{s\} \leq 0$,
(ii) for all real s with $\mathbf{Re}\{s\} > 0$, $G(s) + G^T(s^*) \geq 0$.

Based on the definition of PR, we can give a definition for strict positive realness (SPR).

Definition 3.3. System (3.1) or $G(s)$ is called strictly positive real (SPR) if $G(s - \epsilon)$ is PR for some $\epsilon > 0$.

The basic difference between PR and SPR transfer functions is that PR transfer functions may tolerate poles on the imaginary axis, while SPR transfer functions cannot. In fact, the above two definitions for PR are equivalent.

Lemma 3.1. *Definition 3.1 holds if and only if Definition 3.2 holds.*

The following well-known property of analytic functions, so-called maximum modulus theorem [Titchmarsh (1952)], is needed for the proof of this lemma.

Maximum modulus theorem If $\varphi(s)$ is defined and continuous on a closed-bounded set Ω and analytic on the interior of Ω, then $|\varphi(s)|$ cannot attain the maximum in the interior of Ω unless $\varphi(s)$ is a constant.

This theorem shows that $|\varphi(s)|$ can only achieve its maximum on the boundary Γ of Ω, i.e.,

$$|\varphi(s)| \leq M, \quad \forall s \in \Omega,$$

if $|\varphi(s)| \leq M, \quad \forall s \in \Gamma$.

3.1. Positive real systems

Proof. (Proof of Lemma 3.1) Suppose that Definition 3.1 holds. Let Γ be a simple closed curve shown as in Fig. 3.1 which is composed of a semicircular arc with radius R and the corresponding imaginary axis part if $G(s)$ has no poles on the imaginary axis, otherwise the imaginary axis part near a pole jw_i of $G(s)$ is replaced by a semicircular arc of arbitrarily small radius ρ. Clearly, the region Ω enclosed by Γ is in the right half complex plane. By (ii) and (iii) of Definition 3.1, poles of $G(s)$ on the imaginary axis are simple poles, so $G(s)$ can be written as

$$G(s) = \frac{K_0}{s} + \sum_{i=1}^{l} \left(\frac{K_i}{s - jw_i} + \frac{\overline{K}_i}{s + jw_i} \right) + G_1(s),$$

where $\overline{K}_i$ is the conjugate of K_i and $G_1(s)$ is analytic in the region $\mathbf{Re}(s) \geq 0$. In order to prove the condition (ii) of Definition 3.2, for any s with $\mathbf{Re}(s) > 0$, take suitable scalars R and ρ such that $s \in \Omega$. For any complex vector $x \in \mathbb{C}^m$, let

$$\varphi(s) = x^*(G(s) + G^*(s))x = \varphi_1(s) + \varphi_1^*(s) = 2\mathbf{Re}(\varphi_1(s)),$$

where $\varphi_1(s) = x^* G(s) x$ is analytic in $\Gamma \cup \Omega$. Further, let

$$\psi(s) = e^{-\varphi_1(s)}.$$

Then, $\psi(s)$ is also analytic in $\Gamma \cup \Omega$ and

$$|\psi(s)| = e^{-\mathbf{Re}(\varphi_1(s))}.$$

Therefore, the values of $|\psi(s)|$ on Γ can be divided into three cases:

(i) if jw is not a pole of $G(s)$, then $\mathbf{Re}(\varphi_1(jw)) \geq 0$, so $|\psi(s)| \leq 1$;
(ii) on the small semicircular arc near a pole jw_i of $G(s)$,

$$\varphi_1(s) \approx \frac{x^* K_i x}{s - jw_i} = \frac{x^* K_i x}{\rho} e^{-j\theta}, \quad -\frac{\pi}{2} < \theta < \frac{\pi}{2},$$

since ρ is sufficiently small, then by condition (iii) of Definition 3.1, $\mathbf{Re}(\varphi_1(jw)) \geq 0$, so $|\psi(s)| \leq 1$ on each small semicircular arc corresponding to each pole jw_i of $G(s)$;
(iii) on the large semicircular arc with radius R, $s = Re^{j\theta}$, $-\frac{\pi}{2} < \theta < \frac{\pi}{2}$, at this time

$$\varphi_1(s) \approx x^* G_1(\infty) x \geq 0,$$

so $|\psi(s)| \leq 1$.

By the Maximum Modulus Theorem,

$$|\psi(s)| \leq 1, \quad \forall s, \mathbf{Re}(s) > 0,$$

or

$$x^*(G(s) + G^*(s))x \geq 0, \quad \forall s, \mathbf{Re}(s) > 0, \forall x \in \mathbb{C}^m.$$

Therefore, the conditions of Definition 3.2 hold.

On the other hand, suppose Definition 3.2 holds. The condition (ii) of Definition 3.1 holds directly from Definition 3.2. In addition, suppose jw_i is a pole of order p of any element of $G(s)$. Then, for any m-dimensional complex vector x, the values taken by $x^*G(s)x$ on a semicircular arc of arbitrarily small radius ρ, centered at jw_i, are

$$x^*G(s)x \approx x^*K_i x \rho^{-p} e^{-jp\theta}, \quad -\frac{\pi}{2} \leq \theta \leq \frac{\pi}{2}.$$

Therefore,

$$\rho^p \mathbf{Re}\{x^*G(s)x\} \approx \mathbf{Re}\{x^*K_i x\} \cos p\theta + \mathbf{Im}\{x^*K_i x\} \sin p\theta.$$

For $p > 1$, this expression could have either sign, while condition (ii) of Definition 3.2 implies that it is nonnegative. Hence, p must be limited to one. For $p = 1$, choosing θ as $-\frac{\pi}{2}$, 0, and $\frac{\pi}{2}$ shows that $\mathbf{Im}\{x^*K_i x\} = 0$ and $\mathbf{Re}\{x^*K_i x\} \geq 0$. Hence K_i is positive semidefinite Hermitian. The conditions of Definition 3.1 follow. □

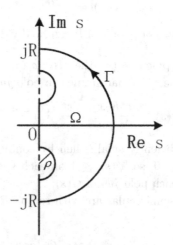

Fig. 3.1 The region $\Omega \cup \Gamma$.

3.1. Positive real systems

From the above definition, one can see that a system is PR if and only if its controllable and observable part is PR. When system (3.1) is an SISO system, the second condition of Definition 3.1 holds when the Nyquist plot of $G(jw)$ lies in the closed right-half complex plane. This is a condition that can be satisfied only if the relative degree of $G(s)$ is zero or one [Khalil (2002)].

The next lemma gives ϵ-free equivalent characterizations of SPR transfer functions for two special cases, which are more computationally tractable [Corless and Shorten (2009); Khalil (2002)].

Lemma 3.2. *Let $G(s)$ be an $m \times m$ proper rational transfer function matrix, and suppose $\det[G(s) + G^T(-s)]$ is not identically zero. Then, if $G(\infty) + G^T(\infty) > 0$, $G(s)$ is SPR if and only if the following two conditions (i) and (ii) hold; if $G(\infty) + G^T(\infty) = 0$, $G(s)$ is SPR if and only if the following three conditions (i)-(iii) hold:*

(i) $G(s)$ is Hurzitz, i.e., poles of all elements of $G(s)$ have negative real parts,
(ii) $G(jw) + G^*(jw) > 0$, $\forall w \in \mathbb{R}$, and
(iii) $\lim_{w \to \infty} w^2(G(jw) + G^*(jw)) > 0$.

Proof. See the following section. □

For the general case of $G(\infty) + G^T(\infty) \geq 0$, ϵ-free equivalent characterizations of SPR was given in [Khalil (2002)] with substituting Lemma 3.2 (iii) by the following condition:

(iv) either $G(\infty) + G^T(\infty) > 0$ or $G(\infty) + G^T(\infty) \geq 0$ and $\lim_{w \to \infty} w^2 M^T(G(jw) + G^*(jw))M > 0$ for any $m \times (m-q)$ full-rank matrix M such that $M^T(G(\infty) + G^T(\infty))M = 0$, where $q = rank(G(\infty) + G^T(\infty))$,

see Lemma 6.1 of [Khalil (2002)]. However, the authors [Corless and Shorten (2009)] found this condition incomplete by illustrating a counter example. And a new side condition was given:

$$\lim_{w \to \infty} w^{2\rho} \det(G(jw) + G^*(jw)) \neq 0, \qquad (3.2)$$

where $\rho = m - q$, m is the order of square matrix $G(\infty)$ and q is the rank of $G(\infty) + G^T(\infty)$. Conditions Lemma 3.2 (i), (ii) and (3.2) equivalently characterize SPR transfer functions, see [Corless and Shorten (2009)] for details. Actually, when $G(\infty) + G^T(\infty) > 0$ or $G(\infty) + G^T(\infty) = 0$, condition (3.2) is equivalent to the condition (iv) given as above, i.e. the condition given in [Khalil (2002)]. However, when $0 < \rho < m$, condition (3.2) is not necessarily the same as (iv), see the following example.

Example 3.1. The transfer function $G(s) = \dfrac{1}{s}$ is PR since it has no poles in $\mathbf{Re}\{s\} > 0$, has a simple pole at $s = 0$ whose residue is 1, and

$$\mathbf{Re}\{G(jw)\} = \mathbf{Re}\left\{\frac{1}{jw}\right\} = 0, \ \forall w \neq 0.$$

It is not SPR since $\dfrac{1}{s-\epsilon}$ has a pole in $\mathbf{Re}\{s\} > 0$ for any $\epsilon > 0$.

The transfer function $G(s) = \dfrac{1}{s+a}$ with $a > 0$ is PR since it has no poles in $\mathbf{Re}\{s\} \geq 0$, and

$$\mathbf{Re}\{G(jw)\} = \frac{a}{w^2 + a^2} > 0, \ \forall w \in \mathbb{R}.$$

Since this is so for every $a > 0$, we see that for any $\epsilon \in (0, a)$ the transfer function $G(s - \epsilon) = \dfrac{1}{s+a-\epsilon}$ will be PR. Hence, $G(s) = \dfrac{1}{s+a}$ is SPR. The same conclusion can also be drawn from Lemma 3.2.

Consider the 2×2 transfer function matrix

$$G(s) = \frac{1}{s+1}\begin{pmatrix} 1 & 1 \\ 1 & 1 \end{pmatrix}.$$

We cannot apply Lemma 3.2 because $\det[G(s) + G^T(-s)] \equiv 0 \ \forall s$. However, $G(s)$ is SPR, as can be seen by checking the conditions of Definition 3.3. Note that, for $0 < \epsilon < 1$, the poles of the elements of $G(s - \epsilon)$ are in $\mathbf{Re}\{s\} < 0$ and

$$G(jw - \epsilon) + G^T(-jw - \epsilon) = \frac{2(1-\epsilon)}{w^2 + (1-\epsilon)^2}\begin{pmatrix} 1 & 1 \\ 1 & 1 \end{pmatrix} \geq 0, \ \forall w \in \mathbb{R}.$$

Similarly, it can be seen that the 2×2 transfer function matrix

$$G(s) = \frac{1}{s+1}\begin{pmatrix} 2s+1 & 1 \\ -1 & 2s+1 \end{pmatrix}$$

is SPR. This time, however, $\det[G(s) + G^T(-s)]$ is not identically zero, and we can apply Lemma 3.2 to arrive at the same conclusion by noting that $G(\infty) + G^T(\infty) > 0$ and

$$G(jw) + G^T(-jw) = \frac{2}{w^2 + 1}\begin{pmatrix} 2w^2+1 & -jw \\ jw & 2w^2+1 \end{pmatrix} > 0, \ \forall w \in \mathbb{R}.$$

Finally, the 2 × 2 transfer function matrix [Corless and Shorten (2009)]

$$G(s) = \begin{pmatrix} 1 & \dfrac{1}{s+1} \\ \dfrac{-1}{s+1} & \dfrac{1}{s+1} \end{pmatrix}$$

has

$$G(\infty) + G^T(\infty) = \begin{pmatrix} 1 & 0 \\ 0 & 0 \end{pmatrix} \geq 0.$$

It can be verified that

$$G(jw - \epsilon) + G^T(-jw - \epsilon) = 2 \begin{pmatrix} 1 & \dfrac{-jw}{(1-\epsilon)^2 + w^2} \\ \dfrac{jw}{(1-\epsilon)^2 + w^2} & \dfrac{1-\epsilon}{(1-\epsilon)^2 + w^2} \end{pmatrix}$$

which has determinant

$$\delta(w, \epsilon) = 4 \frac{(1-\epsilon)^3 - \epsilon w^2}{((1-\epsilon)^2 + w^2)^2}.$$

Consider any $\epsilon > 0$, it can be seen that $\delta(w, \epsilon) < 0$ for large w, and hence $G(jw - \epsilon) + G^T(-jw - \epsilon)$ is not positive definite. We conclude that $G(s)$ is not SPR. However, by taking $M^T = (0, 1)$, it can be verified that

$$\lim_{w \to \infty} w^2 M^T (G(jw) + G^T(-jw)) M = 4 > 0,$$

i.e., the above condition (iv) holds. On the other hand, Lemma 3.2 (i) and (ii) obviously hold. So, $G(s)$ is SPR by Lemma 6.1 of [Khalil (2002)], which contradicts the above conclusion. Therefore, condition (iv) is incomplete for SPR transfer functions.

Remark 3.1. Note that there are different definitions of PR systems which are not necessarily equivalent to each other. For example, given a transfer function $G(s)$, the following definitions of PR and SPR are used in [Sun et al. (1994)]. $G(s)$ is called to be PR if $G(s)$ is analytic in $\mathbf{Re}\{s\} > 0$ and satisfies $G(s) + G^T(s^*) \geq 0$ for $\mathbf{Re}\{s\} > 0$ (Definition 3.2); $G(s)$ is called to be SPR if $G(s)$ is analytic in $\mathbf{Re}\{s\} \geq 0$ and satisfies $G(jw) + G^*(jw) > 0$, $\forall w \in (-\infty, +\infty)$; $G(s)$ is called to be extended strictly positive real (ESPR) if $G(s)$ is SPR and $G(j\infty) + G^*(j\infty) > 0$. By Lemma 3.1, definitions of PR in [Sun et al. (1994)] and Definition 3.1 are equivalent to each other. By Lemma 3.2, obviously when $G(j\infty) + G^*(j\infty) > 0$, SPR in Definition 3.3 is equivalent to ESPR defined above. In addition, SPR in Definition

3.3 implies SPR in [Sun et al. (1994)], but the converse is not true. For example, let $G(s) = \dfrac{s+1}{s^2+s+1}$. Then for $\forall w \in (-\infty, +\infty)$,

$$\mathbf{Re}\{G(jw)\} = \mathbf{Re}\left\{\frac{1+jw}{1-w^2+jw}\right\} = \frac{1}{(1-w^2)^2+w^2} > 0.$$

Hence $G(s)$ is SPR according to the definition in [Sun et al. (1994)]. But on the other hand, $\lim_{w\to\infty} w^2(G(jw) + G^*(jw)) = 0$, so $G(s)$ does not satisfy the conditions of Lemma 3.2. Various conditions for SPR were also discussed in [Wen (1988)]. Throughout this book, we will use Definitions 3.1 and 3.3, unless otherwise indicated.

3.2 Positive real lemma

The properties of PR and SPR can be characterized by algebraic equations, which are known as positive real lemma. In this section, we discuss this remarkable lemma for PR and SPR. The spectral factorization theory is needed for the proof of this lemma [Huang (2003); Khalil (2002); Youla (1961)].

Lemma 3.3. *Let $G(s) = C(sI - A)^{-1}B + D$ be an $m \times m$ transfer function matrix and A be Hurwitz stable. If $G(s)$ is PR, then there exists an $r \times m$ proper rational Hurwitz transfer function matrix $V(s)$ such that*

$$G(s) + G^T(-s) = V^T(-s)V(s),$$

where r is the normal rank of $G(s) + G^T(-s)$, that is, the maximum rank over the field of rational functions of s. Furthermore, $\operatorname{rank}(V(s)) = r$ for $\mathbf{Re}(s) > 0$.

Besides the spectral factorization theory, the following lemma is useful for the proof of positive real lemma [Huang (2003)].

Lemma 3.4. *Given an $m \times m$ proper real rational transfer function matrix $G(s) = C(sI - A)^{-1}B$, and $\left(\begin{array}{c|c} A & B \\ \hline C & 0 \end{array}\right)$ is its minimal realization. Suppose all the eigenvalues of A are on the imaginary axis and $G(s)$ is PR. Then, there exists a positive definite matrix $P = P^T > 0$ such that*

$$PA + A^T P = 0, \quad PB = C^T. \tag{3.3}$$

3.2. Positive real lemma

Proof. By the conditions of the lemma, $G(s)$ can be decomposed as

$$G(s) = \frac{K_0}{s} + \sum_{i=1}^{q} \left(\frac{K_i}{s - jw_i} + \frac{\overline{K}_i}{s + jw_i} \right),$$

where all K_0, K_i and $\overline{K}_i$ are positive semidefinite Hermitian.

First, we prove the conclusion with the transfer function matrix $G_0(s) = \frac{K_0}{s}$. Since K_0 is positive semidefinite, we decompose K_0 as $K_0 = \sum_{i=1}^{r} \lambda_i u_i u_i^T$, $u_i \in \mathbb{R}^m$, and $u_i^T u_i = 1$, $\lambda_i > 0$. Let

$$A_0 = 0 \in \mathbb{R}^{m \times m}, \ C_0 = (\sqrt{\lambda_1} u_1, \cdots, \sqrt{\lambda_r} u_r), \ B_0 = C_0^T,$$

and $P_0 = I_r$. Clearly, $C_0 (sI - A_0)^{-1} B_0 = \frac{K_0}{s}$, and A_0, B_0, C_0 and P_0 satisfy the equation (3.3).

Second, for the transfer function $G_i(s) = \frac{K_i}{s - jw_i} + \frac{\overline{K}_i}{s + jw_i}$, K_i can be decomposed as

$$K_i = \sum_{l=1}^{r_i} x_l x_l^*, \quad x_l \in \mathbb{C}^m,$$

since it is positive semidefinite Hermitian. So $G_i(s)$ can be decomposed as

$$G_i(s) = \sum_{l=1}^{r_i} \left(\frac{x_l x_l^*}{s - jw_i} + \frac{\overline{x}_l x_l^T}{s + jw_i} \right).$$

Let

$$A_{w_i}^l = \begin{pmatrix} 0 & -w_i \\ w_i & 0 \end{pmatrix}, \ B_{w_i}^l = \begin{pmatrix} \frac{x_l^T + \overline{x}_l^T}{\sqrt{2}} \\ \frac{\overline{x}_l^T - x_l^T}{j\sqrt{2}} \end{pmatrix}, \ C_{w_i}^l = (B_{w_i}^l)^T, \ P_{w_i}^l = I_2,$$

$l = 1, \cdots, r_i$, and

$$A_i = \text{diag}(A_{w_i}^1, \cdots, A_{w_i}^{r_i}), \ B_i = ((B_{w_i}^1)^T, \cdots, (B_{w_i}^{r_i})^T)^T, \ C_i = B_i^T,$$

and $P_i = \text{diag}(P_{w_i}^1, \cdots, P_{w_i}^{r_i})$. Then one can get that $G_i(s) = C_i(sI - A_i)^{-1} B_i$ and A_i, B_i, C_i and P_i satisfy the equation (3.3).

Based on the above discussions, for a general transfer function $G(s)$ as given above, we can view $G(s)$ as a parallel connection of $G_0(s)$ and $G_i(s), i = 1, \cdots, q$. Then using the above realizations for $G_0(s)$ and $G_i(s)$, take a realization of $G(s)$ as $A = \text{diag}(A_0, A_1, \cdots, A_q)$, $C = (C_0, C_1, \cdots, C_q)$, $B = C^T$ and $P = \text{diag}(P_0, P_1, \cdots, P_q)$. Obviously, this realization is a minimal realization since all poles of $G(s)$ are of order 1. And A, B, C and P satisfy the equation (3.3). □

With the above preliminaries, we give the following well-known positive real lemma.

Lemma 3.5. (Positive real lemma) Let $G(s) = C(sI - A)^{-1}B + D$ be an $m \times m$ transfer function matrix, where (A, B) is controllable and (A, C) is observable. Then, $G(s)$ is PR if and only if there exist matrices $P = P^T > 0$, L and W such that
$$PA + A^T P = -L^T L,$$
$$PB = C^T - L^T W, \qquad (3.4)$$
$$W^T W = D + D^T.$$

Proof. Sufficiency: Suppose that there exist $P = P^T > 0$, L and W satisfying (3.4). Using $V(x) = x^T P x$ as a Lyapunov function for $\dot{x} = Ax$, (3.4) shows that A has no eigenvalues in $\mathbf{Re}\{s\} > 0$. Let $\Phi(s) = (sI - A)^{-1}$. We have
$$G(s) + G^T(s^*) = D + D^T + C\Phi(s)B + B^T \Phi^T(s^*)C^T.$$
Substituting $B^T P + W^T L$ and $W^T W$ for C and $D + D^T$ respectively by (3.4) gives
$$\begin{aligned} G(s) + G^T(s^*) &= W^T W + (B^T P + W^T L)\Phi(s)B \\ &\quad + B^T \Phi^T(s^*)(PB + L^T W) \\ &= W^T W + W^T L\Phi(s)B + B^T \Phi^T(s^*) L^T W \\ &\quad + B^T \Phi^T(s^*)[(s + s^*)P - A^T P - PA]\Phi(s)B. \end{aligned}$$
Using (3.4) yields
$$\begin{aligned} G(s) + G^T(s^*) &= [W^T + B^T \Phi^T(s^*) L^T][W + L\Phi(s)B] \\ &\quad + (s + s^*)B^T \Phi^T(s^*) P\Phi(s)B \end{aligned} \qquad (3.5)$$
which shows that for all s in $\mathbf{Re}\{s\} > 0$, $G(s) + G^T(s^*) \geq 0$. It follows that $G(s)$ is PR by Definition 3.2.

Necessity: First, we prove the special case that $G(s)$ is Hurwitz. Suppose that $G(s)$ is positive real and Hurwitz, and recall that $\left(\begin{array}{c|c} A & B \\ \hline C & D \end{array}\right)$ is a minimal realization of $G(s)$. By Lemma 3.3, there exists an $r \times m$ proper rational Hurwitz transfer function matrix $V(s)$ such that
$$G(s) + G^T(-s) = V^T(-s)V(s).$$
Let $\left(\begin{array}{c|c} F & K \\ \hline H & J \end{array}\right)$ be a minimal realization of $V(s)$. The matrix F is Hurwitz since $V(s)$ is Hurwitz. It can be easily seen that $\left(\begin{array}{c|c} -F^T & H^T \\ \hline -K^T & J^T \end{array}\right)$ is a minimal

3.2. Positive real lemma

realization of $V^T(-s)$. Therefore,

$$\left(\begin{array}{c|c}\mathcal{A}_1 & \mathcal{B}_1 \\ \hline \mathcal{C}_1 & \mathcal{D}_1\end{array}\right) = \left(\begin{array}{cc|c} F & 0 & K \\ H^T H & -F^T & H^T J \\ \hline J^T H & -K^T & J^T J \end{array}\right)$$

is a realization of the cascade connection $V^T(-s)V(s)$. By checking the controllability and observability and by using the property that $\text{rank}(V(s)) = r$ for $\mathbf{Re}\{s\} > 0$, it can be seen that this realization is minimal [Khalil (2002)].

Consider the Lyapunov equation

$$XF + F^T X = -H^T H.$$

Because the pair (F, H) is observable, there is a unique positive definite solution X. Using the similarity transformation $\begin{pmatrix} I & 0 \\ X & I \end{pmatrix}$, we obtain the following alternative minimal realization of $V^T(-s)V(s)$:

$$\left(\begin{array}{c|c}\mathcal{A}_2 & \mathcal{B}_2 \\ \hline \mathcal{C}_2 & \mathcal{D}_2\end{array}\right) = \left(\begin{array}{cc|c} F & 0 & K \\ 0 & -F^T & XK + H^T J \\ \hline J^T H + K^T X & -K^T & J^T J \end{array}\right).$$

On the other hand, $\left(\begin{array}{c|c} -A^T & C^T \\ \hline -B^T & D^T \end{array}\right)$ is a minimal realization of $G^T(-s)$. Therefore,

$$\left(\begin{array}{c|c}\mathcal{A}_3 & \mathcal{B}_3 \\ \hline \mathcal{C}_3 & \mathcal{D}_3\end{array}\right) = \left(\begin{array}{cc|c} A & 0 & B \\ 0 & -A^T & C^T \\ \hline C & -B^T & D + D^T \end{array}\right)$$

is a realization of the parallel connection $G(s) + G^T(-s)$. Since the eigenvalues of A are in the open left-half plane, while the eigenvalues of $-A^T$ are in the open right-half plane, it can be easily seen that this realization is minimal. Thus, due to the spectral factorization equation, $\left(\begin{array}{c|c}\mathcal{A}_2 & \mathcal{B}_2 \\ \hline \mathcal{C}_2 & \mathcal{D}_2\end{array}\right)$ and $\left(\begin{array}{c|c}\mathcal{A}_3 & \mathcal{B}_3 \\ \hline \mathcal{C}_3 & \mathcal{D}_3\end{array}\right)$ are equivalent minimal realizations of the same transfer function. Therefore, they have the same dimension and there is a nonsingular matrix T such that

$$\mathcal{A}_2 = T\mathcal{A}_3 T^{-1}, \; \mathcal{B}_2 = T\mathcal{B}_3, \; \mathcal{C}_2 = \mathcal{C}_3 T^{-1}, \; J^T J = D + D^T.$$

The matrix T must be a block diagonal matrix. To see this point, partition T compatibly as $\begin{pmatrix} T_{11} & T_{12} \\ T_{21} & T_{22} \end{pmatrix}$. Then, the matrix T_{12} satisfies the equation

$$FT_{12} + T_{12}A^T = 0.$$

Premultiplying by $\exp(Ft)$ and postmultiplying by $\exp(A^T t)$, we obtain
$$0 = \exp(Ft)[FT_{12} + T_{12}A^T]\exp(A^T t) = \frac{d}{dt}[\exp(Ft)T_{12}\exp(A^T t)].$$
Hence, $\exp(Ft)T_{12}\exp(A^T t)$ is constant for all $t \geq 0$. In particular, since $\exp(0) = I$,
$$T_{12} = \exp(Ft)T_{12}\exp(A^T t) \to 0, \quad \text{as} \quad t \to \infty.$$
Therefore, $T_{12} = 0$. Similarly, we can show that $T_{21} = 0$. Consequently, the matrix T_{11} is nonsingular and
$$F = T_{11}AT_{11}^{-1}, \ K = T_{11}B, \ J^T H + K^T X = CT_{11}^{-1}.$$
Define
$$P = T_{11}^T X T_{11}, \ L = HT_{11}, \ W = J.$$
It can be easily verified that P, L, and W satisfy the equations
$$PA + A^T P = -L^T L, \ PB = C^T - L^T W, \ W^T W = D + D^T,$$
which completes the proof when $G(s)$ is Hurwitz.

Next, if $G(s)$ has poles on the imaginary axis, decompose $G(s)$ as $G(s) = G_0(s) + G_1(s)$, where all poles of $G_0(s)$ are on the imaginary axis and $G_1(s)$ is Hurwitz. By Definition 3.1, it follows that $G_0(s)$ and $G_1(s)$ are PR, respectively, if $G(s)$ is PR. By Lemma 3.4 and the above proof for stable transfer functions, take minimal realizations of $G_0(s)$ and $G_1(s)$ as $\left(\begin{array}{c|c} A_0 & B_0 \\ \hline C_0 & 0 \end{array}\right)$ and $\left(\begin{array}{c|c} A_1 & B_1 \\ \hline C_1 & D_1 \end{array}\right)$, respectively. Then there exist matrices $P_0 > 0, P_1 > 0$ and L_1, W such that
$$P_0 A_0 + A_0^T P_0 = 0, \ P_0 B_0 = C_0^T$$
and
$$P_1 A_1 + A_1^T P_1 = -L_1^T L_1, \ P_1 B_1 = C_1^T - L_1^T W, \ W^T W = D_1 + D_1^T$$
hold. Finally, take $A = \text{diag}(A_0, A_1), C = (C_0, C_1), B = (B_0^T, B_1^T)^T, D = D_1, L = (0, L_1)$ and $P = \text{diag}(P_0, P_1)$. Clearly $G(s) = C(sI - A)^{-1}B + D$ and the corresponding equation (3.4) holds for A, B, C, D, L, W and P. □

Lemma 3.6. (**SPR version of positive real lemma**) *Let* $G(s) = C(sI - A)^{-1}B + D$ *be an* $m \times m$ *transfer function matrix where* (A, B) *is controllable and* (A, C) *is observable. Then,* $G(s)$ *is SPR if and only if there exist matrices* $P = P^T > 0$, L, *and* W, *and a positive constant* ϵ *such that*
$$PA + A^T P = -L^T L - \epsilon P,$$
$$PB = C^T - L^T W, \tag{3.6}$$
$$W^T W = D + D^T.$$

3.2. Positive real lemma

Proof. Suppose there exist $P = P^T > 0$, L, W, and $\epsilon > 0$ that satisfy (3.6). Set $\mu = \epsilon/2$ and recall that $G(s - \mu) = C(sI - \mu I - A)^{-1}B + D$. From (3.6), we have

$$P(A + \mu I) + (A + \mu I)^T P = -L^T L.$$

It follows from Lemma 3.5 that $G(s - \mu)$ is PR. Hence, $G(s)$ is SPR. On the other hand, suppose $G(s)$ is SPR. There exists $\mu > 0$ such that $G(s - \mu)$ is PR. It follows from Lemma 3.5 that there are matrices $P = P^T > 0$, L and W, which satisfy (3.4). Setting $\epsilon = 2\mu$ leads to (3.6). □

With Lemma 3.6, we can give a proof for Lemma 3.2.

Proof of Lemma 3.2

Sufficiency: Suppose the conditions of the lemma are satisfied. Since $G(s)$ is Hurwitz, there exist positive constants δ and μ^* such that poles of all elements of $G(s - \mu)$ have real parts less than $-\delta$, for all $\mu < \mu^*$. To show that $G(s)$ is SPR, it is sufficient to show that $G(jw - \mu) + G^T(-jw - \mu) \geq 0$, $\forall w \in \mathbb{R}$. Let (A, B, C, D) be a minimal realization of $G(s)$. Then,

$$\begin{aligned} G(s - \mu) &= D + C(sI - \mu I - A)^{-1}B \\ &= D + C(sI - A)^{-1}(sI - A)(sI - \mu I - A)^{-1}B \\ &= D + C(sI - A)^{-1}(\mu I + sI - \mu I - A)(sI - \mu I - A)^{-1}B \\ &= G(s) + \mu N(s) \end{aligned}$$
(3.7)

where $N(s) = C(sI - A)^{-1}(sI - \mu I - A)^{-1}B$. Since A and $A + \mu I$ are Hurwitz uniformly in μ, there is $k_0 > 0$ such that

$$\sigma_{\max}[N(jw) + N^T(-jw)] \leq k_0, \quad \forall w \in \mathbb{R}. \tag{3.8}$$

Moreover, $\lim_{w \to \infty} w^2 N(jw)$ exists. Hence, there exists $k_1 > 0$ and $w_1 > 0$ such that

$$w^2 \sigma_{\max}[N(jw) + N^T(-jw)] \leq k_1, \quad \forall |w| \geq w_1. \tag{3.9}$$

If $G(\infty) + G^T(\infty) > 0$, there is $\sigma_0 > 0$ such that

$$\sigma_{\min}[G(jw) + G^T(-jw)] \geq \sigma_0, \quad \forall w \in \mathbb{R}.$$

From (3.7), (3.8) and the above inequality,

$$\sigma_{\min}[G(jw - \mu) + G^T(-jw - \mu)] \geq \sigma_0 - \mu k_0, \quad \forall w \in \mathbb{R}.$$

Choosing $\mu < \sigma_0/k_0$ ensures that $G(jw - \mu) + G^T(-jw - \mu) > 0$ for all $w \in \mathbb{R}$. If $G(\infty) + G^T(\infty) = 0$, the third condition of the lemma ensures that there exists $\sigma_1 > 0$ and $w_2 > 0$ such that

$$w^2 \sigma_{\min}[G(jw) + G^T(-jw)] \geq \sigma_1, \quad \forall |w| \geq w_2.$$

From (3.7), (3.9) and the above inequality,
$$w^2 \sigma_{\min}[G(jw - \mu) + G^T(-jw - \mu)] \geq \sigma_1 - \mu k_1, \quad \forall |w| \geq w_3,$$
where $w_3 = \max\{w_1, w_2\}$. On the compact frequency interval $[-w_3, w_3]$, we have
$$\sigma_{\min}[G(jw) + G^T(-jw)] \geq \sigma_2 > 0.$$
Hence, from (3.7), (3.8) and the above inequality,
$$\sigma_{\min}[G(jw - \mu) + G^T(-jw - \mu)] \geq \sigma_2 - \mu k_0, \quad \forall |w| \leq w_3.$$
Choosing $\mu < \min\{\sigma_1/k_1, \sigma_2/k_0\}$ ensures that $G(jw-\mu)+G^T(-jw-\mu) > 0$ for all $w \in \mathbb{R}$.

Necessity: Suppose $G(s)$ is SPR. There is $\mu > 0$ such that $G(s - \mu)$ is PR. It follows that $G(s)$ is Hurwitz and PR. Consequently,
$$G(jw) + G^T(-jw) \geq 0, \quad \forall w \in \mathbb{R}.$$
Therefore,
$$G(\infty) + G^T(\infty) \geq 0.$$
Let $\left(\begin{array}{c|c} A & B \\ \hline C & D \end{array}\right)$ be a minimal realization of $G(s)$. By Lemma 3.6, there exist P, L, W and ϵ that satisfy (3.6). Let $\Phi(s) = (sI - A)^{-1}$. We have
$$G(s) + G^T(-s) = D + D^T + C\Phi(s)B + B^T\Phi^T(-s)C^T.$$
Substitute for C and $D + D^T$ by using (3.6). Then,
$$G(s) + G^T(-s) = W^T W + (B^T P + W^T L)\Phi(s)B$$
$$+ B^T \Phi^T(-s)(PB + L^T W)$$
$$= W^T W + W^T L\Phi(s)B + B^T \Phi^T(-s)L^T W$$
$$+ B^T \Phi^T(-s)[(s + s^*)P - A^T P - PA]\Phi(s)B.$$
Using (3.6) yields
$$G(s)+G^T(-s) = [W^T + B^T\Phi^T(-s)L^T][W + L\Phi(s)B] + \epsilon B^T\Phi^T(-s)P\Phi(s)B.$$
From the last equation, it can be seen that $G(jw)+G^T(-jw) > 0, \forall w \in \mathbb{R}$, for if it were singular at some frequency w, there would exist $0 \neq x \in \mathbb{C}^m$ such that
$$x^*[G(jw) + G^T(-jw)]x = 0, \tag{3.10}$$
which implies
$$x^* B^T \Phi^T(-jw) P \Phi(jw) B x = 0.$$

3.2. Positive real lemma

Consequently, $Bx = 0$. Also (3.10) implies
$$x^*[W^T + B^T\Phi^T(-jw)L^T][W + L\Phi(jw)B]x = 0.$$
Since $Bx = 0$, the preceding equation implies $Wx = 0$. Hence,
$$x^*[G(s) + G^T(-s)] \equiv 0, \ \forall s,$$
which contradicts the assumption that $\det(G(s)+G^T(-s))$ is not identically zero. Now if $G(\infty) + G^T(\infty) > 0$, we are done. If $G(\infty) + G^T(\infty) = 0$, we have $D = 0, W = 0$ and
$$G(jw) + G^T(-jw) = B^T\Phi^T(-jw)(L^TL + \epsilon P)\Phi(jw)B.$$
Note that B must have full column rank; otherwise, there is $x \neq 0$ such that $Bx = 0$. It yields
$$x^T[G(jw) + G^T(-jw)]x = 0, \quad \forall w \in \mathbb{R},$$
which contradicts the positive definiteness of $G(jw) + G^T(-jw)$. Now
$$\lim_{w \to \infty} w^2[G(jw) + G^T(-jw)] = B^T(L^TL + \epsilon P)B.$$
The fact that B has full column rank implies that $B^T(L^TL + \epsilon P)B > 0$. □

Corollary 3.1. *Suppose A is Hurwitz, system (3.1) is strictly proper, i.e., $D = 0$, (A, B) is controllable and (A, C) is observable. Then, $G(s) = C(sI - A)^{-1}B$ is SPR if and only if there exists a matrix $P = P^T > 0$ such that*
$$PA + A^TP < 0, \quad PB - C^T = 0. \tag{3.11}$$
Further, $G(s)$ is PR if and only if there exists a matrix $P = P^T > 0$ such that
$$PA + A^TP \leq 0, \quad PB - C^T = 0. \tag{3.12}$$

Remark 3.2. With the assumptions in Corollary 3.1, if $\det(G(s)+G^T(-s))$ is not identically zero, by Lemma 3.2 $G(s)$ is strictly proper and SPR, i.e.,
$$G(jw) + G^*(jw) > 0, \ \forall w \in \mathbb{R}, \text{ and } \lim_{w \to \infty} w^2(G(jw) + G^*(jw)) > 0. \tag{3.13}$$
Then clearly (3.13) is equivalent to (3.11). In fact, if we only consider the property of $G(s)$ on the imaginary axis such as (3.13), then the Hurwitz stability of A is not necessary for the equivalence of (3.13) and (3.11) [Huang (2003); Leonov et al. (1996)]. Further for the PR case, the property of $G(s)$ on the imaginary axis is described as follows
$$G(jw) + G^*(jw) \geq 0, \ \forall w \in \mathbb{R} \text{ with } |jwI - A| \neq 0. \tag{3.14}$$
Then, (3.12) is equivalent to (3.14) even if A is not Hurwitz stable. This can actually be viewed as a generalized PR lemma for strictly proper transfer functions [Huang (2003); Leonov et al. (1996)].

Remark 3.3. Reconsider the transfer function $G(s) = \dfrac{s+1}{s^2+s+1}$ in Remark 3.1. Let a state-space realization of $G(s)$ be $\left(\begin{array}{c|c} A & B \\ \hline C & D \end{array}\right)$, where $A = \begin{pmatrix} 0 & 1 \\ -1 & -1 \end{pmatrix}$, $B = \begin{pmatrix} 0 \\ 1 \end{pmatrix}$, $C = (1, 1)$, $D = 0$. By Remark 3.1 we know $G(jw) + G^T(-jw) > 0$, $\forall w \in \mathbb{R}$. However, there is no a positive definite matrix P such that $PA + A^T P < 0$, $PB = C^T$. Hence, in the conditions of SPR(Lemma 3.2), the condition $\lim_{w \to \infty} w^2(G(jw) + G^*(jw)) > 0$ is important for strictly proper transfer functions.

The condition $PB = C^T$ was further studied in [Huang et al. (1999)] and all solutions to this equation was given.

Lemma 3.7. *Suppose B and C are full rank. Then there exists a matrix $P = P^T > 0$ such $PB = C^T$ if and only if $CB = B^T C^T > 0$. Further, if this inequality holds, all solutions of $PB = C^T$ are given by*

$$P = C^T(CB)^{-1}C + B^\perp X B^{\perp T}$$

where $B^\perp$ is a matrix that satisfies $B^{\perp T} B = 0$, $B^{\perp T} B^\perp = I$ and $[B, B^\perp]$ nonsingular, and X is an arbitrary positive definite matrix.

Combining with Corollary 1.4 and algebraic Riccati equation theory in Chapter 2, we can also give several equivalent conditions for SPR when $G(\infty) + G^T(\infty) > 0$.

Lemma 3.8. *Consider system (3.1) and define*

$$R_0(X) = A^T X + XA + (C - B^T X)^T (D + D^T)^{-1}(C - B^T X)$$

and

$$S_0(Y) = AY + YA^T + (B - YC^T)(D + D^T)^{-1}(B - YC^T)^T.$$

Suppose A is Hurwitz and $D + D^T > 0$. Then the following statements are equivalent:

(i) System (3.1) is SPR.
(ii) There exists a positive definite solution X such that

$$\begin{pmatrix} XA + A^T X & C^T - XB \\ C - B^T X & -D - D^T \end{pmatrix} < 0.$$

(iii) The algebraic Riccati inequality $R_0(X) < 0$ has a positive definite solution X.

3.2. Positive real lemma

(iv) The algebraic Riccati equation $R_0(X) = 0$ has a symmetric solution $X = X^T$ and $A - B(D + D^T)^{-1}C + B(D + D^T)^{-1}B^T X$ is Hurwitz stable.

(v) There exists a positive definite solution Y such that
$$\begin{pmatrix} YA^T + AY & B - YC^T \\ B^T - CY & -D - D^T \end{pmatrix} < 0.$$

(vi) The algebraic Riccati inequality $S_0(Y) < 0$ has a positive definite solution Y.

(vii) The algebraic Riccati equation $S_0(Y) = 0$ has a symmetric solution $Y = Y^T$ and $A - B(D + D^T)^{-1}C + YC^T(D + D^T)^{-1}C$ is Hurwitz stable.

Proof. See Corollary 1.4 for the equivalence between (i) and (ii). The equivalence between (ii) and (iii) is obvious by Schur complement. In the following, we prove the equivalence between (i) and (iv). If (iv) holds, let $R = D + D^T$ and
$$M(s) = \left[\begin{array}{c|c} A & B \\ \hline R^{-1}(C - B^T X) & I \end{array} \right].$$

Obviously,
$$M^{-1}(s) = \left[\begin{array}{c|c} A - BR^{-1}C + BR^{-1}B^T X & -B \\ \hline R^{-1}(C - B^T X) & I \end{array} \right].$$

Using the Riccati equation $R_0(X) = 0$, by simple algebraic computation we get
$$G(jw) + G^*(jw) = M^*(jw) R M(jw).$$

By stability of A and $A - BR^{-1}C + BR^{-1}B^T X$, $M(s)$ has no zeros and poles on the imaginary axis, i.e., $M(jw)$ is nonsingular for all $w \in \mathbb{R} \cup \{\infty\}$. Therefore, $G(jw) + G^*(jw) > 0$, $\forall w \in \mathbb{R} \cup \{\infty\}$, that is, $G(s)$ is SPR. On the other hand, if $G(s)$ is SPR, let
$$\Phi(s) = G(s) + G^T(-s) = \left(\begin{array}{cc|c} A & 0 & B \\ 0 & -A^T & C^T \\ \hline C & -B^T & R \end{array} \right)$$

and
$$H = \begin{pmatrix} A & 0 \\ 0 & -A^T \end{pmatrix} - \begin{pmatrix} B \\ C^T \end{pmatrix} R^{-1} \begin{pmatrix} C & -B^T \end{pmatrix}$$
$$= \begin{pmatrix} A - BR^{-1}C & BR^{-1}B^T \\ -C^T R^{-1} C & -(A - BR^{-1}C)^T \end{pmatrix}.$$

Then, the state space realization of $\Phi^{-1}(s)$ is

$$\Phi^{-1}(s) = \left(\begin{array}{c|c} H & \hat{B} \\ \hline \hat{C} & R^{-1} \end{array}\right),$$

where

$$\hat{B} = \begin{pmatrix} BR^{-1} \\ C^T R^{-1} \end{pmatrix}, \hat{C} = \begin{pmatrix} -R^{-1}C & R^{-1}B^T \end{pmatrix}.$$

By the property of SPR of $G(s)$, it follows that the Hamiltonian matrix H has no eigenvalues on the imaginary axis. Otherwise, if jw_0 is an eigenvalue of H. Since $\Phi(jw) > 0$ for all $w \in \mathbb{R}$, jw_0 must be an uncontrollable mode of $(H, \hat{B})$ or an unobservable mode of $(H, \hat{C})$. Suppose that it is an unobservable mode. Then there exists $x_0 \neq 0$ such that $Hx_0 = jw_0 x_0$ and $\hat{C}x_0 = 0$. Partitioning x_0 compatibly as $x_0 = (x_1^T, x_2^T)^T$, by the definition of H we have $(jw_0 - A)x_1 = 0$ and $(jw_0 + A^T)x_2 = 0$. Because of SPR of $G(s)$, A has no eigenvalues on the imaginary axis. Then it follows that $x_1 = 0$ and $x_2 = 0$. This is a contradiction with $x_0 \neq 0$. Similarly, we can prove that jw_0 is not an uncontrollable mode of $(H, \hat{B})$. Therefore, H has no eigenvalues on the imaginary axis. By the Riccati equation theory in Chapter 2 (Theorems 2.3 and 2.4), $R_0(X) = 0$ has a symmetric stabilizing solution.

In addition, by pre-multiplying and post-multiplying $T = \mathrm{diag}(-X^{-1}, I)$ and T^T on the two sides of the inequality in (i), we can get (v) equivalently with $X^{-1} = Y$. The other equivalences for (vi) and (vii) can be given similarly as discussed above. □

Remark 3.4. Combining with Lemma 3.8 or Corollary 1.4, we know that if $D + D^T > 0$, the equation (3.6) for SPR is equivalent to a single inequality in (ii) of Lemma 3.8 or

$$\begin{pmatrix} PA + A^T P & PB - C^T \\ B^T P - C & -D - D^T \end{pmatrix} < 0. \tag{3.15}$$

For PR case, the equation (3.4) is equivalent to a single inequality

$$\begin{pmatrix} PA + A^T P & PB - C^T \\ B^T P - C & -D - D^T \end{pmatrix} \leq 0. \tag{3.16}$$

However, we should note that neither the inequality (3.16) nor (3.15) can imply the condition (3.11). Therefore, the KYP lemma (Corollary 1.4) does not include the case of SPR when $G(s)$ is strictly proper (Corollary 3.1), or

$D \neq 0$ but singular. In addition, if we only consider the properties of $G(s)$ on the imaginary axis such as $G(jw) + G^*(jw) > 0$ or $G(jw) + G^*(jw) \geq 0$, the requirement on stability of A is not necessary for the equivalence between frequency and time domain conditions in Corollary 1.4. Also for the case of PR, A can have simple eigenvalues on the imaginary axis as discussed in Lemmas 3.4 and 3.5.

3.3 LMI approach to control of SPR

Consider a generalized plant P given by

$$\begin{aligned}\dot{x} &= Ax + B_1 w + B_2 u, \\ z &= C_1 x + D_{11} w + D_{12} u, \\ y &= C_2 x + D_{21} w + D_{22} u,\end{aligned} \quad (3.17)$$

where x is the state of the plant, w is the exogenous input, u is the control input, z is controlled output, y is measurement output, and $A \in \mathbb{R}^{n \times n}, B_1 \in \mathbb{R}^{n \times m_1}, B_2 \in \mathbb{R}^{n \times m_2}, C_1 \in \mathbb{R}^{p_1 \times n}, D_{11} \in \mathbb{R}^{p_1 \times m_1}, D_{12} \in \mathbb{R}^{p_1 \times m_2}, C_2 \in \mathbb{R}^{p_2 \times n}, D_{21} \in \mathbb{R}^{p_2 \times m_1}$, and $m_2 \leq p_1, p_2 \leq m_1, p_1 = m_1$. The system P can be denoted as

$$P = \begin{pmatrix} P_{11} & P_{12} \\ P_{21} & P_{22} \end{pmatrix} = \left(\begin{array}{c|cc} A & B_1 & B_2 \\ \hline C_1 & D_{11} & D_{12} \\ C_2 & D_{21} & D_{22} \end{array} \right). \quad (3.18)$$

Consider a controller K which is also a linear time-invariant system

$$\begin{aligned}\dot{x}_K &= A_K x_K + B_K y, \\ u &= C_K x_K + D_K y,\end{aligned} \quad (3.19)$$

where x_K is the state of the controller, y is the input to the controller, u is the output of the controller, and $A_K \in \mathbb{R}^{n_K \times n_K}, B_K \in \mathbb{R}^{n_K \times p_2}, C_K \in \mathbb{R}^{m_2 \times n_K}, D_K \in \mathbb{R}^{m_2 \times p_2}$.

The closed-loop system defined by P and K with input w and output z is shown in Fig. 2.6. Its transfer function matrix is equal to the lower fractional transformation $\mathcal{F}_\ell(P, K)$.

The closed-loop system given by Fig. 2.6 has a state space realization

$$\mathcal{F}_\ell(P, K) =$$
$$\left(\begin{array}{cc|c} A + B_2 D_K M C_2 & B_2(I + D_K M D_{22})C_K & B_1 + B_2 D_K M D_{21} \\ B_K M C_2 & A_K + B_K M D_{22} C_K & B_K M D_{21} \\ \hline C_1 + D_{12} D_K M C_2 & D_{12}(I + D_K M D_{22})C_K & D_{11} + D_{12} D_K M D_{21} \end{array} \right),$$
$$(3.20)$$

where $M = (I - D_{22}D_K)^{-1}$.

Definition 3.4. The closed-loop system defined by Fig. 2.6 is said to be internally stable if the matrix

$$\begin{pmatrix} A + B_2 D_K M C_2 & B_2(I + D_K M D_{22})C_K \\ B_K M C_2 & A_K + B_K M D_{22} C_K \end{pmatrix}$$

is stable.

In this section we make assumptions that the pair (A, B_2) is stabilizable and the pair (A, C_2) is detectable which are needed for the closed-loop internal stability.

The SPR control problem is stated as follows.

Given a system P, when does there exist a controller K such that the closed-loop system defined by Fig. 2.6 is internally stable and SPR? When the existence condition is satisfied, provide a formula for one such controller.

For simplicity, we discuss the problem of controller design with two cases. One is for $D_{22} = 0$ and the other is for $D_{22} \neq 0$.

Case I: $D_{22} = 0$.

In this case the LMI method for the SPR control problem is described as follows [Sun et al. (1994)].

Theorem 3.1. *There exists a strictly proper controller K for the system P such that the closed-loop system shown in Fig. 2.6 is internally stable and SPR with $D_{11} + D_{11}^T > 0$ if and only if there exist matrices $W_1 > 0, W_3 > 0$ and W_2, W_4 satisfying*

$$\begin{bmatrix} AW_1 + W_1 A^T + B_2 W_2 + W_2^T B_2^T & W_1 C_1^T + W_2^T D_{12}^T - B_1 \\ C_1 W_1 + D_{12} W_2 - B_1^T & -(D_{11} + D_{11}^T) \end{bmatrix} < 0, \quad (3.21)$$

$$\begin{bmatrix} W_3 A + A^T W_3 + W_4 C_2 + C_2^T W_4^T & W_3 B_1 + W_4 D_{21} - C_1^T \\ B_1^T W_3 + D_{21}^T W_4^T - C_1 & -(D_{11} + D_{11}^T) \end{bmatrix} < 0, \quad (3.22)$$

and the spectral radius $\rho(YX) < 1$, where $X = W_1^{-1}, Y = W_3^{-1}$.

Moreover, when these conditions are satisfied the matrices A_K, B_K and C_K are expressed as

$$\begin{aligned} A_K &= A + B_2 F + (I - YX)^{-1} L C_2 + \Delta, \\ B_K &= -(I - YX)^{-1} L, \quad C_K = F \end{aligned} \quad (3.23)$$

3.3. LMI approach to control of SPR

which achieves the closed-loop internal stability and SPR, where $F = W_2 W_1^{-1}, L = W_3^{-1} W_4$ and

$$\Delta = -(B_1 + (I - YX)^{-1}LD_{21})(D_{11} + D_{11}^T)^{-1}(C_1 - B_1^T X + D_{12}F)$$
$$-(I - YX)^{-1}YF^T(B_2^T X + D_{12}^T(D_{11} + D_{11}^T)^{-1}(C_1 - B_1^T X + D_{12}F))$$
$$+(I - YX)^{-1}YR_F,$$
$$R_F = (A + B_2F)^T X + X(A + B_2F)$$
$$+(C_1 + D_{12}F - B_1^T X)^T(D_{11} + D_{11}^T)^{-1}(C_1 + D_{12}F - B_1^T X).$$

If the states are available for measurement, i.e., $y = x$, then the state feedback controller K: $u = Fx$ can be designed. The corresponding result is stated as the following corollary.

Corollary 3.2. *For the system P, suppose that the pair (A, B_2) is stabilizable. Then there exists a state feedback controller K: $u = Fx$ such that the closed-loop system defined by P and K with the state space realization*

$$\mathcal{F}_\ell(P, K) = \left(\begin{array}{c|c} A + B_2F & B_1 \\ \hline C_1 + D_{12}F & D_{11} \end{array} \right)$$

is internally stable and SPR with $D_{11} + D_{11}^T > 0$ if and only if there exist matrices $W_1 > 0$ and W_2 satisfying (3.21).

Moreover, when such a solution exists the controller $u = Kx = W_2 W_1^{-1} x$ achieves the closed-loop internally stable and SPR.

Case II: $D_{22} \neq 0$.

We will convert the case of $D_{22} \neq 0$ into the first case by a loop shifting. Let

$$\overline{K} = K(I - D_{22}K)^{-1}, \bar{P} = \begin{pmatrix} P_{11} & P_{12} \\ P_{21} & \bar{P}_{22} \end{pmatrix} = \left(\begin{array}{c|cc} A & B_1 & B_2 \\ \hline C_1 & D_{11} & D_{12} \\ C_2 & D_{21} & 0 \end{array} \right). \quad (3.24)$$

Then from the definition of linear fractional transformation and $P_{22} = \overline{P}_{22} + D_{22}$ we have

$$\mathcal{F}_\ell(P, K) = P_{11} + P_{12}K(I - \overline{P}_{22}K - D_{22}K)^{-1}P_{21}$$
$$= P_{11} + P_{12}K(I - D_{22}K)^{-1} \left[I - \overline{P}_{22}K(I - D_{22}K)^{-1} \right]^{-1} P_{21} \quad (3.25)$$
$$= \mathcal{F}_\ell(\overline{P}, \overline{K}).$$

This means that there exists a controller K such that the closed-loop system $\mathcal{F}_\ell(P, K)$ defined by P and K is internally stable and SPR if and only if there exists a controller $\overline{K}$ such that the closed-loop $\mathcal{F}_\ell(\overline{P}, \overline{K})$ defined by $\overline{P}$ and $\overline{K}$ is internally stable and SPR. Note that the problem of designing a controller $\overline{K}$ such that $\mathcal{F}_\ell(\overline{P}, \overline{K})$ is SPR can be realized by the results of case I. Obviously, from (3.24) if a controller

$$\overline{K} = \left(\begin{array}{c|c} \overline{A}_K & \overline{B}_K \\ \hline \overline{C}_K & 0 \end{array} \right)$$

for system $\overline{P}$ is designed in terms of Theorem 3.1 then a controller K for the system P can be expressed as

$$K = (I + \overline{K}D_{22})^{-1}\overline{K} = \left(\begin{array}{c|c} \overline{A}_K - \overline{B}_K D_{22} \overline{C}_K & \overline{B}_K \\ \hline \overline{C}_K & 0 \end{array} \right). \qquad (3.26)$$

3.4 Relationship between SPR control and SBR control

First we introduce the definitions of bounded real (BR) and strictly bounded real (SBR) systems in the network theory [Belevitch (1970); Anderson and Vongpanitlerd (1973)].

Definition 3.5. A stable transfer matrix $S(s)$ satisfying

$$S^T(-j\omega)S(j\omega) \leq I, \quad \forall \omega \in \mathbb{R} \qquad (3.27)$$

is said to be bounded real. If strict inequality prevails in (3.27), then $S(s)$ is said to be strictly bounded real.

The condition (3.27) is equivalent to $\|S(s)\|_\infty \leq 1$. The PR and BR systems play important roles in analyzing the stability and robustness of systems. The relationship between PR and BR systems is connected by Cayley transform.

Let

$$S(s) = (G(s) - I)(G(s) + I)^{-1}. \qquad (3.28)$$

S is said to be Cayley transform of G.

The Cayley transform have the following elementary property [Belevitch (1970); Desoer and Vidyasagar (1975)].

Lemma 3.9. *For any $G(s)$ and $S(s)$ related by the Cayley transform (3.28), $G(s)$ is PR (SPR with $G(\infty) + G^T(\infty) > 0$) if and only if $S(s)$ is BR (SBR).*

From (3.28), $G(s) = (I - S(s))^{-1}(I + S(s)) = 2(I - S(s))^{-1} - I$, it is easy to show the following lemma.

3.4. Relationship between SPR control and SBR control

Lemma 3.10. *If*
$$S(s) = \left(\begin{array}{c|c} A_s & B_s \\ \hline C_s & D_s \end{array} \right), \quad (3.29)$$
then
$$G(s) = \left(\begin{array}{c|c} A_s + B_s(I - D_s)^{-1} C_s & B_s(I - D_s)^{-1} \\ \hline 2(I - D_s)^{-1} C_s & (I - D_s)^{-1}(I + D_s) \end{array} \right). \quad (3.30)$$
The Cayley transform preserves minimality, i.e., (3.29) is minimal realization of $S(s)$ if and only if (3.30) is minimal realization of $G(s)$.

Similarly, we can also establish the relationship between the SBR control and SPR control.

Consider a linear time invariant system:
$$\begin{pmatrix} z \\ y \end{pmatrix} = P \begin{pmatrix} w \\ u \end{pmatrix} \quad (3.31)$$
where P is defined by (3.18). Let K denote the controller. We have the following results.

Theorem 3.2. *Assume that $I - D_{11}$ is invertible and*
$$\det \left(I - D_{11} - D_{12}K(\infty)(I - D_{22}K(\infty))^{-1} D_{21} \right) \neq 0.$$
Then
$$(I - \mathcal{F}_\ell(P, K))^{-1}(I + \mathcal{F}_\ell(P, K)) = \mathcal{F}_\ell(\overline{P}, K), \quad (3.32)$$
where
$$\overline{P} = \begin{pmatrix} I + 2P_{11}(I - P_{11})^{-1} & 2(I - P_{11})^{-1} P_{12} \\ P_{21}(I - P_{11})^{-1} & P_{22} + P_{21}(I - P_{11})^{-1} P_{12} \end{pmatrix}. \quad (3.33)$$
Moreover, K stabilizes P and $\|\mathcal{F}_\ell(P, K)\|_\infty < 1$ if and only if the closed-loop system $\mathcal{F}_\ell(\overline{P}, K)$ defined by $\overline{P}$ and K is internally stable and SPR with $N + N^T - I > 0$, i.e., K is a SBR controller of the system P if and only if K is a SPR controller of the system $\overline{P}$, where
$$N = [I - D_{11} - D_{12}K(\infty)(I - D_{22}K(\infty))^{-1} D_{21}]^{-1}.$$

Proof. In terms of the equality $(I - \Delta)^{-1} = I + \Delta(I - \Delta)^{-1}$ and properties of the Redheffer star product one implies
$$(I - \mathcal{F}_\ell(P, K))^{-1}(I + \mathcal{F}_\ell(P, K)) = \mathcal{F}_\ell \left(\begin{pmatrix} I & 2I \\ I & I \end{pmatrix}, \mathcal{F}_\ell(P, K) \right)$$
$$= \mathcal{F}_\ell \left(S \left(\begin{pmatrix} I & 2I \\ I & I \end{pmatrix}, P \right), K \right) = \mathcal{F}_\ell(\overline{P}, K),$$

where
$$\overline{P} = S\left(\begin{pmatrix} I & 2I \\ I & I \end{pmatrix}, P\right)$$
which can be expressed as (3.33) by the definition of the Redheffer star product. This derives that (3.32) holds. Combining with Lemma 3.9, the results follow. □

The following lemma gives a state space representation of $\overline{P}$ in terms of (3.33) and direct algebra manipulation.

Lemma 3.11. *Assume that P has a state space representation (3.18). Then $\overline{P}$ defined in (3.33) has a state space realization*
$$\overline{P} = \left(\begin{array}{c|cc} \overline{A} & \overline{B}_1 & \overline{B}_2 \\ \hline \overline{C}_1 & \overline{D}_{11} & \overline{D}_{12} \\ \overline{C}_2 & \overline{D}_{21} & \overline{D}_{22} \end{array}\right),$$
where
$$\overline{A} = A + B_1(I - D_{11}^{-1})C_1, \ \overline{B}_1 = B_1(I - D_{11})^{-1},$$
$$\overline{B}_2 = B_2 + B_1(I - D_{11})^{-1}D_{12}, \ \overline{C}_1 = 2(I - D_{11})^{-1}C_1,$$
$$\overline{C}_2 = C_2 + D_{21}(I - D_{11})^{-1}C_1, \ \overline{D}_{11} = (I - D_{11})^{-1}(I + D_{11}), \quad (3.34)$$
$$\overline{D}_{12} = 2(I - D_{11})^{-1}D_{12}, \ \overline{D}_{21} = D_{21}(I - D_{11})^{-1},$$
$$\overline{D}_{22} = D_{22} + D_{21}(I - D_{11})^{-1}D_{12}.$$

Consider the system P defined by (3.17). Assume that the size of the exogenous input w is the same as the size of controlled output z, i.e., $m_1 = p_1$, and $m_2 \leq p_1, p_2 \leq m_1$. The state space realization of P is expressed as (3.18). According to Theorem 3.1 and Theorem 3.2, a different design method from that in chapter 2 for SBR control (H_∞ control with $\gamma = 1$) can be easily derived.

Theorem 3.3. *Assume that (A, B_2) is stabilizable, (A, C_2) is detectable and $\overline{D}_{11} + \overline{D}_{11}^T > 0$. Then there exists a strictly proper controller K such that the closed-loop system defined by Fig. 2.6 is internally stable and $||\mathcal{F}_\ell(P, K)||_\infty < 1$ if and only if there exist matrices $P_1 > 0, P_3 > 0$ and P_2, P_4 satisfying*
$$\begin{pmatrix} \overline{A}P_1 + P_1\overline{A}^T + \overline{B}_2P_2 + P_2^T\overline{B}_2^T & P_1\overline{C}_1^T + P_2^T\overline{D}_{12}^T - \overline{B}_1 \\ \overline{C}_1P_1 + \overline{D}_{12}P_2 - \overline{B}_1^T & -(\overline{D}_{11} + \overline{D}_{11}^T) \end{pmatrix} < 0,$$
$$\begin{pmatrix} P_3\overline{A} + \overline{A}^T P_3 + P_4\overline{C}_2 + \overline{C}_2^T P_4^T & P_3\overline{B}_1 + P_4\overline{D}_{21} - \overline{C}_1^T \\ \overline{B}_1^T P_3 + \overline{D}_{21}^T P_4^T - \overline{C}_1 & -(\overline{D}_{11} + \overline{D}_{11}^T) \end{pmatrix} < 0,$$

and the spectral radius $\rho(YX) < 1$, where $X = P_1^{-1}$ and $Y = P_3^{-1}$.
Moreover, when these conditions are satisfied a strictly proper controller K is expressed as

$$K = \left(\begin{array}{c|c} \overline{A}_K - \overline{B}_K \overline{D}_{22} \overline{C}_K & \overline{B}_K \\ \hline \overline{C}_K & 0 \end{array} \right),$$

where

$$\overline{A}_K = \overline{A} + \overline{B}_2 F + (I - YX)^{-1} L \overline{C}_2 + \overline{\Delta},$$
$$\overline{B}_K = -(I - YX)^{-1} L, \ \overline{C}_K = F, \ F = P_2 P_1^{-1}, \ L = P_3^{-1} P_4,$$
$$\overline{\Delta} = -\left[\overline{B}_1 + (I - YX)^{-1} L \overline{D}_{21} \right] (\overline{D}_{11} + \overline{D}_{11}^T)^{-1} (\overline{C}_1 - \overline{B}_1^T X + \overline{D}_{12} F)$$
$$-(I - YX)^{-1} Y F^T \left(\overline{B}_2^T X + \overline{D}_{12}^T (\overline{D}_{11} + \overline{D}_{11}^T)^{-1} (\overline{C}_1 - \overline{B}_1^T X + \overline{D}_{12} F) \right)$$
$$+(I - YX)^{-1} Y \overline{R}_F,$$
$$\overline{R}_F = (\overline{A} + \overline{B}_2 F)^T X + X(\overline{A} + \overline{B}_2 F)$$
$$+(\overline{C}_1 + \overline{D}_{12} F - \overline{B}_1^T X)^T (\overline{D}_{11} + \overline{D}_{11}^T)^{-1} (\overline{C}_1 + \overline{D}_{12} F - \overline{B}_1^T X).$$

and $\overline{A}, \overline{B}_1, \overline{B}_2, \overline{C}_1, \overline{C}_2, \overline{D}_{11}, \overline{D}_{12}, \overline{D}_{21}, \overline{D}_{22}$ are defined by (3.34).

Remark 3.5. Compared with the results of H_∞ control given in Chapter 2, the differences are that the controller obtained by Theorem 3.3 is strictly proper and D_{22} is not necessary to be zero matrix. But it should be pointed out that the conditions $P_1 = m_1, m_2 \leq p_1$ and $m_2 \leq p_1$ are required in Theorem 3.3 while these limitation are not needed in the results of H_∞ control in Chapter 2.

3.5 Multiplier design for SPR

This section considers the problem that given a real square transfer function matrix $G_1(s) = \left(\begin{array}{c|c} A_1 & B_1 \\ \hline C_1 & D_1 \end{array} \right)$ to design a transfer function $G_2(s) = \left(\begin{array}{c|c} A_2 & B_2 \\ \hline C_2 & D_2 \end{array} \right)$ such that $G_1(s)G_2(s)$ is SPR. Multiplier design often appears in robust control and filtering problems. By the state space realization $G_1(s)G_2(s) = \left(\begin{array}{c|c} A & B \\ \hline C & D \end{array} \right) = \left(\begin{array}{cc|c} A_1 & B_1 C_2 & B_1 D_2 \\ 0 & A_2 & B_2 \\ \hline C_1 & D_1 C_2 & D_1 D_2 \end{array} \right)$ and Corollary 1.4, one can get the following result.

Theorem 3.4. Given $G_1(s)$ as above, A_1 stable. There exists a stable multiplier $G_2(s)$ such that $D + D^T > 0$ and $G_1(s)G_2(s)$ is SPR if and only if $G_1(s)$ invertible and $G_1^{-1}(s) \in \mathcal{RH}_\infty$.

Proof. Since $D+D^T > 0$, by Lemma 3.8 $G_1(s)G_2(s)$ is SPR, if and only if there exists $Q > 0$ such that

$$\begin{pmatrix} AQ + QA^T & B - QC^T \\ B^T - CQ & -D - D^T \end{pmatrix} < 0. \tag{3.35}$$

Substitute the state space realization of $G_1(s)G_2(s)$ into the inequality above, set $Q = \begin{pmatrix} Q_1 & Q_{12} \\ Q_{12}^T & Q_2 \end{pmatrix}$ and rewrite (3.35) as

$$M + FKH + (FKH)^T < 0, \tag{3.36}$$

where $M = \begin{pmatrix} A_1Q_1 + Q_1A_1^T & A_1Q_{12} & -Q_1C_1^T \\ Q_{12}^T A_1^T & 0 & -Q_{12}^T C_1^T \\ -C_1Q_1 & -C_1Q_{12} & 0 \end{pmatrix}$, $F = \begin{pmatrix} B_1 & 0 \\ 0 & I \\ -D_1 & 0 \end{pmatrix}$, $K = \begin{pmatrix} D_2 & C_2 \\ B_2 & A_2 \end{pmatrix}$, $H = \begin{pmatrix} 0 & 0 & I \\ Q_{12}^T & Q_2 & 0 \end{pmatrix}$. By the projection lemma in Chapter 1, there exists K such that (3.36) holds, if and only if

$$F^\perp M F^{\perp T} < 0 \tag{3.37}$$

and

$$H^{T\perp} M H^{T\perp T} < 0. \tag{3.38}$$

Further, take $F^\perp = \left(\begin{pmatrix} B_1 \\ -D_1 \end{pmatrix}^\perp \ 0 \right) \begin{pmatrix} I & 0 & 0 \\ 0 & 0 & I \\ 0 & I & 0 \end{pmatrix}$, then (3.37) holds, if and only if there exists $\lambda > 0$ such that

$$\begin{pmatrix} A_1Q_1 + Q_1A_1^T & -Q_1C_1^T \\ -C_1Q_1 & 0 \end{pmatrix} - \lambda \begin{pmatrix} B_1 \\ -D_1 \end{pmatrix} \begin{pmatrix} B_1 \\ -D_1 \end{pmatrix}^T < 0. \tag{3.39}$$

And using Schur complement, (3.39) holds if and only if D_1 is nonsingular and

$$A_1Q_1 + Q_1A_1^T + Q_1C_1^T(\lambda D_1 D_1^T)^{-1} C_1 Q_1 - Q_1 C_1^T D_1^{-T} B_1^T - B_1 D_1^{-1} C_1 Q_1 < 0.$$

Take λ sufficiently large, the inequality above holds if and only if

$$A_1Q_1 + Q_1 A_1^T - Q_1 C_1^T D_1^{-T} B_1^T - B_1 D_1^{-1} C_1 Q_1 < 0. \tag{3.40}$$

On the other hand, (3.38) holds if and only there exists a scalar $\beta > 0$ such that

$$\begin{pmatrix} A_1Q_1 + Q_1A_1^T & A_1Q_{12} \\ Q_{12}^T A_1^T & 0 \end{pmatrix} - \beta \begin{pmatrix} Q_{12} \\ Q_2 \end{pmatrix} (Q_{12}^T \ Q_2) < 0. \tag{3.41}$$

3.5. Multiplier design for SPR   71

Set $P = Q^{-1} = \begin{pmatrix} P_1 & P_{12} \\ P_{12}^T & P_2 \end{pmatrix}$, multiply by P on the two sides of (3.41), one can get

$$\begin{pmatrix} P_1 A_1 + A_1^T P_1 & A_1^T P_{12} \\ P_{12}^T A_1 & 0 \end{pmatrix} - \beta \begin{pmatrix} 0 & 0 \\ 0 & I \end{pmatrix} < 0.$$

Clearly, there exists a scalar $\beta > 0$ such that the inequality above holds if and only if

$$P_1 A_1 + A_1^T P_1 < 0. \quad (3.42)$$

And clearly, P_1 and Q_1 satisfy

$$\begin{pmatrix} P_1 & I \\ I & Q_1 \end{pmatrix} \geq 0. \quad (3.43)$$

Clearly, that (3.40), (3.42) and (3.43) hold simultaneously is equivalent to that (3.40) and (3.42) hold. That is, A_1 and $A_1 - B_1 D_1^{-1} C_1$ are stable, i.e., $G_1(s)$, $G_1^{-1}(s) \in \mathcal{RH}_\infty$. This completes the proof. □

Remark 3.6. For single-input single-output(SISO) case, the result in Theorem 3.4 holds obviously [Rantzer and Megretski (1994); Geng (2004)]. In addition, Theorem 3.4 shows that the stability of $G_1(s)$ and $G_1^{-1}(s)$ is also the necessary and sufficient condition for the existence of a muliplier transfer function matrix such that the resulting plant is SPR for multi-input multi-output(MIMO) case. Therefore, all such kinds of multipliers $G_2(s)$ can be parameterized as follows

$$\mathcal{G} = \{G_2(s) = G_1^{-1}(s) E(s) \mid E(s) \text{ is SPR}\}.$$

In this parameterization, the order of $G_2(s)$ might be very high. One can also parameterize all multipliers $G_2(s)$ with the same order of $G_1(s)$ by the solutions to LMIs as in the next remark.

Remark 3.7. Given a stable square plant $G_1(s)$, by the proof of Theorem 3.4, multiplier design procedure (for example, multiplier design with the same order of $G_1(s)$) can be given as follows.

(i) Take $P_1 > 0, Q_1 > 0$ such that (3.40) and (3.42) hold and $P_1 - Q_1^{-1} > 0$;

(ii) Take P_{12}, Q_{12} such that $P_1 Q_1 + P_{12} Q_{12}^T = I$;

(iii) Take $Q = \begin{pmatrix} Q_1 & Q_{12} \\ Q_{12}^T & Q_2 \end{pmatrix} > 0$ such that (3.41) holds;

(iv) Then by the H_∞ method in Chapter 2, all matrices K (i.e., A_2, B_2, C_2, D_2) that solve (3.36) can be parameterized as follows (noticing that $HH^T > 0$)

$$K = -\rho F^T X H^T Y + Z^{1/2} L Y^{1/2},$$

where the scalar ρ and matrix L are the free parameters subject to

$$X = (FF^T - \frac{1}{\rho}M)^{-1} > 0, \quad \|L\|_2 < \rho$$

and the matrices Z and Y are defined by

$$Z = I - F^T(X - XH^T YHX)F, \quad Y = (HXH^T)^{-1}.$$

Remark 3.8. Obviously, inequalities (3.40), (3.42) and (3.43) are similar to the linear matrix inequalities in filtering theory [Duan et al. (2006c)] which are easier than the inequalities in H_∞ control theory [Iwasaki and Skelton (1994)]. By the method in [Duan et al. (2006c); Iwasaki and Skelton (1994)] and inequality (3.43), multiplier reduction design problem can be studied similarly as H_∞ controller and filter reduction problem.

Corollary 3.3. *The following two statements are equivalent:*

(i) there exists a real constant matrix D_2 such that $D_1 D_2 + D_2^T D_1^T > 0$ and $G_1(s)D_2$ is SPR;

(ii) there is a common Lyapunov matrix $Q_1 > 0$ such that (3.40) holds and $Q_1 A_1^T + A_1 Q_1 < 0$.

Proof. By Lemma 3.8, there exists a real constant matrix D_2 such that $G_1(s)D_2$ is SPR if and only if there exists $Q_1 > 0$ such that

$$\begin{pmatrix} Q_1 A_1^T + A_1 Q_1 & -Q_1 C_1^T \\ -C_1 Q_1 & 0 \end{pmatrix} + \begin{pmatrix} B_1 \\ -D_1 \end{pmatrix} D_2 (0 \quad I) + \begin{pmatrix} 0 \\ I \end{pmatrix} D_2^T (B_1^T \quad -D_1^T) < 0.$$

Then by the method in Theorem 3.4, one can complete the proof easily. □

Remark 3.9. Corollary 3.3 shows that when there is a common Lyapunov matrix for A_1 and $A_1 - B_1 D_1^{-1} C_1$, multiplier reduction problem can be solved completely. In fact, at this time, the multiplier design is closely related to the input transformation problem [Duan et al. (2005a)]. Compared with various results on feedback controller design methods, the input transformation has been paid much less attention in linear system theory except that it played an important role in input and output decoupling control. In fact, input transformation can also be used to improve system performances. Given a linear stable system

$$\begin{cases} \dot{x} = A_1 x + B_1 u, \\ y = C_1 x + D_1 u, \end{cases} \quad (3.44)$$

where A_1, B_1, C_1 and D_1 are matrices with compatible dimensions. Take an input transformation $u = Lv$, then system (3.44) becomes

$$\begin{cases} \dot{x} = A_1 x + B_1 Lv, \\ y = C_1 x + D_1 Lv. \end{cases} \quad (3.45)$$

Obviously, by Corollary 3.3, there exists an input transformation $u = Lv$ such that $D_1 L + L^T D_1^T > 0$ and system (10.47) is SPR if and only if D_1 is nonsingular, A_1 and $A_1 - B_1 D_1^{-1} C_1$ have a common Lyapuov matrix. Therefore, Corollary 3.3 establishes a relationship between common Lyapunov function and SPR problems.

Remark 3.10. For any two stable matrices A_1 and A_2, write A_2 as $A_1 - BC$, then there is a common Lyapunov matrix for A_1 and A_2, if and only if there is a constant nonsingular matrix D such that $D + D^T > 0$ and $(C(sI - A_1)^{-1}B + I)D$ is SPR.

3.6 Notes and references

Positive real systems have many applications to passivity analysis and control [Desoer and Vidyasagar (1975); Lozano et al. (2000); Slotine (1991)], quadratic optimal control [Anderson and Moore (1970); Anderson and Vongpanitlerd (1973)], adaptive system theory [Landau (1979)], robust control [Rantzer and Megretski (1994); Geng (2004)] and nonlinear systems [Narendra and Taylor (1973); Popov (1973); Khalil (2002)]. The celebrated positive real lemma establishes the equivalent relationship between frequency-domain and time-domain conditions for PR and SPR [Khalil (2002); Popov (1973); Rantzer (1996)]. Generally, SPR transfer functions are defined by PR transfer functions through an ϵ parallel shift of the imaginary axis. Recently, a complete ϵ-free equivalent characterization of SPR transfer functions was given in [Corless and Shorten (2009)]. The SPR control problems were discussed in [Huang et al. (1999); Sun et al. (1994)]. Multiplier design and input transportation problems were discussed in [Duan et al. (2004a, 2005a)].

Chapter 4

Absolute Stability and Dichotomy of Lur'e Systems

The absolute stability of Lur'e systems has been well-studied in many references. This chapter studies absolute stability and dichotomy of Lur'e systems in a unified framework. Sections 1 and 2 introduce the canonical circle and Popov criteria for absolute stability of SISO Lur'e systems without proofs. Section 3 introduces Aizerman and Kalman conjectures. Section 4 gives a unified proof for the stability criterion of MIMO Lur'e systems which covers the circle and Popov criteria. Sections 5 and 6 introduce the dichotomy property of Lur'e systems with sector and bounded derivative conditions. Section 7 concludes the chapter.

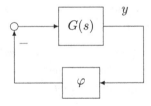

Fig. 4.1 System structure in absolute stability problems.

4.1 Circle criterion of SISO Lur'e systems

Consider the feedback structure shown in Fig. 4.1. The forward path is a linear time-invariant system, and the feedback part is a nonlinear function.

The equations for such systems can be written as

$$\begin{cases} \dot{x} = Ax + bu, \\ y = cx, \\ u = -\varphi(t,y), \end{cases} \quad (4.1)$$

where $G(s) = c(sI - A)^{-1}b$ and $\varphi(t,y)$ is a memoryless, possibly time-varying, nonlinearity, which is piecewise continuous in t and locally Lipschitz in y, and it satisfies the sector condition:

$$\varphi(t,0) \equiv 0; \quad \mu_1 \leq \frac{\varphi(t,y)}{y} \leq \mu_2, \ \mu_1 \leq \mu_2 \quad (t \in \mathbb{R}_+, y \neq 0). \quad (4.2)$$

Sometimes, the condition (4.2) is also written as

$$\mu_1 y^2 \leq \varphi(t,y)y \leq \mu_2 y^2.$$

Definition 4.1. System (4.1) is called absolutely stable if the origin is globally uniformly asymptotically stable for any nonlinearity in the given sector.

Similar to the Nyquist criterion for stability of linear systems, a well known stability condition for Lur'e systems is the circle criterion, whose basic version can be stated as follows [Khalil (2002); Leonov et al. (1996); Huang (2003)]:

Theorem 4.1. *(circle criterion) Suppose that (A,b) is controllable, and the following conditions hold:*

(i) the matrix A has no eigenvalues with zero real parts;
(ii) there is $\mu_0 \in [\mu_1, \mu_2]$ such that $A - \mu_0 bc$ is Hurwitz;
(iii) the following frequency-domain inequality is valid:

$$\mathbf{Re}\{[1 + \mu_1 G(jw)]^*[1 + \mu_2 G(jw)]\} > 0, \quad \forall w \in \mathbb{R} \cup \{\infty\}. \quad (4.3)$$

Then system (4.1) is absolutely stable for all nonlinear functions with sector condition (4.2).

Remark 4.1. The inequality (4.3) is sometimes equivalently written as

$$\mathbf{Re}\left\{\frac{1 + \mu_2 G(jw)}{1 + \mu_1 G(jw)}\right\} > 0, \quad \forall w \in \mathbb{R} \cup \{\infty\}.$$

The frequency-domain condition (4.3) can also be explained by the Nyquist plot of $G(s)$ according to the different cases of μ_1 and μ_2 being larger than zero or smaller than zero [Khalil (2002)]. In addition, by Corollary 1.4 the

4.1. Circle criterion of SISO Lur'e systems

frequency-domain inequality (4.3) is equivalent to the existence of a matrix $P = P^T$ such that the following matrix inequality hold

$$\begin{pmatrix} PA + A^T P - \mu_1\mu_2 c^T c & Pb - \frac{\mu_1+\mu_2}{2} c^T \\ b^T P - \frac{\mu_1+\mu_2}{2} c & -1 \end{pmatrix} < 0. \quad (4.4)$$

Corollary 4.1. *Conditions (ii) and (iii) of Theorem 4.1 guarantee that P in (4.4) is positive definite, i.e., $P > 0$, further, for any $\mu \in [\mu_1, \mu_2]$, $A - \mu bc$ is Hurwitz stable.*

Proof. Take $N = \begin{pmatrix} I & -\mu_0 c^T \\ 0 & 1 \end{pmatrix}$. Multiplying the left and right sides of (4.4) by N and N^T gives

$$\begin{pmatrix} P(A - \mu_0 bc) + (A - \mu_0 bc)^T P + \alpha c^T c & Pb - \frac{\mu_1+\mu_2}{2} c^T + \mu_0 c^T \\ b^T P - \frac{\mu_1+\mu_2}{2} c + \mu_0 c & -1 \end{pmatrix} < 0, \quad (4.5)$$

where $\alpha = (\mu_1 - \mu_0)(\mu_0 - \mu_2)$. Clearly, $\alpha \geq 0$, so the above inequality and the stability of $A - \mu_0 bc$ guarantee the positive definiteness of P. Replacing μ_0 by any $\mu \in [\mu_1, \mu_2]$, the positive definiteness of P guarantees the stability of $A - \mu bc$. □

Corollary 4.2. *Conditions (ii) and (iii) of Theorem 4.1 are equivalent to that $1 + (\mu_2 - \mu_1)c(sI - A + \mu_1 bc)^{-1}b$ is SPR.*

Proof. By Corollary 4.1, under condition (iii), condition (ii) of Theorem 4.1 is equivalent to that $A - \mu_1 bc$ is Hurwitz. Replacing μ_0 in (4.5) by μ_1, then by Corollary 1.4 (4.5) is equivalent to SPR property of $1 + (\mu_2 - \mu_1)c(sI - A + \mu_1 bc)^{-1}b$. □

Remark 4.2. Due to its graphical interpretation, control engineers prefer the frequency-domain condition (4.3). Let $G(jw) = U(w) + jV(w)$. Condition (4.3) is equivalent to either

$$1 + \mu_2 U(w) > 0 \quad \forall w \in \mathbb{R} \cup \{\infty\}, \quad (4.6)$$

when $\mu_1 = 0$, or for the case $\mu_1 \neq 0$,

$$\mu_1\mu_2 \left[V^2(w) + U^2(w) + \frac{1}{\mu_1\mu_2} + \left(\frac{1}{\mu_1} + \frac{1}{\mu_2}\right) U(w) \right] > 0 \quad \forall w \in \mathbb{R} \cup \{\infty\},$$

which is equivalent to

$$\left[U(w) + \frac{1}{2}\left(\frac{1}{\mu_1} + \frac{1}{\mu_2}\right) \right]^2 + V^2(w) > \frac{1}{4}\left(\frac{1}{\mu_1} - \frac{1}{\mu_2}\right)^2 \quad \forall w \in \mathbb{R} \cup \{\infty\}, \quad (4.7)$$

when $\mu_1 > 0$ or $\mu_2 < 0$, or

$$\left[U(w) + \frac{1}{2}\left(\frac{1}{\mu_1} + \frac{1}{\mu_2}\right)\right]^2 + V^2(w) < \frac{1}{4}\left(\frac{1}{\mu_1} - \frac{1}{\mu_2}\right)^2 \quad \forall w \in \mathbb{R} \cup \{\infty\}, \tag{4.8}$$

when $\mu_1 < 0$ and $\mu_2 > 0$. In the complex plane (U, V), condition (4.6) represents the half plane to the right of the line $1 + \mu_2 U = 0$ as shown in Fig. 4.2 (a), and condition (4.7) or (4.8) represents the outside part (Fig. 4.2 (b)) or the inside part (Fig. 4.2 (c)) of the circle $\mathbf{S}$,

$$\mathbf{S} = \{z \mid |z - z_0| = \rho\}, \quad z_0 = -\frac{1}{2}\left(\frac{1}{\mu_1} + \frac{1}{\mu_2}\right), \quad \rho = \frac{1}{2}\left|\frac{1}{\mu_1} - \frac{1}{\mu_2}\right|.$$

Based on the above geometric interpretations, the test of condition (4.3) becomes to test if the curve $G(jw)$ $(w \in \mathbb{R})$ falls in the shaded part of Figures 4.2 (a)-(c) for different cases of μ_1 and μ_2, so the condition (4.3) can be graphically tested. Reference [Khalil (2002)]) also gives a Nyquist plot description of the circle criterion.

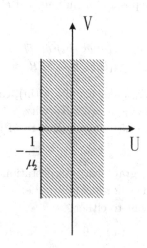

Fig. 4.2 (a).

Remark 4.3. In this remark, we discuss the relationship between the frequency-domain condition (4.3) and the sector condition (4.2). The sector condition (4.2) can be equivalently written as

$$(\varphi(t, y) - \mu_1 y)(\varphi(t, y) - \mu_2 y) \leq 0.$$

4.1. Circle criterion of SISO Lur'e systems

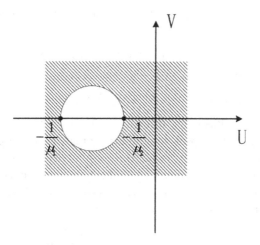

Fig. 4.2 (b).

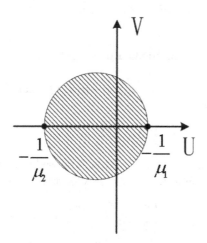

Fig. 4.2 (c).

Fig. 4.2 Geometric interpretation of the circle criterion.

Considering the case of complex parameters and $y = cx$, the above inequality can be rewritten as

$$\mathbf{Re}\{(\varphi(t,y) - \mu_1 cx)^*(\varphi(t,y) - \mu_2 cx)\} \leq 0.$$

Based on the system (4.1), representing x in the above inequality by $-(jwI - A)^{-1}b$ gives

$$\mathbf{Re}\{\varphi^*(t,y)(1+\mu_1 G(jw))^*(1+\mu_2 G(jw))\varphi(t,y)\} \leq 0, \quad \forall w \in \mathbb{R} \cup \{\infty\}.$$

Considering the arbitrariness of $\varphi(t,y)$, this is equivalent to

$$\mathbf{Re}\{(1+\mu_1 G(jw))^*(1+\mu_2 G(jw))\} \leq 0, \quad \forall w \in \mathbb{R} \cup \{\infty\}, \quad (4.9)$$

which has a contrary inequality direction with the circle criterion inequality (4.3) in Theorem 4.1. Absolute stability implies that the corresponding linear system is stable after replacing the nonlinearity by the linear function in the sector. Hence the inequality (4.9) obtained from the sector constraint condition by substituting into the linear system characteristic should not hold for absolutely stable systems. This implies that there is not a process of harmonic wave input and output in the closed-loop linear system after replacing the nonlinear function $\varphi(t,y)$ by the corresponding linear function $\varphi(t,y) = \mu y$, $\mu \in [\mu_1, \mu_2]$ when system (4.1) is absolutely stable, which is an intrinsic relationship between conditions (4.2) and (4.3) for absolute stability of Lur'e systems [Huang (2003)]).

4.2 Popov criterion of SISO Lur'e systems

In condition (4.2), let $\varphi_1(t,y) = \varphi(t,y) - \mu_1 y$. One can get a new constraint condition for $\varphi_1(t,y)$:

$$\varphi_1(t,0) \equiv 0; \quad 0 \leq \frac{\varphi_1(t,y)}{y} \leq \mu = \mu_2 - \mu_1.$$

Hence, the sector condition (4.2) is often supposed to be in the above form.

Consider the following Lur'e system with a time-invariant nonlinearity

$$\begin{cases} \dot{x} = Ax - b\varphi(y), \\ y = cx, \end{cases} \quad (4.10)$$

where $\varphi(y)$ satisfies the sector condition:

$$\varphi(0) \equiv 0; \quad 0 \leq \frac{\varphi(y)}{y} \leq \mu \quad (y \neq 0, \mu \geq 0). \quad (4.11)$$

Sometimes, the condition (4.11) is equivalently written as

$$0 \leq \varphi(y)y \leq \mu y^2 \quad (\mu \geq 0).$$

For absolute stability of Lur'e system (4.10), the following version of Popov criterion is very popular [Popov (1973); Leonov et al. (1996); Khalil (2002); Huang (2003)].

4.2. Popov criterion of SISO Lur'e systems

Theorem 4.2. *(Popov criterion) Suppose that A is Hurwitz and (A, b) is controllable. If there is a real parameter ν such that*

$$\frac{1}{\mu} + \text{Re}\{G(jw) + \nu jwG(jw)\} > 0, \quad \forall w \in \mathbb{R} \cup \{\infty\}, \tag{4.12}$$

then system (4.10) is absolutely stable for all nonlinear functions $\varphi(y)$ with condition (4.11).

Remark 4.4. Noticing the stability of A, condition (4.12) means that $\frac{1}{\mu} + G(s) + \nu s G(s)$ is SPR. When $\mu = +\infty$, the condition (4.12) naturally reduces to

$$\text{Re}\{G(jw) + \nu jwG(jw)\} > 0, \quad \forall w \in \mathbb{R} \cup \{\infty\},$$

which means that $G(s) + \nu s G(s)$ is SPR.

Remark 4.5. By the transfer function $G(s) = c(sI - A)^{-1}b$, one knows that

$$sG(s) = cA(sI - A)^{-1}b + cb.$$

Then similar to matrix inequality (4.4), by Corollary 1.4 condition (4.12) is equivalent to the existence of a matrix $P = P^T$ such that the following matrix inequality holds

$$\begin{pmatrix} PA + A^T P & Pb - \frac{1}{2}(c^T + \nu A^T c^T) \\ b^T P - \frac{1}{2}(c + \nu c A) & -\frac{1}{\mu} - \nu c b \end{pmatrix} < 0. \tag{4.13}$$

When A is Hurwitz, condition (4.12) or (4.13) guarantees that $A - \mu_0 bc$ is Hurwitz for any $\mu_0 \in [0, \mu]$, see the following section for the proof of this conclusion.

Remark 4.6. In this remark, we give a geometric interpretation of Popov criterion. Let $U(w) = \text{Re}\{G(jw)\}$, $V(w) = w\text{Im}\{G(jw)\}$ and $\hat{G}(jw) = U(w) + jV(w)$. $\hat{G}(jw)$ is generally called the modified frequency response of $G(s)$. Then condition (4.12) can be rewritten as

$$\frac{1}{\mu} + U(w) - \nu V(w) > 0, \quad \forall w \in \mathbb{R} \cup \{\infty\}. \tag{4.14}$$

In the (ξ, η) plane, $\Gamma : \xi - \nu\eta + \frac{1}{\mu} = 0$ represents a line (Figure 4.3). Condition (4.14) means that the frequency response of $\hat{G}(jw)$ falls in the right part of the line Γ, so the Popov criterion can also be graphically tested.

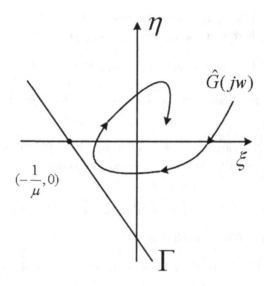

Fig. 4.3 Geometric interpretation of Popov criterion.

Remark 4.7. With the sector constraint (4.11), the corresponding frequency-domain condition in circle criterion becomes $1+\mathbf{Re}\{\mu G(jw)\} > 0$ or $\frac{1}{\mu} + \mathbf{Re}\{G(jw)\} > 0$, which is consistent with Corollary 4.2. Comparing with condition (4.12), one more degree of freedom $\mathbf{Re}\{\nu jwG(jw)\}$ has been added to improve the frequency-domain condition in Popov criterion. Hence, generally Popov criterion is less conservative than circle criterion. However, circle criterion in Theorem 4.1 is suitable for Lur'e systems with time-varying nonlinearities as in system (4.1), while Popov criterion is not suitable for system (4.1). And see [Khalil (2002); Huang (2003); Leonov et al. (1996)] for more details.

4.3 Aizerman and Kalman conjectures

The circle criterion or the Popov criterion are sufficient conditions for absolute stability of Lur'e systems. The question of necessary and sufficient conditions for absolute stability is still open, despite having been studied for many years.

In system (4.10) and (4.11), representing the nonlinearity by a linear function $\varphi(y) = \sigma y$ gives a linear system

$$\dot{x} = (A - \sigma bc)x, \quad \sigma \in [0, \mu]. \tag{4.15}$$

Obviously, if system (4.10) is absolutely stable with nonlinearities (4.11), then linear system (4.15) is stable for all $\sigma \in [0, \mu]$. However, how about the reverse of this statement? In 1949, M. A. Aizerman stated the following conjecture: if linear system (4.15) is stable with $\sigma \in [0, \mu]$, then system (4.10) is absolutely stable with nonlinearities (4.11). This conjecture motivated numerous investigations. At least, it can be right for some second-order Lur'e systems. For example, consider the following second-order Lur'e system

$$\begin{cases} \dot{x}_1 = x_2, \\ \dot{x}_2 = -x_2 - h(y), \\ y = x_1, \end{cases}$$

where $h(y)$ is a nonlinear function with some sector condition. Note that the linear part in the above system is not stable. Defining the transformation $\varphi(y) = h(y) - \alpha y$ leads to the system in the form (4.10) with matrices

$$A = \begin{pmatrix} 0 & 1 \\ -\alpha & -1 \end{pmatrix}, \quad b = \begin{pmatrix} 0 \\ 1 \end{pmatrix}, \quad c = (1\ 0).$$

Assume $h(y)$ belongs to a sector $[\alpha, \beta]$, where $\beta > \alpha > 0$. Then, $\varphi(y)$ belongs to the sector $[0, \mu]$, where $\mu = \beta - \alpha$. Obviously, the corresponding linear system (4.15) in this example is stable for all $\sigma \in [0, \mu]$. On the other hand, condition (4.12) in this example takes the form

$$\frac{1}{\mu} + \frac{\alpha - w^2 + \nu w^2}{(\alpha - w^2)^2 + w^2} > 0, \quad \forall w \in \mathbb{R}.$$

For all finite positive values of α and μ, this inequality is satisfied by choosing $\nu > 1$. Hence, the absolute stability coincides with the stability of the corresponding linear system in this example. This statement even holds for the case $\mu = \infty$, as detailed in [Khalil (2002)]. Another second-order example was also given in [Huang (2003); Leonov et al. (1996)] to show that the region decided by condition (4.12) is the same as the stable region of the corresponding linear system. However, Aizerman conjecture does not hold for general Lur'e systems, see [Krasovsky (1952); Pliss (1964); Yakubovich (1958)] for counterexamples.

Since the stability of linear systems proved to be insufficient for system (4.10) and (4.11) to be absolutely stable, another conjecture with more restrictions on $\varphi(y)$ was brought forward in [Kalman (1957)]. It is called Kalman conjecture.

Suppose that the nonlinear function $\varphi(y)$ is continuously differentiable and $\alpha < \varphi'(y) < \beta$, and the corresponding linear system (4.15) is stable for $\sigma \in (\alpha, \beta)$, then Kalman conjecture states that system (4.10) is absolutely stable with sector (α, β).

In [Barabanov (1988)], it was proved that the Kalman conjecture is true for the case $n = 3$. But for $n \geq 4$, the examples are known when the conjecture fails [Barabanov (1988); Leonov et al. (1996)]. In addition, with bounded derivative conditions on $\varphi(y)$, the Popov criterion can be strengthened [Brockett and Willems (1965a,b); Huang (2003); Leonov et al. (1996); Yakubovich (1965)].

4.4 MIMO Lur'e systems

This section considers MIMO Lur'e systems described by the differential equation

$$\begin{aligned} \dot{x} &= Ax + Bu, \\ y &= Cx, \\ u &= -f(y), \end{aligned} \qquad (4.16)$$

where $x = (x_1, \cdots, x_n)^T \in \mathbb{R}^n$ is the system state vector, $y = (y_1, \cdots, y_m)^T \in \mathbb{R}^m$ is the output vector, $f(y) = (f_1(y_1), \cdots, f_m(y_m))^T$, $f_i : \mathbb{R} \to \mathbb{R}$ is time-invariant nonlinear function which satisfies the sector condition given below,

$$\gamma_i \tau^2 \leq f_i(\tau)\tau \leq \delta_i \tau^2, \ \delta_i \geq \gamma_i, \ f_i(0) = 0, \ i = 1, \cdots, m. \qquad (4.17)$$

Let $\Gamma = diag(\gamma_1, \cdots, \gamma_m)$, $\Delta = diag(\delta_1, \cdots, \delta_m)$. Obviously $\Delta \geq \Gamma$. In the following, we say f belongs to the sector $[\Gamma, \Delta]$, denoted by $f \in [\Gamma, \Delta]$, it means that $f_i, i = 1, \cdots, m$ satisfies the condition (4.17). The canonical Lyapunov method leads to the following theorem for absolute stability of system (4.16).

Theorem 4.3. *System (4.16) is absolutely stable for all $f \in [\Gamma, \Delta]$, if $A - B\Gamma C$ is stable and there exist diagonal matrices P and Q with $Q \geq 0$, and a symmetric matrix W such that the following LMI is feasible*

$$\begin{pmatrix} WA + A^T W - C^T \Gamma \Delta QC & WB + A^T C^T P - \frac{1}{2}C^T Q(\Gamma + \Delta) \\ B^T W + PCA - \frac{1}{2}(\Gamma + \Delta)QC & PCB + B^T C^T P - Q \end{pmatrix} < 0. \qquad (4.18)$$

4.4. MIMO Lur'e systems

Before proving the theorem, two lemmas and some discussions on frequency-domain inequalities are needed.

Lemma 4.1. W and P are given as in (4.18), then the stability of $A - B\Gamma C$ is equivalent to $W - C^T P \Gamma C > 0$.

Proof. Defining
$$\Psi := \begin{pmatrix} WA + A^T W - C^T \Gamma \Delta QC & WB + A^T C^T P - \frac{1}{2} C^T Q(\Gamma + \Delta) \\ B^T W + PCA - \frac{1}{2}(\Gamma + \Delta)QC & PCB + B^T C^T P - Q \end{pmatrix}$$
and
$$T := \begin{pmatrix} I_n & -C^T \Gamma \\ 0 & I_m \end{pmatrix},$$
by (4.18) a congruence transformation on Ψ leads to $T\Psi T^T < 0$, i.e.,
$$\begin{pmatrix} YA_\Gamma + A_\Gamma^T Y & YB + A_\Gamma^T C^T P - \frac{1}{2} C^T Q(\Delta - \Gamma) \\ B^T Y + PCA_\Gamma - \frac{1}{2}(\Delta - \Gamma)QC & PCB + B^T C^T P - Q \end{pmatrix} < 0, \quad (4.19)$$
where $Y = W - C^T P \Gamma C$, $A_\Gamma = A - B\Gamma C$. By this inequality, the lemma obviously holds. □

Suppose the transfer function of system (4.16) from u to y is
$$G(s) = C(sI - A)^{-1} B,$$
then
$$sG(s) = CA(sI - A)^{-1} B + CB.$$
Then, we have

Theorem 4.4. The existence of W such that (4.18) holds is equivalent to
$$\mathbf{Re}\{jwPG(jw) - (I + \Gamma G(jw))^* Q(I + \Delta G(jw))\} < 0, \; \forall w \in \mathbb{R} \cup \{\infty\}. \quad (4.20)$$

Proof. Let
$$M = \begin{pmatrix} -C^T \Gamma \Delta QC & A^T C^T P - \frac{1}{2} C^T Q(\Gamma + \Delta) \\ PCA - \frac{1}{2}(\Gamma + \Delta)QC & PCB + B^T C^T P - Q \end{pmatrix}.$$
The inequality (4.18) can be rewritten as
$$M + \begin{pmatrix} WA + A^T W & WB \\ B^T W & 0 \end{pmatrix} < 0.$$
Then by Corollary 1.4, (4.18) holds if and only if
$$\begin{pmatrix} (jwI - A)^{-1} B \\ I \end{pmatrix}^* M \begin{pmatrix} (jwI - A)^{-1} B \\ I \end{pmatrix} < 0, \; \forall w \in \mathbb{R} \cup \{\infty\}.$$
Noticing the arbitrariness of P, the above inequality is equivalent to (4.20).
□

Theorem 4.5. *The existence of Y such that (4.19) holds is equivalent to*
$$\mathbf{Re}\{-(\Delta - \Gamma)QG_\Gamma(jw) - Q + jwPG_\Gamma(jw)\} < 0, \, \forall w \in \mathbb{R} \cup \{\infty\}, \quad (4.21)$$
where $G_\Gamma(s) = C(sI - A + B\Gamma C)^{-1}B$.

Proof. Similarly to the proof of Theorem 4.4. □

Obviously, (4.18), (4.19), (4.20) and (4.21) are equivalent to each other.

Remark 4.8. With the stability of $A - B\Gamma C$, condition (4.21) means that $Q(\Delta - \Gamma)G_\Gamma(s) + Q - sPG_\Gamma(s)$ is SPR, which is consistent with Popov criterion of SISO Lur'e system (Remark 4.4).

Lemma 4.2. *If $A - B\Gamma C$ is stable, then (4.18) guarantees that $A - B\tilde{\Gamma}C$ is stable for all diagonal matrix $\tilde{\Gamma}$ with $\Gamma \leq \tilde{\Gamma} \leq \Delta$.*

Proof. We prove this lemma by reduction to absurdity. Assume that there exists a $\tilde{\Gamma}$ such that $A - B\tilde{\Gamma}C$ is not stable. Then there exists a $\tilde{\Delta}$, $0 \leq \tilde{\Delta} \leq \Delta - \Gamma$, such that $A - B\tilde{\Gamma}C = A - B\Gamma C - B\tilde{\Delta}C$, which is not stable. So, there exists an $\alpha \in (0, 1]$ such that $A - B\Gamma C - \alpha B\tilde{\Delta}C$ has an imaginary eigenvalue, that is, there exists a $w_0 \in \mathbf{R}$ such that
$$det(jw_0 I - A + B\Gamma C + \alpha B\tilde{\Delta}C) = 0.$$
By the stability of $A - B\Gamma C$, we have
$$det(I + \alpha(jw_0 I - A + B\Gamma C)^{-1}B\tilde{\Delta}C) = 0$$
which is equivalent to
$$det(I + \alpha C(jw_0 I - A + B\Gamma C)^{-1}B\tilde{\Delta}) = 0, \text{ i.e. } det(I + \alpha G_\Gamma(jw_0)\tilde{\Delta}) = 0,$$
where $G_\Gamma(s) = C(sI - A + B\Gamma C)^{-1}B$. Hence, there exists a non-zero vector η such that
$$\eta^*(I + \alpha \tilde{\Delta} G_\Gamma^*(jw_0)) = 0$$
which also provides $\eta^*\tilde{\Delta} \neq 0$. In this case
$$\begin{aligned}\eta^*\tilde{\Delta}\left(-2Q - (\Delta - \Gamma)QG_\Gamma(jw_0) - G_\Gamma^*(jw_0)Q(\Delta - \Gamma)\right)\tilde{\Delta}\eta \\ = \eta^*(-2\tilde{\Delta}Q\tilde{\Delta} + \tfrac{2}{\alpha}(\Delta - \Gamma)Q\tilde{\Delta})\eta \geq 0.\end{aligned} \quad (4.22)$$
On the other hand, by (4.21) it holds that
$$\begin{aligned}0 > \eta^*\tilde{\Delta}(-2Q - (\Delta - \Gamma)QG_\Gamma(jw_0) - G_\Gamma^*(jw_0)Q(\Delta - \Gamma) \\ + jw_0 PG_\Gamma(jw_0) - jw_0 G_\Gamma^*(jw_0)P)\tilde{\Delta}\eta \\ = \eta^*(-2\tilde{\Delta}Q\tilde{\Delta} + \tfrac{2}{\alpha}(\Delta - \Gamma)Q\tilde{\Delta})\eta + \tfrac{jw_0}{\alpha}\eta^*\tilde{\Delta}P\eta - \tfrac{jw_0}{\alpha}\eta^*P\tilde{\Delta}\eta \\ = \eta^*(-2\tilde{\Delta}Q\tilde{\Delta} + \tfrac{2}{\alpha}(\Delta - \Gamma)Q\tilde{\Delta})\eta,\end{aligned}$$
which contradicts (4.22). Therefore, $A - B\Gamma C - B\tilde{\Delta}C$ is stable. □

4.4. MIMO Lur'e systems

Of course, this lemma is also suitable for SISO Lur'e systems (Remark 4.5).

Proof. (*proof of Theorem 4.3*) Take for Lyapunov function candidate

$$v(x) =: x^T W x - 2 \sum_{i=1}^{m} p_i \int_0^{y_i} f_i(\tau) d\tau. \quad (4.23)$$

The time derivative of $v(x)$ along any trajectory of system (4.16) is given by

$$\dot{v}(x) = 2x^T W (Ax - Bf(y)) - 2f^T(y) PC(Ax - Bf(y)), \quad (4.24)$$

where $P =: diag(p_1, \cdots, p_m)$. Note that for all $y_i \neq 0$ the sector condition of f_i given in (4.17) is equivalent to

$$s_i(x_i) := (f_i(y_i) - \gamma_i y_i)(f_i(y_i) - \delta_i y_i) \leq 0.$$

Then for any $q_i \geq 0$, $i = 1, \cdots, m$, we have

$$\sum_{i=1}^{m} q_i s_i(x_i) = f^T(y) Q f(y) - f^T(y)(\Gamma + \Delta) Q C x + x^T C^T \Gamma \Delta Q C x \leq 0, \quad (4.25)$$

where $Q = diag(q_1, \cdots, q_m)$. Considering sector condition (4.17) and inequality (4.25), we know that $\dot{v}(x) < 0$ if

$$2x^T W (Ax - Bf(y)) - 2f^T(y) PC(Ax - Bf(y))$$
$$< f^T(y) Q f(y) - f^T(y)(\Gamma + \Delta) Q C x + x^T C^T \Gamma \Delta Q C x, \quad \forall (x, f(y)) \neq 0.$$

Obviously, the inequality above can be written as

$$\xi \begin{pmatrix} WA + A^T W - C^T \Gamma \Delta Q C & R \\ R^T & PCB + B^T C^T P - Q \end{pmatrix} \xi^T < 0,$$

for any $\xi = (x^T \; f^T(y)) \neq 0$, which is equivalent to the existence of a feasible solution to the LMI (4.18), where $R = -WB - A^T C^T P + \frac{1}{2} C^T Q (\Gamma + \Delta)$. Then (4.18) guarantees that $\dot{v}(x) < 0$. In fact, (4.18) also guarantees that the Lyapunov candidate $v(x)$ defined in (4.23) is positive definite. First rewrite $v(x)$ in (4.23) as

$$v(x) = x^T (W - C^T P \Gamma C) x - 2 \sum_{i=0}^{m} p_i \int_0^{y_i} (f_i(\tau) - \gamma_i \tau) d\tau.$$

By Lemma 4.1, $W - C^T P \Gamma C$ is positive definite. And $\int_0^{y_i} (f_i(\tau) - \gamma_i \tau) d\tau \geq 0$ because of the sector condition (4.17). So $v(x) > 0$ holds if $p_i \leq 0, i = 1, \cdots, m$. If there exists $p_i > 0$, we suppose that $p_1 > 0, \cdots, p_k > 0$ and $p_r \leq 0, k+1 \leq r \leq m$ without loss of generality.

Let $P_k = diag(p_1, \cdots, p_k, 0, \cdots, 0)$, $\Gamma_k = diag(\gamma_1, \cdots, \gamma_k, 0, \cdots, 0)$, and $\Delta_k = diag(\delta_1, \cdots, \delta_k, 0, \cdots, 0)$. Then we have

$$\begin{aligned} v(x) &\geq x^T(W - C^T P\Gamma C)x - 2\sum_{i=1}^{k} p_i \int_0^{y_i}(\delta_i\tau - \gamma_i\tau)d\tau \\ &\quad -2\sum_{i=k+1}^{m} p_i \int_0^{y_i}(f_i(\tau) - \gamma_i\tau)d\tau \\ &= x^T(W - C^T P\Gamma C - C^T P_k(\Delta_k - \Gamma_k)C)x \\ &\quad -2\sum_{i=k+1}^{m} p_i \int_0^{y_i}(f_i(\tau) - \gamma_i\tau)d\tau. \end{aligned} \quad (4.26)$$

Taking

$$T_k =: \begin{pmatrix} I_n & -C^T(\Delta_k - \Gamma_k) \\ 0 & I_m \end{pmatrix}$$

and making a congruence transformation on the inequality (4.19) as in the proof of Lemma 4.1 leads to the positive definiteness of $W - C^T P\Gamma C - C^T P(\Delta_k - \Gamma_k)C = W - C^T P\Gamma C - C^T P_k(\Delta_k - \Gamma_k)C$ by the stability of $A - B\Gamma C - B(\Delta_k - \Gamma_k)C$ (Lemma 4.2). Therefore, (4.26) implies $v(x) > 0$. Then by the canonical Lyapunov theory, system (4.16) is absolutely stable. □

Remark 4.9. Similarly to the canonical circle and Popov criteria for SISO Lur'e systems, Theorems 4.4 and 4.5 also give frequency domain interpretations for the LMIs (4.18) and (4.19). Obviously, (4.20) becomes a simple extension of the canonical circle criterion for MIMO Lur'e systems when $P = 0$ and (4.21) becomes a simple extension of the canonical Popov criterion when $\Gamma = 0$ [Huang (2003); Khalil (2002)]. Therefore, (4.20) unifies Circle and Popov criteria together. In addition, it is worth to point out that it is also proved that the diagonal matrix P in Theorem 4 needs not to be positive in [Park (1997)] by the choice of a special Lyapunov function. By using the technique of [Bernstein et al. (1994); Park (1997)], we also proved this point by choosing the canonical Lur'e-Postnikov Lyapunov function. Compared with the result of [Park (1997)], Q is another degree of freedom introduced here. Generally, this additional variable would render a less conservative result. In addition, when $P = 0$, Theorem 4.3 is also suitable for Lur'e system (4.16) with time-varying nonlinearity.

For the case $\Gamma = 0$ and $\Delta > 0$, the sector condition (4.17) reduces to

$$f_i(y_i)(f_i(y_i) - \delta_i y_i) \leq 0, \quad 'i = 1, \cdots, m,$$

which can be rewritten as

$$f_i(y_i)(\delta_i^{-1} f_i(y_i) - y_i) \leq 0, \quad, i = 1, \cdots, m.$$

4.5. Dichotomy of Lur'e systems

Correspondingly, the condition (4.25) reduces to

$$f^T(y)\Delta^{-1}Qf(y) - f^T(y)QCx \leq 0.$$

Then Theorem 4.3 can be simplified as follows.

Corollary 4.3. *System (4.16) is absolutely stable for all* $f \in [0, \Delta]$, $\Delta > 0$, *if A is stable and there exist diagonal matrices P and Q with* $Q \geq 0$, *and a symmetric matrix W such that the following LMI is feasible*

$$\begin{pmatrix} WA + A^TW & WB + A^TC^TP - \frac{1}{2}C^TQ \\ B^TW + PCA - \frac{1}{2}QC & PCB + B^TC^TP - \Delta^{-1}Q \end{pmatrix} < 0. \quad (4.27)$$

Corresponding to Theorem 4.5, we have

Corollary 4.4. *The existence of W such that (4.27) holds is equivalent to*

$$\mathbf{Re}\{-QG(jw) - \Delta^{-1}Q + jwPG(jw)\} < 0, \forall w \in \mathbb{R} \cup \{\infty\}. \quad (4.28)$$

Remark 4.10. Corollary 4.4 is consistent with Theorem 4.2 for SISO Lur'e systems. In fact, with the stability of A the condition (4.28) means that $\Delta^{-1}Q + QG(s) - sPG(s)$ is SPR. Obviously, Corollaries 4.3 and 4.4 are suitable for the case of infinite sector, i.e., $\Delta = +\infty$, in which case $\Delta^{-1} = 0$, condition (4.28) reduces to $\mathbf{Re}\{-QG(jw) + jwPG(jw)\} < 0, \forall w \in \mathbb{R} \cup \{\infty\}$, i.e., $QG(s) - sPG(s)$ is SPR.

Remark 4.11. Noticing that in the canonical circle and Popov criteria for Lur'e systems, (A, B) is generally supposed to be controllable [Huang (2003); Khalil (2002)]. In fact, with the assumption on stability of $A - B\Gamma C$ in time-domain models, we can see from Theorem 4.3 that the controllability of (A, B) is not necessary for absolute stability. And in KYP lemma (Corollary 1.4), the controllability of (A, B) is not necessary too for the case of strict inequality. In order to be consistent with the traditional literature, in Sections 4.1 and 4.2, (A, b) is supposed to be controllable for circle and Popov criteria of SISO Lur'e systems. Actually, this controllability condition can be removed.

4.5 Dichotomy of Lur'e systems

The absolute stability implies the global stability of all solutions for the system. Besides global stability, the property of dichotomy is another interesting property for nonlinear systems, which shows the convergence of bounded solutions. The study of dichotomy is important in chaotic systems when one cares about the nonexistence of chaotic or periodic solutions.

Definition 4.2. System (4.16) is called to be dichotomous if every bounded solution is convergent to an equilibrium of (4.16).

Definition 4.3. A point $\hat{x}$ is called an w-limit point of the solution $x(t, x_0)$ to system (4.16) if there exists an increasing sequence $\{t_n\}$, $t_n \to +\infty$, such that $x(t_n, x_0) \to \hat{x}$ as $n \to +\infty$. The set of w-limit points is called w-limit set and is denoted by Ω.

Using the method of [Leonov et al. (1996)], we can obtain a basic LMI-based result for dichotomy of system (4.16). The next lemma [Leonov et al. (1996)] is useful for proving the main result.

Lemma 4.3. *If $f : \mathbb{R}_+ \to \mathbb{R}$ is uniformly continuous and $f \in L^2[0, +\infty)$, then*
$$\lim_{t \to +\infty} f(t) = 0.$$

Theorem 4.6. *Suppose that A has no eigenvalues on the imaginary axis and (A, B) is controllable and (A, C) is observable. And suppose that nonlinear function f_i, $i = 1, \cdots, m$, is piecewise continuously differentiable and $f'_i(\tau)$ is bounded. If system (4.16) has isolated equilibria and there exist diagonal matrices P and Q with $Q \geq 0$, a scalar $\epsilon > 0$ and a symmetric matrix W such that the following LMI is feasible*
$$\begin{pmatrix} WA + A^T W - C^T \Gamma \Delta QC + \epsilon A^T C^T CA \\ B^T W + PCA + \epsilon B^T C^T CA - \frac{1}{2}(\Gamma + \Delta)QC \\ WB + A^T C^T P + \epsilon A^T C^T CB - \frac{1}{2}C^T Q(\Gamma + \Delta) \\ PCB + B^T C^T P - Q + \epsilon B^T C^T CB \end{pmatrix} \leq 0,$$
(4.29)
then system (4.16) is dichotomous for all $f \in [\Gamma, \Delta]$.

Proof. For a given symmetric matrix W, take $W(t) = x^T(t) W x(t)$ and
$$V(x) = \frac{dW(t)}{dt} - f^T(y) P \dot{y} + \epsilon \dot{y}^T \dot{y},$$
where P and ϵ are required as in the theorem. Note that
$$V(x) = 2x^T W(Ax - Bf(y)) - f^T(y) P(CAx - CBf(y)) \\ + \epsilon (CAx - CBf(y))^T (CAx - CBf(y)).$$
(4.30)

Similarly to the proof of Theorem 4.3, (4.29) guarantees
$$V(x) \leq f^T(y) Q f(y) - f^T(y)(\Gamma + \Delta) QCx + x^T C^T \Gamma \Delta QCx,$$

$\forall (x, f(y)) \neq 0$. By (4.25), we know $V(x) \leq 0$. From 0 to $t(t \geq 0)$, integrating $V(t)$ gives

$$W(t) - W(0) \leq \int_0^t f^T(y) P \dot{y} dt - \epsilon \int_0^t \dot{y}^T \dot{y} dt$$

or

$$\epsilon \int_0^t \dot{y}^T \dot{y} dt \leq \sum_{i=1}^m p_i \int_{y_i(0)}^{y_i(t)} f_i(y_i) dy_i - W(t) + W(0),$$

where p_i is the i-th diagonal element of P. For any bounded solution of (4.16), $y(t)$ and $W(t)$ are also bounded. Therefore, it follows from the above inequality that

$$\dot{y}_i \in L^2[0, +\infty), \quad i = 1, \cdots, m. \tag{4.31}$$

Since $f_i(y_i), i = 1, \cdots, m$, has a bounded derivative for almost all $t \geq 0$, $\dot{y}_i$ has a bounded derivative too. Hence, $\dot{y}$ is uniformly continuous on $[0, +\infty)$. Further by lemma 4.3, we have

$$\dot{y} \to 0, \quad \text{as} \quad t \to +\infty. \tag{4.32}$$

Since the trajectory $y(t)$ is bounded, its w-limit point set Ω is nonempty. Then for any trajectory belonging to Ω it is true that

$$\dot{y}(t) = 0$$

and consequently

$$y(t) = y_0, \quad y_0 \text{ is a constant vector.}$$

From system (4.16) we have that for trajectories belonging to Ω it is true that

$$CAx(t) - CBf(y(t)) = 0, \quad f(y(t)) = f(y_0).$$

Thus for trajectories belonging to Ω we have

$$CAx(t) = CBf(y_0) \quad \text{and consequently} \quad CA\dot{x}(t) = 0.$$

Combining with system (4.16) we can get the following algebraic equations

$$\begin{aligned} CAx &= CBf(y_0), \\ CA^2 x &= CABf(y_0), \\ CA^3 x &= CA^2 B f(y_0), \\ &\cdots \cdots \cdots \\ CA^n x &= CA^{n-1} B f(y_0). \end{aligned} \tag{4.33}$$

By the observability of (A, C) and the nonsingularity of A, we know that (4.33) has a unique solution x_0 for given $f(y_0)$. And if x_0 is a solution of (4.33), then we have that
$$Ax_0 - Bf(y_0) = 0, \quad \text{i.e.,} \quad x_0 = A^{-1}Bf(y_0).$$
Noticing the equilibrium set of (4.16) is $\Lambda = \{x_{eq} \mid x_{eq} = A^{-1}Bf(y_{eq}), y_{eq} = Cx_{eq}\}$. Therefore $\Lambda = \Omega$. Then according to (4.32), we have that every bounded solution of (4.16) tends to Λ as $t \to +\infty$. In addition (4.16) has isolated equilibria, so every bounded solution tends to a certain equilibrium of (4.16). □

The condition for dichotomy of system (4.16) is given by LMI (4.29) in Theorem 4.6. This LMI can be easily solved by the powerful LMI toolbox and can also be used to consider the controller design problem. By Corollary 1.4, we can also establish frequency domain condition in which we can see the effects of the sector condition (4.2).

Theorem 4.7. *Suppose that (A, B) is controllable and A has no eigenvalues on the imaginary axis. Then, the LMI (4.29) is feasible if and only if the following frequency domain inequality holds*
$$\mathbf{Re}\{jwPG(jw) - (I + \Gamma G(jw))^*Q(I + \Delta G(jw))\} + \epsilon w^2 G^*(jw)G(jw) \leq 0, \tag{4.34}$$
$\forall w \in \mathbb{R} \cup \{\infty\}$.

Proof. Similarly to the proof of Theorem 4.4, by Corollary 1.4, (4.29) holds if and only if
$$\mathbf{Re}\{2jwPG(jw) - (I + \Gamma G(jw))^*Q(I + \Delta G(jw))\} + \epsilon w^2 G^*(jw)G(jw) \leq 0,$$
$\forall w \in \mathbb{R} \cup \{\infty\}$. Obviously, by the choice of P, the coefficient 2 can be deleted. This completes the proof. □

From the frequency domain inequality (4.34), we can see that the sector condition (4.17) plays the same role for dichotomy as for the canonical Circle and Popov criteria for absolute stability.

Remark 4.12. From Theorems 4.4, 4.5, 4.6 and 4.7, we can see some differences between the conditions of dichotomy and absolute stability. First we know that if (4.20) holds, then there exists a sufficiently small scalar $\epsilon > 0$ such that
$$\mathbf{Re}\{jwPG(jw) - (I + \Gamma G(jw))^*Q(I + \Delta G(jw))\} + \epsilon w^2 G^*(jw)G(jw) < 0. \tag{4.35}$$

The inequality (4.34) for dichotomy is not strict, but the inequality (4.35) for absolute stability is a strict inequality. In Theorems 4.6 and 4.7, $A - B\Gamma C$ is not necessarily stable. And Theorem 4.6 allows the existence of multiple equilibria. In contrast, absolute stability means global stability of the unique zero equilibrium, and the stability of the corresponding linear system is a necessary result. Of course, if system (4.16) is absolutely stable, it is also dichotomous. On the other hand, in order to establish the condition of dichotomy in Theorem 4.6, we required that f has bounded derivatives for almost all $t \geq 0$. Generally, this additional assumption should lead to a less conservative condition (see the following section).

Remark 4.13. Compared with the property of dichotomy, the absolute stability is a strong property for global convergence of all solutions to Lur'e systems. When we only care about the convergence of bounded solutions, the dichotomy is a very important property for nonlinear systems, which has many applications to nonlinear analysis and control [Leonov et al. (1996)]. In the study of chaos, it is very hard to know the existence of chaotic attractors in theory [Chen and Dong (1998); Madan (1993); Stewart (1988)]. Hence, it is interesting to know the nonexistence of chaotic solutions by the study of dichotomy (obviously, dichotomous systems have no chaotic solutions). And the dichotomy property is an important aspect in view of the existence and nonexistence of complex behaviors in dynamical systems [Chua (1998)]. In fact, for a linear system $\dot{x} = Ax$, the dichotomy is equivalent to that A has no eigenvalues on the imaginary axis. Therefore, for nonlinear system (4.16), that A has no eigenvalues on the imaginary axis is also important, then an LMI condition (4.29) or a frequency-domain condition (4.34) guarantees the dichotomy of (4.16).

4.6 Bounded derivative conditions

In Kalman conjecture, the nonlinear functions are supposed to have bounded derivatives. Generally, the more we know about the nonlinear functions, the more are expected from the the criteria for stability/dichotomy. The slope condition has been used to reduce the conservativeness of stability criteria for a long time, see [Park (2002); Suykens et al. (1998); Leonov et al. (1996)] and references therein. In the following, we discuss the sector characteristic and bounded derivative condition for dichotomy in a unified framework.

Suppose that the nonlinear function in system (4.16) is continuously differentiable for almost all $\tau \in \mathbb{R}$ and satisfies

$$\theta_i \leq \frac{df_i(\tau)}{d\tau} \leq \lambda_i, \; i = 1, \cdots, m. \tag{4.36}$$

Let

$$z = \begin{pmatrix} x \\ f(y) \end{pmatrix}, \; \tilde{A} = \begin{pmatrix} A & -B \\ 0 & 0 \end{pmatrix}, \; L = \begin{pmatrix} 0 \\ I_m \end{pmatrix}, \; C_0^T = \begin{pmatrix} C^T \\ 0 \end{pmatrix}.$$

Then system (4.16) can be rewritten as

$$\dot{z} = \tilde{A}z + L\psi(y), \; y = C_0 z \tag{4.37}$$

where $\psi(y) = f'(y)\dot{y}$.

Let $\Theta = diag(\theta_1, \cdots, \theta_m)$, $\Lambda = diag(\lambda_1, \cdots, \lambda_m)$, $C_\Gamma = (-\Gamma C \;\; I)$, $C_\Delta = (-\Delta C \;\; I)$, $\tilde{C} = C_0 \tilde{A} = (CA \;\; -CB)$. Comparing conditions (4.17) and (4.36), we know that in general $\Gamma \geq \Theta$ and $\Lambda \geq \Delta$. By these notations and the method above, we can improve the condition for dichotomy with both conditions (4.17) and (4.36).

Theorem 4.8. *Suppose that A has no pure imaginary eigenvalues and (A, B) is controllable and (A, C) is observable. And suppose that system (4.16) has isolated equilibria. Then system (4.16) is dichotomous for all f with conditions (4.17) and (4.36), if there exist diagonal matrices P, Q and R with $Q \geq 0$ and $R \geq 0$, and a symmetric matrix W and a scalar $\epsilon > 0$ such that the following LMI is feasible*

$$\begin{pmatrix} W\tilde{A} + \tilde{A}^T W - \mathbf{Re}\{C_\Gamma^T Q C_\Delta\} - \tilde{C}^T \Theta R \Lambda \tilde{C} - LP\tilde{C} - \tilde{C}^T PL^T + \epsilon \tilde{C}^T \tilde{C} \\ L^T W + \frac{1}{2}(\Theta + \Lambda)R\tilde{C} \\ WL + \frac{1}{2}\tilde{C}^T R(\Theta + \Lambda) \\ -R \end{pmatrix} \leq 0,$$

(4.38)

where $\mathbf{Re}\{C_\Gamma^T Q C_\Delta\} = \frac{1}{2} C_\Gamma^T Q C_\Delta + \frac{1}{2} C_\Delta^T Q C_\Gamma$.

Proof. Similar to the proof of Theorem 4.6, take $W(t) = z^T(t)Wz(t)$ and

$$V(z) = \frac{dW(t)}{dt} - 2f^T(y)P\dot{y} + \epsilon \dot{y}^T \dot{y}.$$

Note that

$$V(z) = 2z^T W(\tilde{A}z + L\psi(y)) - 2z^T LP\tilde{C}z + \epsilon z^T \tilde{C}^T \tilde{C} z.$$

For all $z_i \neq 0$, similar to the sector condition, the bounded derivative condition (4.36) is equivalent to

$$u_i(x_i) =: (\psi_i(y_i) - \theta_i y_i)(\psi_i(y_i) - \lambda_i y_i) \leq 0,$$

4.6. Bounded derivative conditions

where $\psi_i(y_i) = df_i(y_i)/dt$. Then noticing that for $r_i \geq 0$, $i = 1, \cdots, m$, we have

$$\sum_{i=1}^{m} r_i u_i(x_i) = \psi^T(y) R \psi(y) - \psi^T(y)(\Theta + \Lambda) R \tilde{C} z + z^T \tilde{C}^T \Theta \Lambda R \tilde{C} z \leq 0, \quad (4.39)$$

where $R = diag(r_1, \cdots, r_m)$. Noticing the notations of C_Γ and C_Δ, (4.25) can be rewritten as

$$\sum_{i=1}^{m} q_i s_i(x_i) = z^T C_\Gamma^T Q C_\Delta z \leq 0. \quad (4.40)$$

In addition, (4.38) guarantees that

$$V(z) \leq z^T C_\Gamma^T Q C_\Delta z + \psi^T(y) R \psi(y) - \psi^T(y)(\Theta + \Lambda) R \tilde{C} z + z^T \tilde{C}^T \Theta \Lambda R \tilde{C} z,$$

$\forall (z, \psi(y)) \neq 0$. Then by (4.39) and (4.40), we know that $V(z) \leq 0$. Similarly to the proof of Theorem 4.6, we know that system (4.16) is dichotomous. $\square$

As discussed in Theorem 4.7, we can also give a frequency domain interpretation for the LMI (4.38).

Theorem 4.9. *Suppose that (A, B) is controllable and (A, C) is observable, and A has no eigenvalues on the imaginary axis. Then, the LMI (4.38) is feasible if and only if the following frequency domain inequality holds*

$$\mathbf{Re}\{jw PG(jw) - (I + \Gamma G(jw))^* Q(I + \Delta G(jw)) + \epsilon w^2 G^*(jw) G(jw)$$
$$-w^2(I + \Theta G(jw))^* R(I + \Lambda G(jw))\} \leq 0, \quad \forall w \in \mathbb{R} \cup \{\infty\}. \quad (4.41)$$

Proof. Obviously, we have

$$C_\Gamma(sI - \tilde{A})^{-1} L = \frac{1}{s} I + \frac{1}{s} \Gamma G(s), \quad C_\Delta(sI - \tilde{A})^{-1} L = \frac{1}{s} I + \frac{1}{s} \Delta G(s),$$

$$\tilde{C}(sI - \tilde{A})^{-1} L = -G(s), \quad L^T(sI - \tilde{A})^{-1} L = \frac{1}{s} I.$$

Then similarly to the proof of Theorem 4.4, by Corollary 1.4 and Remark 3.4, (4.38) holds if and only if the following inequality holds

$$\mathbf{Re}\{\tfrac{2}{jw} PG(jw) - \tfrac{1}{w^2}(I + \Gamma G(jw))^* Q(I + \Delta G(jw)) + \epsilon G^*(jw) G(jw)$$
$$-(I + \Theta G(jw))^* R(I + \Lambda G(jw))\} \leq 0, \quad \forall w \in \mathbb{R} \cup \{\infty\}.$$

The equivalence between the inequality above and (4.41) completes the proof. $\square$

Remark 4.14. Similarly to Remark 4.3, from the inequality (4.41) we can see that the term $(I+\Gamma G(jw))^*Q(I+\Delta G(jw))$ is related to the sector condition (4.25) and $(I+\Theta G(jw))^*R(I+\Lambda G(jw))$ is related to the bounded derivative condition (4.36). The effects of (4.25) and (4.36) appear in a unified framework. Comparing with Theorem 4.7, we know that if we do not use the condition of bounded derivatives, Theorem 4.9 reduces to Theorem 4.8 ($R = 0$). Generally, we can see from the forthcoming example that Theorem 4.9 is less conservative than Theorem 4.8.

Example 4.1. Consider the following system
$$\dot{x}_1 = x_1 - x_3 + \alpha h(x_3),$$
$$\dot{x}_2 = -x_2 - g(x_2) + 2x_3 + 2h(x_3), \quad (4.42)$$
$$\dot{x}_3 = 5x_1 - x_2 - g(x_2) - 2x_3 + \beta h(x_3).$$

This system can be written as a system in the form of (4.16) with

$$A = \begin{pmatrix} 1 & 0 & -1 \\ 0 & -1 & 2 \\ 5 & -1 & -2 \end{pmatrix}, B = \begin{pmatrix} 0 & -\alpha \\ 1 & -2 \\ 1 & -\beta \end{pmatrix}, C = \begin{pmatrix} 0 & 1 & 0 \\ 0 & 0 & 1 \end{pmatrix}, f(y) = \begin{pmatrix} g(x_2) \\ h(x_3) \end{pmatrix}.$$
(4.43)

In what follows, we discuss this system with different values of α and β.

Case 1 $\alpha = -1$ and $\beta = -2$. By Theorem 4.3, we know that at this time system (4.42) is absolutely stable for the sectors
$$x_2^2 \leq x_2 g(x_2) \leq 500 x_2^2, \quad x_3^2 \leq x_3 h(x_3) \leq 500 x_3^2. \quad (4.44)$$

Case 2 $\alpha = 0$ and $\beta = -2$. In this case, $A - B\Gamma C$ is not stable. Of course, system (4.42) is not absolutely stable. However, by Theorem 4.6, we know that this system is dichotomous for the nonlinearities in the sectors (4.44) with bounded derivatives.

Case 3 $\alpha = 0.1$ and $\beta = 1.2$. In this case, $A - B\Gamma C$ is not stable too. By Theorem 4.6, we know that the system above is dichotomous for the nonlinearities in the sectors
$$x_2^2 \leq x_2 g(x_2) \leq 20 x_2^2, \quad x_3^2 \leq x_3 h(x_3) \leq 2.6 x_3^2.$$
with bounded derivatives. However, Theorem 4.6 does not hold for the sectors
$$x_2^2 \leq x_2 g(x_2) \leq 20 x_2^2, \quad x_3^2 \leq x_3 h(x_3) \leq 2.7 x_3^2. \quad (4.45)$$
But by Theorem 4.8, we can still know that at this time the above system is dichotomous with the sectors (4.45) and the condition of bounded derivatives
$$0.8 \leq \frac{dg(x_2)}{dx_2} < 50, \quad 0.8 \leq \frac{dh(x_3)}{dx_3} < 50.$$
Therefore, generally Theorem 4.8 is less conservative than Theorem 4.6.

The methods established in this chapter can also be used to study the canonical Chua's circuit, see Chapter 11 for details.

4.7 Notes and references

The absolute stability of Lur'e systems which are composed of a linear system and a feedback nonlinear loop has been extensively studied for several decades and different criteria have been established, see [Lur'e and Postnikov (1944); Popov (1973); Kalman (1957); Yakubovich (1962); Krasovsky (1952); Khalil (2002)] and references therein. The frequency-domain condition based on transfer functions for Lur'e systems is popular in engineering since it can be effectively graphically tested [Narendra and Taylor (1973); Huang (2003); Leonov et al. (1996)]. An important concept in control theory related to such a method is the notion of positive realness which has many applications to passivity analysis and control [Desoer and Vidyasagar (1975); Lozano et al. (2000)], quadratic optimal control [Huang (1964); Anderson and Moore (1970)], and adaptive system theory [Landau (1979)]. On the other hand, the time-domain method based on Lyapunov functions is effective in computation and controller design [Yakubovich (1962); Arcak et al. (2003); de Oliveira et al. (2002a)] because of the development of linear matrix inequality (LMI) theory [Boyd et al. (1994); Gahinet et al. (1995)]. And the remarkable Kalman-Yakubovich-Popov (KYP) lemma [Yakubovich (1962); Leonov et al. (1996, 1992b); Rantzer (1996)] established the equivalence between frequency-domain conditions (e.g., circle and Popov criteria) and time-domain conditions (LMIs) for absolute stability of Lur'e systems. This lemma has been known as one of the most important results in systems and control theory. At the beginning study of Lur'e systems, the stability conditions were applicable to single-input and single-output (SISO) systems. Later on, they were generalized to multi-input and multi-output (MIMO) systems, e.g., the Popov criterion was extended to multiple sector restricted Lur'e systems in [Park (1997)] and it was shown that the diagonal parameter matrix in Popov multiplier does not need to be positive. And more exact criteria were established for absolute stability with sector and slope conditions, see [Park (2002); Suykens et al. (1998); Leonov et al. (1996); Huang (2003)] and references therein. The boundedness without absolute stability for a special class of Lur'e systems was studied in [Arcak et al. (2002)]. The study of absolute stability is of fundamental importance in many problems of control and electrical circuits, e.g.,

Lur'e systems play important roles in neural networks [Guzelis and Chua (1993); Kaszkurewicz and Bhaya (1994); Liao (2000)] and synchronization theory [Curran and Chua (1997); Wu and Chua (1994); Wu (2002)]. In the latter case, Chua's circuit [Chua (1994); Madan (1993)] which is representable in Lur'e form, is often employed in the synchronization schemes. Besides absolute stability, dichotomy characteristic of Lur'e systems was studied in [Leonov et al. (1996)], which is important for the test of the nonexistence of chaotic and periodic solutions. The dichotomy property was used to study chaos control in Chua's circuits [Leonov et al. (1996); Duan et al. (2005b); Wang et al. (2006d)].

Chapter 5

Pendulum-like Feedback Systems

This chapter introduces two kinds of special form of pendulum-like feedback systems and discusses their global properties including dichotomy, gradient-like property, Lagrange stability and Bakaev stability as well as various stability test criteria. The arrangement of this chapter is as follows: Section 5.1 presents several examples which can be written in a unified system form with a linear part and a periodic nonlinearity. Section 5.2 introduces the concept and two forms of pendulum-like feedback systems and discusses the relationship of the two forms. Section 5.3 gives the time-domain conditions related to linear matrix inequality (LMI) and frequency-domain conditions to test dichotomy of pendulum-like feedback systems. Sections 5.4, 5.5 and 5.6, respectively, discuss the time-domain and frequency-domain conditions guaranteeing gradient-like property, Lagrange stability and Bakaev stability.

5.1 Several examples

Example 5.1. Consider the canonical mathematical pendulum equation

$$\ddot{\sigma} + a\dot{\sigma} + \sin\sigma - \gamma = 0, \quad a > 0, \gamma \in (0, 1). \tag{5.1}$$

A simple generalization of (5.1) gives the equation

$$\ddot{\sigma} + a\dot{\sigma} + \varphi(\sigma) = 0, \quad a > 0, \tag{5.2}$$

where $\varphi(\sigma)$ is a Δ-periodic continuous function. Let $\eta = \dot{\sigma}$. Then (5.2) can be rewritten as

$$\begin{aligned}\dot{\sigma} &= \eta \\ \dot{\eta} &= -a\eta - \varphi(\sigma).\end{aligned} \tag{5.3}$$

The above second-order systems were studied in detail in [Leonov et al. (1996)]. Obviously, system (5.3) has infinite equilibria and includes a linear part and a nonlinearity part as in the Lur'e system studied in Chapter 4.

Example 5.2. Consider the equation which describes the work of synchronous single-phase motor with pulsing vibrating moment [Morary (1970); Leonov et al. (1996)]

$$\ddot{\zeta} + a\dot{\zeta} + \beta \cos t \cos \zeta + \gamma \sin 2\zeta + \delta \cos 2t \sin 2\zeta = 0, \qquad (5.4)$$

where $a > 0$, β, γ and δ are numbers. With the help of $\sigma = \zeta + \pi/2$, we have

$$\ddot{\sigma} + a\dot{\sigma} + \varphi(t, \sigma) = 0, \qquad (5.5)$$

where $\varphi(t, \sigma) = \beta \cos t \sin \sigma - \gamma \sin 2\sigma - \delta \cos 2t \sin 2\sigma$. The boundedness of all solutions of (5.4) was studied in [Leonov et al. (1996)]. Obviously, (5.5) can also be written in the form (5.3) with a time-varying nonlinearity.

Example 5.3. Let us consider an example of concrete systems studied in the theory of phase-locked loops (PLL). The locking-in of an arbitrary solution to an equilibrium point is just a synonym of the global asymptotic stability for PLL. The locking-in property is extremely significant for PLL. Many papers and monographs are devoted to it, for example [Gardner (1979); Abramovitch (1990); Wu (2002)].

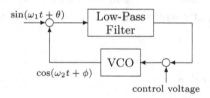

Fig. 5.1   Phase-locked loop.

The block diagram shown in Fig. 5.1 is close to the actual implementation of a PLL. Then with some assumptions the model in Fig. 5.1 can be simplified to the one in Fig. 5.2 and the voltage controlled oscillator (VCO) can be considered as an integrator [Abramovitch (1990)]. The model shown in Fig. 5.2 is rearranged so that all linear components appear in the forward path and the nonlinear component appears in the feedback path as shown in Fig. 5.3. Both the reference phase signal and the VCO control signal are set to zero in Fig. 5.3.

5.1. Several examples

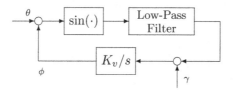

Fig. 5.2 Model of Phase-locked loop.

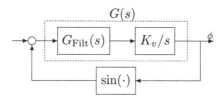

Fig. 5.3 Rearrangement into a linear component with a feedback nonlinear component.

From Fig. 5.3, the transfer function from the input $\sin(\cdot)$ to the output $-\phi$ is

$$G(s) = \frac{1}{s}K(s)$$

where $K(s)$ is the transfer function from $\sin(\cdot)$ to $-\dot{\phi}$ and described by state-space model of the form

$$\begin{aligned}
\dot{x} &= \begin{pmatrix} -63 & -20 \\ 32 & 0 \end{pmatrix} x + \begin{pmatrix} 8 \\ 0 \end{pmatrix} \xi \\
\dot{\phi} &= \begin{pmatrix} 2 & -2 \end{pmatrix} x - 0.5\xi \\
\xi &= \sin\phi
\end{aligned} \qquad (5.6)$$

System (5.6) describes the dynamics of an autonomous phase-locked loop with a second order filter of type "2/2" [Leonov et al. (1996)] and a 2π-periodic input nonlinearity.

A large class of practical engineering systems as described above can be written in a system form including a linear part and a periodic nonlinearity, which has infinite equilibria. Such kind of systems were called pendulum-like systems in [Leonov et al. (1996)]. The global properties of pendulum-like systems will be studied in detail in this chapter.

5.2 Pendulum-like feedback systems

Consider the ordinary differential equation

$$\dot{x} = f(t, x) \quad (f : \mathbb{R}_+ \times \mathbb{R}^n \to \mathbb{R}^n) \tag{5.7}$$

where $f : \mathbb{R}_+ \times \mathbb{R}^n \to \mathbb{R}^n$ is continuous, locally Lipschitz continuous in the second argument. Suppose that every solution $x(t, t_0, x_0)$ of (5.7) with $x(t_0, t_0, x_0) = x_0$, $t_0 \geq 0$ and $x_0 \in \mathbb{R}^n$ may be continued to $[t_0, +\infty]$.

Let $\Gamma := \left\{ \sum_{j=1}^{m} k_j d_j \mid k_j \in \mathbb{Z}, 1 \leq j \leq m \right\}$, where $d_j \in \mathbb{R}^n$ are supposed to be linearly independent ($m \leq n$).

Definition 5.1. We say that (5.7) is pendulum-like with respect to Γ if for any solution $x(t, t_0, x_0)$ of (5.7) we have

$$x(t, t_0, x_0 + d) = x(t, t_0, x_0) + d \tag{5.8}$$

for all $t \geq t_0$ and $d \in \Gamma$.

The property (5.8) is just like the characteristic of mathematical pendulum equation, so system (5.7) with the property (5.8) is called pendulum-like in [Leonov et al. (1996)].

Theorem 5.1. *System (5.7) is pendulum-like with respect to Γ if and only if*

$$f(t, x + d) = f(t, x) \tag{5.9}$$

for all $t \geq 0, x \in \mathbb{R}^n$ and $d \in \Gamma$.

Proof. Suppose that (5.9) holds. Let $y(t) = x(t, t_0, x_0) + d$. Then $y(t_0) = x_0 + d$ and

$$\dot{y}(t) = \dot{x}(t, t_0, x_0) = f(t, x(t, t_0, x_0)) = f(t, x(t, t_0, x_0) + d) = f(t, y)$$

which implies that $y(t)$ is a solution of (5.7). Thus $x(t, t_0, x_0 + d) = y(t)$ by uniqueness of solution. This derives $x(t, t_0, x_0 + d) = x(t, t_0, x_0) + d$.

Conversely, suppose that (5.7) is a pendulum-like system with respect to Γ. That is, $x(t, t_0, x_0 + d) = x(t, t_0, x_0) + d$ for arbitrary $(t_0, x_0) \in \mathbb{R}_+ \times \mathbb{R}^n$ and $d \in \Gamma$. Then $\dot{x}(t, t_0, x_0 + d) = \dot{x}(t, t_0, x_0)$ and thus $f(t, x(t, t_0, x_0)) = f(t, x(t, t_0, x_0 + d))$. Let $t = t_0$. Then $f(t_0, x_0) = f(t_0, x_0 + d)$ and thus the result follows by arbitrariness of t_0 and x_0. □

Remark 5.1. If the equilibrium point $x(t) = x_{eq}$ is a solution of (5.7) then $x(t) = x_{eq} + d$ ($d \in \Gamma$) is also a solution of (5.7). So the set of equilibria of (5.7) is either empty or infinite.

5.2.1 The first canonical form of pendulum-like feedback system

Consider a feedback control system

$$\dot{x} = Ax + B\varphi(t,y), \quad y = Cx, \tag{5.10}$$

where $A \in \mathbb{R}^{n \times n}$, $B \in \mathbb{R}^{n \times m}$ and $C \in \mathbb{R}^{m \times n}$, and

$$\begin{aligned}\varphi(t,y) &= (\varphi_1(t,y_1), \varphi_2(t,y_2), \cdots, \varphi_m(t,y_m))^T, \\ \varphi_i(t,y_i) &\text{ is } \Delta_i - \text{periodic with respect to } y_i, \, i = 1,2,\cdots,m.\end{aligned} \tag{5.11}$$

Theorem 5.2. *Assume that system (5.10) is a pendulum-like with respect to* $\Gamma = \{jd, j \in \mathbf{Z}\}$ *and* (A, C) *is observable. Then without loss of generality it can be assumed that* $Ad = 0$ *and* $\varphi_i(t, y_i)$ *is* $e_i^T Cd$-*periodic with respect to* y_i *for* $i = 1, 2, \cdots, m$, *where* e_i *is a vector with the* i-*th element being* 1 *and others being zero.*

Proof. Since system (5.10) is pendulum-like with respect to $\Gamma = \{jd, j \in \mathbf{Z}\}$, then by Theorem 5.1 we have

$$Ad + B\varphi(t, y + Cd) = B\varphi(t,y) \tag{5.12}$$

for all $(t,y) \in \mathbb{R}_+ \times \mathbb{R}^m$. In what follows, we show that there exists i, $1 \leq i \leq m$ such that $e_i^T Cd \neq 0$. Suppose to the contrary, that is, $e_i^T Cd = 0$ for all $i = 1, 2, \cdots, m$. Then it follows from (5.12) that

$$CA^k d = 0, \quad k = 0, 1, \cdots, n-1. \tag{5.13}$$

Let $N^T = [C^T \quad A^T C^T \quad \cdots \quad (A^{n-1})^T C^T]$. Then $Nd = 0$ and the matrix N is with full column rank by observability of (A, C). This implies that $d = 0$, which contradicts that (5.10) is pendulum-like.

Without loss of generality, we suppose that $e_i^T Cd \neq 0$ for $1 \leq i \leq k$ and $e_i^T Cd = 0$ for $k < i \leq m$. Rewriting system (5.10) in the form

$$\begin{aligned}\dot{x} &= (A - TC)x + \bar{\varphi}(t,y) \\ y &= Cx\end{aligned} \tag{5.14}$$

where

$$T = \frac{1}{k} Ad \left(\frac{1}{e_1^T Cd} \quad \cdots \quad \frac{1}{e_k^T Cd} \quad 0 \quad \cdots \quad 0 \right)$$

and

$$\bar{\varphi}(t,y) = B\varphi(t,y) + Ty \tag{5.15}$$

Obviously, it can be derived that $(A - TC)d = 0$ and thus

$$\bar{\varphi}(t, y + Cd) = B\varphi(t, y + Cd) + T(y + Cd)$$
$$= B\varphi(t, y) - Ad + T(y + Cd)$$
$$= \bar{\varphi}(t, y) - Ad + TCd = \bar{\varphi}(t, y)$$

from (5.12) and (5.15). That is, $\bar{\varphi}_i(t, y_i)$ is $e_i^T Cd$-periodic with respect to y_i for $i = 1, 2, \cdots, m$. Therefore, we have shown that the new system (5.14) has the properties of Theorem 5.2. $\square$

Theorem 5.3. *Suppose that*

(i) A has m zero eigenvalues and the multiplicity of elementary factor is 1 corresponding to each zero eigenvalue.
(ii) (5.11) holds.
(iii) (A, B) is controllable and (A, C) is observable.

Then system (5.10) is pendulum-like.

Proof. By assumption (i) we can choose a nonsingular matrix T such that

$$T^{-1}AT = \begin{pmatrix} A_1 & 0 \\ A_2 & 0 \end{pmatrix} = \bar{A}$$

where $A_1 \in \mathbb{R}^{(n-m)\times(n-m)}$. Let $x = Tz$. Then system (5.10) can be converted into

$$\begin{aligned} \dot{z} &= \bar{A}z + T^{-1}B\varphi(t, y) \\ y &= [C_1 \ C_2]z \end{aligned} \quad (5.16)$$

where $[C_1 \ C_2] = CT$, $C_1 \in \mathbb{R}^{m\times(n-m)}$ and $C_2 \in \mathbb{R}^{m\times m}$. By observability of (A, C), C_2 is nonsingular. Let

$$\Gamma = \left\{ d = \begin{pmatrix} 0_{(n-m)\times m} \\ C_2^{-1} D_\Delta \end{pmatrix} \begin{pmatrix} k_1 \\ \vdots \\ k_m \end{pmatrix} \middle| k_i \in \mathbf{Z} \right\} \quad (5.17)$$

where $D_\Delta = diag(\Delta_1, \cdots, \Delta_m)$. By condition (ii),

$$\bar{A}(z + d) + T^{-1}B\varphi(t, (C_1 \ C_2)(z + d))$$
$$= \bar{A}z + T^{-1}B\varphi\left(t, (C_1 \ C_2)z + D_\Delta \begin{pmatrix} k_1 \\ \vdots \\ k_m \end{pmatrix}\right)$$
$$= \bar{A}z + T^{-1}B\varphi\left(t, (C_1, C_2)z\right)$$

implies that system (5.16) is pendulum-like with respect to Γ by Theorem 5.1.

Further, let $\Gamma_0 = \{Td, d \in \Gamma\}$. Then by $\tilde{A}d = T^{-1}ATd = 0$, we have
$$A(x+Td) + B\varphi(t, Cx + CTd)$$
$$= Ax + B\varphi\left(t, Cx + D_\Delta \begin{pmatrix} k_1 \\ \vdots \\ k_m \end{pmatrix}\right)$$
$$= Ax + B\varphi(t, Cx),$$
which implies that system (5.10) is pendulum-like with respect to Γ_0. $\square$

System (5.10) satisfying the conditions of Theorem 5.3 is called the first canonical form of pendulum-like feedback system. From (5.17), we can see the relationship between Γ and Δ_i (the period of nonlinear function $\varphi_i(t, y_i)$).

5.2.2 The second canonical form of pendulum-like feedback system

Consider the following multi-input and multi-output system,
$$\begin{cases} \dfrac{dx}{dt} = Ax + B\varphi(t, y) \\ \dfrac{dy}{dt} = Cx + D\varphi(t, y) \end{cases} \quad (5.18)$$
where $A \in \mathbb{R}^{n \times n}, B \in \mathbb{R}^{n \times m}, C \in \mathbb{R}^{m \times n}, D \in \mathbb{R}^{m \times m}, \varphi(t, y) = (\varphi_1(t, y_1), \cdots, \varphi_m(t, y_m))^T$. Suppose that $\varphi_i : \mathbb{R}_+ \times \mathbb{R} \to \mathbb{R}$ is Δ_i−periodic with the second argument and local Lipschitz continuous, where $i = 1, \cdots, m$. Viewing $\dot{y}$ as the output of the system, then the transfer function from $\varphi(t, y)$ to $\dot{y}$ is
$$G(s) = C(sI - A)^{-1}B + D.$$
Let
$$\Gamma = \{d = (0, \cdots, 0, k_1\Delta_1, \cdots, k_m\Delta_m)^T \mid k_i \in \mathbf{Z}\}$$
and $\eta = \begin{pmatrix} x \\ y \end{pmatrix}$. Viewing (5.18) as
$$\dot{\eta} = f(t, \eta), \quad (5.19)$$
where $f : \mathbb{R}_+ \times \mathbb{R}^{n+m} \to \mathbb{R}^{n+m}$ is continuous and locally Lipschitz continuous in the second argument, by the periodicity of φ_i system (5.18) satisfies
$$f(t, \eta + d) = f(t, \eta), \quad t \geq 0, \ d \in \Gamma.$$
Therefore (5.18) is a pendulum-like system by Theorem 5.1.

System (5.18) is called the second canonical form of a pendulum-like feedback system.

5.2.3 The relationship between the first and the second forms of pendulum-like feedback systems

From the following discussions, it can be viewed that the two forms of pendulum-like feedback system discussed above are equivalent to each other. First, the first form (5.10) can be transformed into the second form (5.18). In fact, for the first form of pendulum-like feedback system (5.10), since the matrix A has m zero eigenvalues and the multiplicity of elementary factor is 1 corresponding to each eigenvalue. We can find a nonsingular linear transformation $x = T \begin{pmatrix} \xi \\ \eta \end{pmatrix}$ such that system (5.10) can be transformed into

$$\begin{cases} \dot{\xi} = A_1 \xi + B_1 \varphi(t, y), \\ \dot{\eta} = C_1 \xi + B_2 \varphi(t, y), \\ y = C_2 \xi + D_1 \eta, \end{cases} \quad (5.20)$$

where $A_1 \in \mathbb{R}^{(n-m) \times (n-m)}$, B_1, B_2, C_1, C_2 and D_1 are matrices with compatible dimensions. By the observability of (A, C) in (5.10), D_1 is nonsingular. Differentiating y, (5.20) can be rewritten as

$$\begin{cases} \dot{\xi} = A_1 \xi + B_1 \varphi(t, y), \\ \dot{y} = (C_2 A_1 + D_1 C_1) \xi + (C_2 B_1 + D_1 B_2) \varphi(t, y), \end{cases} \quad (5.21)$$

System (5.21) is in the form of (5.18), which is the second form of pendulum-like feedback system. Let the transfer function from $\varphi(t, y)$ to y in (5.10) be

$$\chi(s) = C(sI - A)^{-1} B$$

From (5.20), we know that

$$\chi(s) = \frac{1}{s} G(s), \quad G(s) = (C_2 A_1 + D_1 C_1)(sI - A_1)^{-1} B_1 + C_2 B_1 + D_1 B_2. \quad (5.22)$$

In addition, from the form (5.18), by defining $\eta = y - Cx$ we can get a new system

$$\begin{cases} \dot{x} = Ax + B\varphi(t, y) \\ \dot{\eta} = (C - CA)x + (D - CB)\varphi(t, y) \\ y = Cx + \eta \end{cases}$$

which is just in the form of (5.10).

Remark 5.2. By the above discussions, two canonical forms of pendulum-like systems can be transformed to each other. But for convergent solutions, the convergent equilibrium may be different since the differentiating operator is involved during the transformations.

5.3. Dichotomy of pendulum-like feedback systems

We introduce the following basic concepts for system (5.7) which will be used in subsequent sections. For systems with multiple equilibria, these concepts are important for the study of global properties of the system.

Definition 5.2. If all the solutions of (5.7) are bounded we say that the equation (5.7) is Lagrange stable.

Definition 5.3. Equation (5.7) is said to be dichotomous if its every bounded solution is convergent.

Definition 5.4. Equation (5.7) is said to be gradient-like if its every solution converges to an equilibrium.

5.3 Dichotomy of pendulum-like feedback systems

5.3.1 Dichotomy of the second form of autonomous pendulum-like feedback systems

Consider the second canonical form of autonomous pendulum-like feedback system

$$\begin{cases} \dfrac{dx}{dt} = Ax + B\varphi(y) \\ \dfrac{dy}{dt} = Cx + D\varphi(y) \end{cases} \quad (5.23)$$

where $A \in \mathbb{R}^{n\times n}, B \in \mathbb{R}^{n\times m}, C \in \mathbb{R}^{m\times n}, D \in \mathbb{R}^{m\times m}, \varphi(y) = (\varphi_1(y_1), \cdots, \varphi_m(y_m))^T$. Let $G(s) = C(sI - A)^{-1}B + D$.

Throughout this chapter, The following assumptions for the above system are made.

Assumption 5.1 A has no eigenvalues on the imaginary axis, (A, B) is controllable, (A, C) is observable and $G(0)$ is nonsingular.

Assumption 5.2 $\varphi_i : \mathbb{R} \to \mathbb{R}$ is Δ_i-periodic, local Lipschitz continuous, where $i = 1, \cdots, m$.

Assumption 5.3 φ_i possesses a finite number of isolated zeroes on $[0, \Delta_i)$, where $i = 1, \cdots, m$.

With the Assumptions above, we discuss the set of equilibria of (5.23). Suppose that (x_{eq}, y_{eq}) is an equilibrium of (5.23). Then,

$$Ax_{eq} = -B\varphi(y_{eq}), \quad Cx_{eq} = -D\varphi(y_{eq}).$$

By the non-singularity of A, we get

$$(-CA^{-1}B + D)\varphi(y_{eq}) = 0.$$

By assumption 5.1, $G(0) = D - CA^{-1}B$ is nonsingular which implies $\varphi(y_{eq}) = 0$ and consequently $x_{eq} = 0$. Therefore, the equilibrium set Π of (5.23) is

$$\Pi = \{(x_{eq}, y_{eq}) \mid x_{eq} = 0, \varphi(y_{eq}) = 0\}.$$

Generally, according to the periodicity of φ system (5.23) has infinitely many equilibria. For the case that φ_i has a finite number of isolated zeros on $[0, \Delta_i)$, $i = 1, \cdots, m$, we shall need the following auxiliary result which was introduced in [Leonov et al. (1996)].

Lemma 5.1. *Let $\psi : \mathbb{R} \to \mathbb{R}$ be a continuous Δ-periodic function with a finite number of isolated zeros on $[0, \Delta)$ and $\alpha : \mathbb{R}_+ \to \mathbb{R}$ be a continuous function. If*

$$\lim_{t \to +\infty} \psi(\alpha(t)) = 0,$$

then

$$\lim_{t \to +\infty} \alpha(t) = \hat{\alpha},$$

where $\hat{\alpha}$ is a zero of the function ψ.

In this section we suppose that φ_i is continuously differentiable and

$$\theta_i \leq \frac{d\varphi_i(\tau)}{d\tau} \leq \lambda_i, \ \lambda_i < +\infty, \ \theta_i > -\infty, \ i = 1, \cdots, m. \tag{5.24}$$

Let $\Theta = \text{diag}\{\theta_1, \cdots, \theta_m\}$, $\Lambda = \text{diag}\{\lambda_1, \cdots, \lambda_m\}$, and

$$z = \begin{pmatrix} x \\ \varphi(y) \end{pmatrix}, \ \tilde{A} = \begin{pmatrix} A & B \\ 0 & 0 \end{pmatrix}, \ L = \begin{pmatrix} 0 \\ I_m \end{pmatrix}, \ \tilde{C}^T = \begin{pmatrix} C^T \\ D^T \end{pmatrix},$$

then system (5.23) can be rewritten as

$$\dot{z} = \tilde{A}z + L\psi(y), \ \dot{y} = \tilde{C}z \tag{5.25}$$

where $\psi(y) = \varphi'(y)\dot{y}$.

Theorem 5.4. *Under Assumptions 5.1, 5.2 and 5.3, system (5.23) is dichotomous for all φ with condition (5.24), if there exist diagonal matrices P and R with $R \geq 0$, a symmetric matrix W and a scalar $\epsilon > 0$ such that the following LMI is feasible*

$$\begin{pmatrix} W\tilde{A} + \tilde{A}^T W - \tilde{C}^T \Theta R \Lambda \tilde{C} + S & WL + \frac{1}{2}\tilde{C}^T R(\Theta + \Lambda) \\ L^T W + \frac{1}{2}(\Theta + \Lambda)R\tilde{C} & -R \end{pmatrix} \leq 0, \tag{5.26}$$

where $S = LP\tilde{C} + \tilde{C}^T PL^T + \epsilon \tilde{C}^T \tilde{C}$.

5.3. Dichotomy of pendulum-like feedback systems

Proof. Similarly to the proofs of Theorems 4.6 and 4.8, take $W(t) = z^T(t)Wz(t)$ and

$$V(z) = \frac{dW(t)}{dt}\bigg|_{(5.25)} + 2\varphi^T(y)P\dot{y} + \epsilon \dot{y}^T \dot{y}.$$

Note that

$$V(z) = 2z^T W(\tilde{A}z + L\psi(y)) + 2z^T LP\tilde{C}z + \epsilon z^T \tilde{C}^T \tilde{C}z.$$

For all $z_i \neq 0$, the bounded derivative condition (5.24) is equivalent to

$$u_i(x_i) =: (\psi_i(y_i) - \theta_i \dot{y}_i)(\psi_i(y_i) - \lambda_i \dot{y}_i) \leq 0, \ i = 1, \cdots, m.$$

Then noticing that for any $r_i \geq 0$, $i = 1, \cdots, m$, we have

$$\sum_{i=1}^{m} r_i u_i(x_i) = \psi^T(y)R\psi(y) - \psi^T(y)(\Theta + \Lambda)R\tilde{C}z + z^T \tilde{C}^T \Theta \Lambda R\tilde{C}z \leq 0, \tag{5.27}$$

where $R = \text{diag}\{r_1, \cdots, r_m\}$. In fact, (5.26) guarantees that

$$V(z) \leq \psi^T(y)R\psi(y) - \psi^T(y)(\Theta+\Lambda)R\tilde{C}z + z^T \tilde{C}^T \Theta \Lambda R\tilde{C}z, \quad \forall (z, \psi(y)) \neq 0.$$

By (5.27), it follows that $V(z) \leq 0$. Similarly to the proof of Theorem 4.6, we can get

$$\dot{y}_i \in L^2[0, +\infty), \quad i = 1, \cdots, m. \tag{5.28}$$

Since $\varphi_i(y_i), i = 1, \cdots, m$, has a bounded derivative for almost all $t \geq 0$, $\dot{y}_i$ has a bounded derivative too. Hence, $\dot{y}$ is uniformly continuous on $[0, +\infty)$. Further by Lemma 4.3, we have

$$\dot{y} \to 0, \quad \text{as} \quad t \to +\infty. \tag{5.29}$$

Since the trajectory $y(t)$ is bounded, its ω-limit point set Ω is nonempty. Then for any trajectory belonging to Ω it is true that

$$\dot{y}(t) = 0$$

and consequently

$$y(t) = y_0, \quad y_0 \text{ is a constant vector.}$$

From system (5.23) we have that for trajectories belonging to Ω it is true that

$$Cx(t) + D\varphi(y(t)) = 0, \quad \varphi(y(t)) = \varphi(y_0).$$

Thus for trajectories belonging to Ω we have

$$Cx(t) = -D\varphi(y_0) \quad \text{and consequently} \quad C\dot{x}(t) = 0.$$

Combining with system (5.23) we can get the following algebraic equations
$$Cx = -D\varphi(y_0),$$
$$CAx = -CB\varphi(y_0),$$
$$CA^2x = -CAB\varphi(y_0), \qquad (5.30)$$
$$\cdots \qquad \cdots\cdots\cdots$$
$$CA^{n-1}x = -CA^{n-2}B\varphi(y_0).$$
By the observability of (A, C), we know that (5.30) has a unique solution x_0 for given $\varphi(y_0)$. And if x_0 is a solution of (5.30), then we have that
$$Ax_0 + B\varphi(y_0) = 0, \quad \text{and} \quad Cx_0 + D\varphi(y_0) = 0$$
Hence, from the nonsingularity of $K(0)$ we conclude that $\varphi(y_0) = 0$ and $x_0 = 0$. Noticing the equilibrium set of (5.23) is $\Pi = \{(x_{eq}, y_{eq}) \mid x_{eq} = 0, \varphi(y_{eq}) = 0\}$. Therefore $\Pi = \Omega$. Then according to (5.29), we have that every bounded solution of (5.23) tends to Π as $t \to +\infty$. In addition equation (5.23) has isolated equilibria by Assumption 5.3, so every bounded solution tends to an equilibrium of (5.23) from Lemma 5.1. □

Theorem 5.5. *With Assumption 5.1, the LMI (5.26) is feasible if and only if the following frequency domain inequality holds*
$$\mathbf{Re}\{PG(jw) - (jwI - \Theta G(jw))^*R(jwI - \Lambda G(jw))\} + \epsilon G^*(jw)G(jw) \leq 0, \qquad (5.31)$$
$\forall w \in \mathbb{R} \cup \{\infty\}$.

Proof. Similarly to the proof of Theorem 4.4, by Corollary 1.4 and Remark 3.4, (5.26) holds if and only if
$$\mathbf{Re}\{\tfrac{2}{w^2}PG(jw) - \tfrac{1}{w^2}(jwI - \Theta G(jw))^*R(jwI - \Lambda G(jw))\}$$
$$+ \tfrac{\epsilon}{w^2}G^*(jw)G(jw) \leq 0, \forall w \in \mathbb{R} \cup \{\infty\}.$$
Obviously, by the choice of P, the coefficient 2 can be deleted. This completes the proof. □

Obviously, when $R = 0$, the frequency-domain inequality in Theorem 5.5 reduces to
$$\mathbf{Re}\{PG(jw)\} + \epsilon G^H(jw)G(jw) \leq 0, \ \forall w \in \mathbb{R} \cup \{\infty\}.$$
And correspondingly, the LMI (5.26) reduces to a simple inequality as shown in the following corollary.

Corollary 5.1. *Under Assumptions 5.1, 5.2, and 5.3, system (5.23) is dichotomous, if there exist a diagonal matrix P, a symmetric matrix W and a scalar $\epsilon > 0$ such that the following LMI is feasible*
$$\begin{pmatrix} WA + A^TW + \epsilon C^TC & WB + C^TP + \epsilon C^TD \\ B^TW + PC + \epsilon D^TC & \epsilon D^TD + PD + D^TP \end{pmatrix} \leq 0. \qquad (5.32)$$

5.3. Dichotomy of pendulum-like feedback systems

Clearly, in Corollary 5.1, the condition (5.24) is not necessary.

Example 5.4. Consider the system in the form of (5.23) with the following data

$$A = \begin{pmatrix} -0.4 & 3 \\ -1 & -0.5 \end{pmatrix}, B = \begin{pmatrix} 1 & 0 \\ -1.4 & 1 \end{pmatrix}, C = \begin{pmatrix} 2 & -1 \\ 0 & 1 \end{pmatrix}, D = \begin{pmatrix} \alpha & 1.2 \\ -2 & 1 \end{pmatrix},$$
(5.33)

and $\varphi_1(y_1) = \sin(y_1) - 0.2$, $\varphi_2(y_2) = \sin(2y_2) - 0.1$. By Theorem 5.4, we know that this system is dichotomous when $\alpha \geq 1.9$. If we take $\alpha = -1.9$, the condition (5.26) is not satisfied. At this time, we cannot test the dichotomy of the system. Computer simulation shows the dynamical behaviors of phase space as shown in Fig. 5.4 and Fig. 5.5 at the initial value $x_0 = (1, -0.5)^T$, $y_0 = (0.1, -5)^T$. From this solution, we can see that the solution x is something like a chaotic solution and y is unbounded when $t \to \infty$ which shows the complexity of solutions of pendulum-like systems. The chaotic phenomenon in pendulum-like systems is still a new topic which is worth to be studied further. In addition, we should mention that for this example, if we take $R = 0$ in the condition (5.26) or (5.31), we can also know that this system is dichotomous when $\alpha \geq 1.9$. This means that the condition (5.24) for this example does not play any role in the test of dichotomy. Please refer to the following example for the effects of (5.24).

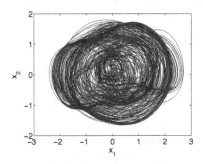

Fig. 5.4 The phase portrait of x.

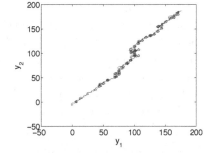

Fig. 5.5 The phase portrait of y.

Example 5.5. Consider the system in the form of (5.23) with the following data

$$A = \begin{pmatrix} -5 & 3 \\ -3 & 1 \end{pmatrix}, B = \begin{pmatrix} 5 & 0 \\ 0 & 1 \end{pmatrix}, C = \begin{pmatrix} 0.5 & 0 \\ 0 & 0.7 \end{pmatrix}, D = \begin{pmatrix} 0.7 & -0.4 \\ 1 & 0.3 \end{pmatrix}, \quad (5.34)$$

and $\varphi_1(y_1) = \sin(y_1) - 0.3$, $\varphi_2(y_2) = \sin(2y_2) - 0.2$. We know that this system is dichotomous by Theorem 5.4. However for this example, if we take $R = 0$ in Theorem 5.4, LMI (5.26) is infeasible. That is, we cannot test its dichotomy. This shows that the condition (5.24) can reduce the conservativeness generally.

5.3.2 Dichotomy of the first form of pendulum-like feedback systems

Consider the first form of autonomous pendulum-like feedback systems
$$\begin{cases} \dfrac{dx}{dt} = \hat{A}x + \hat{B}\varphi(y) \\ y = \hat{C}x \end{cases} \tag{5.35}$$
where $\hat{A} \in \mathbb{R}^{n \times n}$, $\hat{B} \in \mathbb{R}^{n \times m}$, $\hat{C} \in \mathbb{R}^{m \times n}$, $\varphi(y) = (\varphi_1(y_1), \cdots, \varphi_m(y_m))^T$. Let the transfer function from $\varphi(y)$ to y in (5.35) be
$$\chi(s) = \hat{C}(sI - \hat{A})^{-1}\hat{B}.$$
Assumption 5.4 $(\hat{A}, \hat{B})$ is controllable and $(\hat{A}, \hat{C})$ is observable. $\hat{A}$ has m zero eigenvalues with the multiplicity of elementary factor being 1 corresponding to each zero eigenvalue and $\hat{A}$ has no eigenvalues on the imaginary axis other than zero, and $G(0)$ is nonsingular, where $\chi(s) = \dfrac{1}{s}G(s)$.

If Assumption 5.4 holds, according to the discussions in Section 5.1.3, system (5.35) can be transformed into the second form of pendulum-like system as follows
$$\dot{\xi} = A\xi + B\varphi(y), \quad \dot{y} = C\xi + D\varphi(y) \tag{5.36}$$
where $A \in \mathbb{R}^{(n-m) \times (n-m)}$ and B, C and D are matrices with compatible dimensions, and A has no eigenvalues on the imaginary axis. Obviously, system (5.35) is dichotomous if and only if system (5.36) is dichotomous. Let the transfer function from $\varphi(y)$ to $\dot{y}$ in (5.36) is $G(s) = C(sI - A)^{-1}B + D$. Then we know that
$$\chi(s) = \hat{C}(sI - \hat{A})^{-1}\hat{B} = \frac{1}{s}G(s) \tag{5.37}$$
By Theorem 5.5, we can establish the following frequency domain criterion of dichotomy expressed in $\chi(s)$ for system (5.35).

Theorem 5.6. *Under Assumptions 5.2, 5.3 and 5.4, system (5.35) is dichotomous for all φ with condition (5.24), if there exist diagonal matrices P and R with $R \geq 0$, and a scalar $\epsilon > 0$ such that $\forall w \in \mathbb{R} \cup \{\infty\}$,*
$$\operatorname{Re}\left\{\frac{1}{-jw}P\chi(jw) - (I - \Theta\chi(jw))^*R(I - \Lambda\chi(jw))\right\} + \epsilon\chi^*(jw)\chi(jw) \leq 0. \tag{5.38}$$

5.3. Dichotomy of pendulum-like feedback systems

Since $\hat{A}$ has m zero eigenvalues, we can take a nonsingular matrix T such that

$$T\hat{A}T^{-1} = \begin{pmatrix} A & 0 \\ 0 & 0 \end{pmatrix}, \quad T\hat{B} = \begin{pmatrix} B \\ B_2 \end{pmatrix} \quad (5.39)$$

where A and B are given as in (5.36), $B_2 \in \mathbb{R}^{m \times m}$. By the controllability of $(\hat{A}, \hat{B})$, B_2 is nonsingular. Let

$$C_0 = (0_{m \times (n-m)} \quad B_2^{-1}) \quad (5.40)$$

Then similarly to Theorem 5.4, we can also establish LMI-based criterion of dichotomy expressed in $(\hat{A}, \hat{B}, \hat{C})$.

Theorem 5.7. *Under Assumptions 5.2, 5.3 and 5.4, system (5.35) is dichotomous for all φ with condition (5.24), if there exist diagonal matrices P and R with $R \geq 0$, and a symmetric matrix W and a scalar $\epsilon > 0$ such that the following LMI is feasible*

$$\begin{pmatrix} W\hat{A} + \hat{A}^T W + \Psi & W\hat{B} + \frac{1}{2}\hat{C}^T R(\Theta + \Lambda) \\ \hat{B}^T W + \frac{1}{2}(\Theta + \Lambda)R\hat{C} & -R \end{pmatrix} \leq 0, \quad (5.41)$$

where $\Psi = -\hat{C}^T \Theta R \Lambda \hat{C} + T^T C_0^T P \hat{C} + \hat{C}^T P C_0 T + \epsilon \hat{C}^T \hat{C}$.

Proof. By (5.39), it follows that

$$C_0 T(sI - \hat{A})^{-1}\hat{B} = C_0 T(sI - \hat{A})^{-1}T^{-1}T\hat{B} = C_0(sI - T\hat{A}T^{-1})^{-1}T\hat{B} = \frac{1}{s}I.$$

Then similarly to the proof of Theorem 4.4, we know that the LMI (5.41) is equivalent to (5.38). □

From Theorems 5.6 and 5.7 we can easily get the following corollary.

Corollary 5.2. *Theorem 5.6 is still true with $R = 0$ in (5.38) if the system (5.35) has not properties (5.24).*

If $R = 0$ in (5.38), correspondingly, inequality (5.41) will reduce to a simple inequality as discussed in Corollary 5.1. For single input and single output systems (5.35) ($m = 1$), if it has not properties (5.24), then we have the following corollary.

Corollary 5.3. *Under Assumptions 5.2, 5.3 and 5.4, system (5.35) is dichotomous if*

$$\mathbf{Re}\{jw\chi(jw)\} \neq 0 \quad \text{for all} \quad w \in R \cup \{\infty\}. \quad (5.42)$$

Remark 5.3. Theorems 5.5 and 5.6 are still true with $R = 0$ no matter if there are properties (5.24) for φ_i, $i = 1, 2, \cdots, m$. If (5.24) holds for a system we can use Theorem 5.4 and Corollary 5.1 to test dichotomy of system with the second form. Similarly, we can use Theorem 5.7 and Corollary 5.2 to test dichotomy of system with the first form.

Remark 5.4. Obviously, if the nonlinear functions φ_i in a pendulum-like system satisfy the sector conditions discussed in Chapter 4, then the pendulum-like system (5.35) can also be viewed as a Lur'e system. Then theorems similar to Theorems 4.8 and 4.9 can also be established for dichotomy of pendulum-like systems.

5.4 Gradient-like property of pendulum-like feedback systems

5.4.1 Gradient-like property of the second form of pendulum-like feedback systems

In this section we discuss the global convergence of system (5.23).
Let

$$v_i = \int_0^{\Delta_i} \varphi_i(\sigma)d\sigma / \int_0^{\Delta_i} |\varphi_i(\sigma)|d\sigma, \ V = diag(v_1, \cdots, v_m),$$
$$F_i(\sigma) = \varphi_i(\sigma) - v_i|\varphi_i(\sigma)| \quad (5.43)$$

Note that $\int_0^{\Delta_i} F_i(\sigma)d\sigma = 0, i = 1, \cdots, m$, i.e., $F_i(\sigma)$ has mean value zero. Using the method of the above section and notations given in Theorem 5.4 and (5.43), we can get the following result.

Theorem 5.8. *Under Assumptions 5.1, 5.2 and 5.3, system (5.23) for all φ with condition (5.24) is gradient-like, if A is stable and there exist diagonal matrices P, Q, E and R with $R \geq 0$, $E > 0$ and $Q > 0$, and a symmetric matrix W such that the following LMIs are feasible*

(i) $\begin{pmatrix} W\tilde{A} + \tilde{A}^T W + \Psi & WL + \frac{1}{2}\tilde{C}^T R(\Theta + \Lambda) \\ L^T W + \frac{1}{2}(\Theta + \Lambda)R\tilde{C} & -R \end{pmatrix} \leq 0,$
where $\Psi = -\tilde{C}^T \Theta R \Lambda \tilde{C} + \frac{1}{2}LP\tilde{C} + \frac{1}{2}\tilde{C}^T PL^T + \tilde{C}^T Q\tilde{C} + LEL^T,$

(ii) $\begin{pmatrix} 2E & PV \\ VP & 2Q \end{pmatrix} > 0.$

Proof. System (5.23) can be rewritten as (5.25). Take Lyapunov function

5.4. Gradient-like property of pendulum-like feedback systems

candidate

$$v(z) =: z^T W z + \sum_{i=1}^{m} p_i \int_0^{y_i} F_i(\sigma) d\sigma. \tag{5.44}$$

The time derivative of $v(z)$ along any trajectory of system (5.25) is given by

$$\dot{v}(z) = 2z^T W(\tilde{A} z + L\psi(y)) + z^T L P \tilde{C} z - |\varphi(y)|^T P V \tilde{C} z, \tag{5.45}$$

where $|\varphi(y)|^T =: (|\varphi_1(y_1)|, \cdots, |\varphi_m(y_m)|)$ and $P =: \text{diag}(p_1, \cdots, p_m)$. Furthermore,

$$\dot{v}(z) = 2z^T W(\tilde{A} z + L\psi(y)) + z^T L P \tilde{C} z - |\varphi(y)|^T P V \tilde{C} z - \varphi^T(y) E \varphi(y)$$
$$- z^T \tilde{C}^T Q \tilde{C} z + \varphi^T(y) E \varphi(y) + z^T \tilde{C}^T Q \tilde{C} z,$$

where $Q =: \text{diag}(q_1, \cdots, q_m)$ and $E =: \text{diag}(e_1, \cdots, e_m)$. By the condition (ii), we know that there exist $Q_0 =: \text{diag}\{q_{01}, \cdots, q_{0m}\}$ and $E_0 =: \text{diag}\{e_{01}, \cdots, e_{0m}\}$ such that

$$|\varphi(y)|^T P V \tilde{C} z + \varphi^T(y) E \varphi(y) + z^T \tilde{C}^T Q \tilde{C} z \geq \varphi^T(y) E_0 \varphi(y) + z^T \tilde{C}^T Q_0 \tilde{C} z.$$

Hence, we have

$$\dot{v}(z) + \varphi^T(y) E_0 \varphi(y) + z^T \tilde{C}^T Q_0 \tilde{C} z \leq 2z^T W(\tilde{A} z + L\psi(y))$$
$$+ z^T L P \tilde{C} z + \varphi^T(y) E \varphi(y) + z^T \tilde{C}^T Q \tilde{C} z, \tag{5.46}$$

On the other hand, condition (i) of the theorem guarantees that

$$2z^T W(\tilde{A} z + L\psi(y)) + z^T L P \tilde{C} z + \varphi^T(y) E \varphi(y) + z^T \tilde{C}^T Q \tilde{C} z$$
$$\leq \psi^T(y) R \psi(y) - \psi^T(y)(\Theta + \Lambda) R \tilde{C} z + z^T \tilde{C}^T \Theta \Lambda R \tilde{C} z, \quad \forall (z, \psi(y)) \neq 0. \tag{5.47}$$

Combining (5.46), (5.47) and (5.27), we know that

$$\dot{v}(z) + \varphi^T(y) E_0 \varphi(y) + z^T \tilde{C}^T Q_0 \tilde{C} z \leq 0.$$

Integrating the two sides of the inequality above from 0 to t gives

$$v(t) - v(0) \leq -\sum_{i=1}^{m} \int_0^t (e_{0i} \varphi_i^2(y_i(t)) + q_{0i} \dot{y}_i^2(t)) dt, \quad \forall t \geq 0. \tag{5.48}$$

In addition, $v(z)$ is bounded. This property follows from the facts that A is stable (and consequently $z(t)$ is bounded) and the functions $F_i(\sigma)$ have mean value zero (and consequently $\int_0^{y_i(t)} F_i(\sigma) d\sigma$ are bounded). Then it follows that

$$\int_0^{+\infty} \varphi_i^2(y_i(t)) dt < +\infty, \quad i = 0, \cdots, m.$$

Together with the fact that $\varphi_i(y_i(t))$ are uniformly continuous, by Lemma 4.3 we have
$$\lim_{t\to+\infty} \varphi_i(y_i(t)) = 0, \quad i = 0, \cdots, m. \tag{5.49}$$
Since φ_i are periodic functions and have finite number of zeros on $[0, \Delta_i)$, it follows from (5.49) and Lemma 5.1 that
$$\lim_{t\to+\infty} y_i(t) = \hat{y}_i, \quad i = 0, \cdots, m, \tag{5.50}$$
where $\varphi_i(\hat{y}_i) = 0$. In addition, by the first equation of (5.23) we have
$$x(t) = e^{At}x(0) + \int_0^t e^{A(t-\tau)} B\varphi(y(\tau))d\tau.$$
Combining with the stability of A and the property (5.49), we get
$$\lim_{t\to+\infty} x(t) = 0. \tag{5.51}$$
The validity of (5.50) and (5.51) means that every solution of system (5.23) converges to a certain equilibrium. □

Similarly to Theorem 5.5, a frequency domain interpretation for the condition (i) of Theorem 5.8 can be given as follows.

Theorem 5.9. *With Assumption 5.1, condition (i) of Theorem 5.8 is feasible if and only if the following frequency domain inequality holds*
$$\begin{aligned}&\mathrm{Re}\{PG(jw) - (jwI - \Theta G(jw))^* R(jwI - \Lambda G(jw))\} \\ &+ G^*(jw)QG(jw) + E \leq 0, \forall w \in \mathbb{R} \cup \{\infty\}.\end{aligned} \tag{5.52}$$

Proof. Similarly to the proof of Theorem 4.4, LMI (i) holds if and only if
$$\begin{aligned}&\mathrm{Re}\{\tfrac{1}{w^2}PG(jw) - \tfrac{1}{w^2}(jwI - \Theta G(jw))^* R(jwI - \Lambda G(jw))\} \\ &+ \tfrac{1}{w^2}G^*(jw)QG(jw) + \tfrac{1}{w^2}E \leq 0, \forall w \in \mathbb{R} \cup \{\infty\}.\end{aligned}$$
Multiplying w^2 on the both sides of the inequality above completes the proof. □

Obviously, if $v_i = 0, i = 1, \cdots, m$, then the condition (ii) ca be removed in Theorem 5.8. Hence, we can get

Corollary 5.4. *Under Assumptions 5.1, 5.2 and 5.3, if A is stable and $\int_0^{\Delta_i} \varphi_i(\sigma)d\sigma = 0$, $i = 1, \cdots, m$, and there exist diagonal matrices P and R with $R \geq 0$ and a scalar $\epsilon > 0$ such that the following frequency domain inequality holds*
$$\begin{aligned}&\mathrm{Re}\{PG(jw) - (jwI - \Theta G(jw))^* R(jwI - \Lambda G(jw))\} \\ &+ \epsilon G^*(jw)G(jw) \leq 0, \forall w \in \mathbb{R} \cup \{\infty\},\end{aligned} \tag{5.53}$$
then system (5.23) is gradient-like for all φ with condition (5.24).

5.4. Gradient-like property of pendulum-like feedback systems

Similarly to the discussions in the above section, when $R = 0$ in Theorem 5.8, condition (i) of Theorem 5.8 can be simplified as follows.

Corollary 5.5. *Under Assumptions 5.1, 5.2 and 5.3, system (5.23) is gradient-like, if A is stable and there exist diagonal matrices P, Q, E, and a symmetric matrix W such that (ii) of Theorem 5.8 and the following LMI are feasible*

$$\begin{pmatrix} WA + A^TW + C^TQC & WB + C^TP + C^TQD \\ B^TW + PC + D^TQC & \Psi \end{pmatrix} \leq 0, \quad (5.54)$$

where $\Psi = PD + D^TP + D^TQD + E$.

Similarly to Theorem 5.9, we have

Corollary 5.6. *With Assumption 5.1, the condition (5.54) holds if and only if*

$$\text{Re}\{PG(jw)\} + G^*(jw)QG(jw) + E \leq 0, \quad \forall w \in \mathbb{R} \cup \{\infty\}.$$

Example 5.6. Consider the system in the form of (5.23) with the following data

$$A = \begin{pmatrix} -0.9 & 0 & 0.8 \\ 0 & 0 & 1.1 \\ 0 & -2.5 & -1 \end{pmatrix}, \quad B = \begin{pmatrix} 1 & 0 \\ 0 & 0 \\ 0 & -1 \end{pmatrix},$$
$$C = \begin{pmatrix} 0.3 & 0 & 0 \\ 0 & 0.2 & 0 \end{pmatrix}, \quad D = \begin{pmatrix} 0.5 & -0.1 \\ -0.1 & 1 \end{pmatrix}, \quad (5.55)$$

and $\varphi_1(y_1) = \sin(y_1) - r$, $\varphi_2(y_2) = \sin(2y_2) - r$, r is a parameter to be determined. By Theorem 5.8, we know that this system is gradient-like when $r \leq 0.82$. However, if we take $R = 0$ in the condition (i) of Theorem 5.8, we can only get that the system is gradient-like when $r \leq 0.77$. This shows the effects of (5.24) in reducing the conservativeness. Please refer to Fig. 5.6 and Fig. 5.7 for the gradient-like behavior of its solutions with $r = 0.82$ at three initial values $x_0 = (1, 0.5, 0.1)^T$, $y_0 = (2.1, 2)^T$; $x_0 = (0.2, -2.5, 5)^T$, $y_0 = (3, 1)^T$; $x_0 = (0.5, -1.5, 4.2)^T$, $y_0 = (3.5, 0.9)^T$.

5.4.2 Gradient-like property of the first form of pendulum-like feedback systems

Similar to Theorems 5.6 and 5.7, we can also establish frequency and time domain criteria of gardient-like behavior expressed in $\chi(s)$ or $(\hat{A}, \hat{B}, \hat{C})$ for system (5.35).

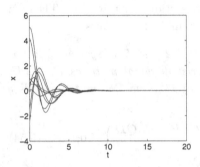
Fig. 5.6 The solution x.

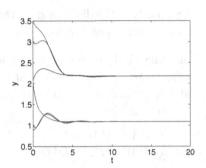
Fig. 5.7 The solution y.

Theorem 5.10. *Under Assumptions 5.2, 5.3 and 5.4, if A in (5.39) is stable and there exist diagonal matrices P, Q, E and R with $Q > 0$, $E > 0$ and $R \geq 0$ such that the following frequency domain inequality*

$$\mathbf{Re}\{\tfrac{1}{-jw}P\chi(jw) - (I - \Theta\chi(jw))^*R(I - \Lambda\chi(jw))\} \\ + \chi^*(jw)Q\chi(jw) + \tfrac{1}{w^2}E \leq 0, \forall w \in \mathbb{R} \cup \{\infty\} \quad (5.56)$$

and (ii) of Theorem 5.8 are feasible, then system (5.35) is gradient-like for all φ with condition (5.24).

Theorem 5.11. *Under Assumptions 5.2, 5.3 and 5.4, if A in (5.39) is stable and there exist diagonal matrices P, Q, E and R with $R \geq 0$, $E > 0$ and $Q > 0$, and a symmetric matrix W such that the following LMI*

$$\begin{pmatrix} W\hat{A} + \hat{A}^T W + \Psi & W\hat{B} + \tfrac{1}{2}\hat{C}^T R(\Theta + \Lambda) \\ \hat{B}^T W + \tfrac{1}{2}(\Theta + \Lambda)R\hat{C} & -R \end{pmatrix} \leq 0 \quad (5.57)$$

and (ii) of Theorem 5.8 are feasible, where $\Psi = -\hat{C}^T\Theta R\Lambda\hat{C} + \tfrac{1}{2}T^T C_0^T P\hat{C} + \tfrac{1}{2}\hat{C}^T PC_0 T + \hat{C}^T Q\hat{C} + T^T C_0^T EC_0 T$, C_0 and T are given as in (5.39) and (5.40), then system (5.35) is gradient-like for all φ with condition (5.24).

5.5 Lagrange stability of pendulum-like feedback systems

Consider the pendulum-like system in the first canonical form:

$$\dot{x} = Ax + b\xi, \quad \sigma = cx, \quad \xi = \varphi(t, \sigma) \quad (5.58)$$

where $A \in \mathbb{R}^{n \times n}$ is a constant real matrix, $b \in \mathbb{R}^{n \times 1}$ and $c^T \in \mathbb{R}^{n \times 1}$ are real vectors and $\det A = 0$. $\varphi : \mathbb{R}_+ \times \mathbb{R} \to \mathbb{R}$ is continuous and locally Lipchitz continuous in the second argument,

$$\varphi(t, \sigma + \Delta) = \varphi(t, \sigma), \quad t \in \mathbb{R}_+, \sigma \in \mathbb{R} \quad (5.59)$$

5.5. Lagrange stability of pendulum-like feedback systems

Furthermore we assume that φ belongs to the sector $M[\mu_1, \mu_2]$, i.e.,

$$\mu_1 \leq \frac{\varphi(t,\sigma)}{\sigma} \leq \mu_2, \tag{5.60}$$

where μ_1, μ_2 are two finite numbers with $\mu_1 < 0$ and $\mu_2 > 0$. In this section we also assume that $\chi(s) = c(A - sI)^{-1}b$ is non-degenerate, i.e., (A, b) is controllable and (A, c) is observable. Before providing the criterion of Lagrange stability, we first introduce the following lemma [Leonov et al. (1996)].

Lemma 5.2. *Suppose that* $v : [t_0, +\infty) \to \mathbb{R}$ *is absolutely continuous,* $v(t_0) \leq 0$ *(resp.* $v(t_0) < 0$*) and there exists a number* γ *such that*

$$\dot{v}(t) \leq \gamma v(t), \quad t \geq t_0. \tag{5.61}$$

Then $v(t) \leq 0$ *(resp.* $v(t) < 0$*) for all* $t \geq t_0$.

Proof. By (5.61), we have $e^{-\gamma t}\dot{v}(t) - \gamma e^{-\gamma t}v(t) \leq 0$ for $t \geq t_0$. Therefore

$$\frac{d[e^{-\gamma t}v(t)]}{dt} \leq 0,$$

for $t \geq t_0$. Integrating the inequality above from t_0 to t we get $v(t) \leq e^{\gamma(t-t_0)}v(t_0) \leq 0$ for all $t \geq t_0$. This completes the proof. □

Theorem 5.12. *Suppose* (A, b) *is controllable,* (A, c) *is observable and* $\det(A) = 0$. *Then system (5.58) is Lagrange stable for all* φ *satisfying (5.59) and (5.60), if there exist a symmetric matrix* W *and a scalar* $\lambda > 0$ *such that* $A + \lambda I$ *has* $n - 1$ *eigenvalues with negative real parts and the following LMI is feasible*

$$\begin{pmatrix} W(A + \lambda I) + (A + \lambda I)^T W - \mu_1\mu_2 c^T c & Wb + \frac{1}{2}(\mu_1 + \mu_2)c^T \\ b^T W + \frac{1}{2}(\mu_1 + \mu_2)c & -1 \end{pmatrix} \leq 0. \tag{5.62}$$

Proof. Suppose $x(t, t_0, x_0)$ is a solution of (5.58). Let d be an eigenvector of A corresponding to its zero eigenvalue, such that $cd = \Delta$. For this d, (5.58) is also pendulum-like with respect to $\{jd, j \in \mathbf{Z}\}$. It follows that

$$x(t, t_0, x_0) - jd = x(t, t_0, x_0 - jd), \quad t \geq t_0, \quad j \in \mathbf{Z}. \tag{5.63}$$

On the other hand, the condition (5.62) guarantees that

$$2x^*W((A + \lambda I)x + b\xi) \leq \mu_1\mu_2 x^* c^T cx - \xi^* \frac{1}{2}(\mu_1 + \mu_2)cx$$

$$-x^* \frac{1}{2}(\mu_1 + \mu_2)c^T \xi + \xi^*\xi = (\xi - \mu_1\sigma)^*(\xi - \mu_2\sigma)$$

holds for any vector $(x^T \ \xi) \neq 0$ where $\xi = \varphi(t, \sigma)$. And by the sector condition (5.60), we know that $(\xi - \mu_1\sigma)^*(\xi - \mu_2\sigma) \leq 0$. Hence,

$$2x^*W((A + \lambda I)x + b\xi) \leq 0.$$

Let $v(x(t)) = x^*(t)Wx(t)$. Then differentiating $v(t)$ along the trajectory of (5.58) gives

$$\dot{v}(x(t)) = 2x^*W(Ax + b\xi) \leq 2\lambda v(x(t)).$$

Combining with Lemma 5.2, we know that the set $\{x \mid x^*Wx < 0\}$ is positively invariant for (5.58). Hence the set

$$\Omega_j = \{x \mid (x - jd)^*W(x - jd) < 0\}$$

is also positively invariant for (5.58). Indeed, for an arbitrary $x_0 \in \Omega_j$ it follows that $x_0 - jd \in \Omega_0$. Then in virtue of positive invariance of Ω_0 we have

$$x(t, t_0, x_0 - jd) \in \Omega_0 \quad (\forall t \geq t_0, t_0 \in \mathbb{R}).$$

Because of (5.63) we have that

$$(x(t, t_0, x_0) - jd)^*W(x(t, t_0, x_0) - jd) < 0 \quad (\forall t \geq t_0, t_0 \in \mathbb{R}).$$

Then by (5.62), we have

$$W(A + \lambda I) + (A + \lambda I)^*W \leq \mu_1\mu_2 c^*c.$$

It follows from Theorem 1.8 that W has one negative eigenvalue and $n - 1$ positive eigenvalues. Combining with the definition of d in the beginning of this proof and the nonsingular transformation discussed in the section above, we know that

$$d^*Wd < 0. \tag{5.64}$$

For an arbitrary integer j, define the set

$$\Gamma_j := \Omega_j \cap \Omega_{-j}.$$

Obviously, the set Γ_j is also positively invariant. Noticing the spectrum of W, there exists a vector h such that

$$\{x \mid h^*x = 0, x \neq 0, i = 1, \cdots, m\} \subset \{x \mid x^*Wx > 0\}. \tag{5.65}$$

Indeed it is sufficient to reduce W to the form $W = \text{diag}\{I_{n-1}, -1\}$ by means of a non-singular linear transformation and to choose $h^* = (0, \cdots, 0, 1)$ in this case. It follows from (5.64) and (5.65) that $h^*d \neq 0$.

5.5. Lagrange stability of pendulum-like feedback systems

Consider an arbitrary solution $x(t, t_0, x_0)$. By the fact that $h^*d \neq 0$ and $d^*Wd < 0$ we can find a number j large enough such that

$$|h^*x_0| < j|h^*d|,$$
$$x_0^*Hx_0 + 2jx_0^*Hd + j^2 d^*Hd < 0, \qquad (5.66)$$
$$x_0^*Hx_0 - 2jx_0^*Hd + j^2 d^*Hd < 0.$$

From above it follows that

$$x(t, t_0, x_0) \in \Gamma_j \quad \forall t \geq t_0. \qquad (5.67)$$

In what follows, we show that

$$|h^*x(t, t_0, x_0)| < j|h^*d| \quad \forall t \geq t_0. \qquad (5.68)$$

Indeed if (5.68) was not valid, there would exist a $\bar{t} > t_0$ such that $|h^*x(\bar{t}, t_0, x_0)| = j|h^*d|$. By (5.65), this would mean that either $(x(\bar{t}, t_0, x_0) + jd)^* H(x(\bar{t}, t_0, x_0) + jd) \geq 0$ or $(x(\bar{t}, t_0, x_0) - jd)^* H(x(\bar{t}, t_0, x_0) - jd) \geq 0$ which is impossible because of (5.67).

With the above discussions, now we can show the boundedness of $x(t, t_0, x_0)$. It follows from (5.65) that the matrix W can be represented in the form $W = M - \tau hh^*$ with a positive definite matrix M and a positive number τ. Let ϵ be a positive number such that $M > \epsilon I_n$. For a solution $x(t) = x(t, t_0, x_0)$ that satisfies (5.67) and (5.68) it is true that

$$\begin{aligned}\epsilon|x(t) - jd|^2 &\leq (x(t) - jd)^* M(x(t) - jd) \\ &= (x(t) - jd)^* W(x(t) - jd) + \tau|h^*(x(t) - jd)|^2 \\ &< \tau[2(h^*x)^2 + 2j^2(h^*d)^2]^2 \\ &\leq 4\tau j^2 (h^*d)^2\end{aligned}$$

for all $t \geq t_0$, which means that $x(t, t_0, x_0)$ is bounded on $[t_0, +\infty)$. $\square$

Combining with the Riccati equation theory in Chapter 2, for the case of strict inequality corresponding to (5.62) in Theorem 5.12, we can get some other equivalent conditions.

Theorem 5.13. *Assume that $\chi(s) = c(A - sI)^{-1}b$ is non-degenerate, and $\det(j\omega I - \lambda I - A) \neq 0$ for $\omega \in R$. Let*

$$H = \begin{pmatrix} \lambda I + A - \alpha\beta bc & \beta bb^T \\ -(1 + \beta\alpha^2)c^T c & -(\lambda I + A - \alpha\beta bc)^T \end{pmatrix}$$

where $\beta = -\mu_1\mu_2$ and $\alpha = \dfrac{\mu_1^{-1} + \mu_2^{-1}}{2}$. Then the following statements are equivalent:

(i) there exists a matrix $W = W^T$ such that the strict inequality corresponding to (5.62) holds;

(ii) there exists a matrix $X = X^T$ such that

$$\begin{pmatrix} (\lambda I + A)^T X + X(\lambda I + A) + c^T c & Xb - \alpha c^T \\ b^T X - \alpha c & \mu_1^{-1}\mu_2^{-1} \end{pmatrix} < 0; \quad (5.69)$$

(iii)

$$\mu_1^{-1}\mu_2^{-1} + (\mu_1^{-1} + \mu_2^{-1})\text{Re}\chi(i\omega - \lambda) + |\chi(i\omega - \lambda)|^2 < 0 \quad \forall \omega \in \mathbb{R} \cup \{\infty\}; \quad (5.70)$$

(iv) there exists a matrix $Y = Y^T$ such that

$$\begin{pmatrix} Y(\lambda I + A)^T + (\lambda I + A)Y & b - \alpha Y c^T & Y c^T \\ b^T - \alpha cY & \mu_1^{-1}\mu_2^{-1} & 0 \\ cY & 0 & -I \end{pmatrix} < 0; \quad (5.71)$$

(v) there exists a matrix $X = X^T$ satisfying algebraic Riccati inequality (ARI):

$$X(\lambda I + A - \alpha\beta bc) + (\lambda I + A - \alpha\beta bc)^T X + \beta X bb^T X + (1 + \alpha^2\beta)c^T c < 0;$$

(vi) there exists a matrix $Y = Y^T$ satisfying ARI:

$$(\lambda I + A - \alpha\beta bc)Y + Y(\lambda I + A - \alpha\beta bc)^T + (1 + \alpha^2\beta)Y c^T cY + \beta bb^T < 0;$$

(vii) H has no eigenvalues on the imaginary axis;

(viii) algebraic Riccati equation:

$$X(\lambda I + A - \alpha\beta bc) + (\lambda I + A - \alpha\beta bc)^T X + \beta X bb^T X + (1 + \alpha^2\beta)c^T c = 0$$

has a stabilizing solution $X = X^T$;

(ix) there exists an $X = X^T$ such that

$$X(\lambda I + A - \alpha\beta bc) + (\lambda I + A - \alpha\beta bc)^T X + \beta X bb^T X + (1 + \alpha^2\beta)c^T c = 0$$

and $\lambda I + A - \alpha\beta bc + \beta bb^T X$ has no eigenvalues on the imaginary axis;

(x) algebraic Riccati equation:

$$(\lambda I + A - \alpha\beta bc)Y + Y(\lambda I + A - \alpha\beta bc)^T + (1 + \alpha^2\beta)Y c^T cY + \beta bb^T = 0$$

has a stabilizing solution $Y = Y^T$.

Proof. Since $\mu_1 < 0$ and $\mu_2 > 0$ it is easy to show that (i) holds if and only if (ii) holds. The frequency-domain inequality of (iii) can be rewritten as

$$\left(b^T((j\omega I - \lambda I - A)^{-1})^* \; I\right)\begin{pmatrix} c^T c & -\alpha c^T \\ -\alpha c & \mu_1^{-1}\mu_2^{-1} \end{pmatrix}\begin{pmatrix} (j\omega I - \lambda I - A)^{-1}b \\ I \end{pmatrix} < 0,$$

5.5. Lagrange stability of pendulum-like feedback systems

which is equivalent to (ii) in terms of Corollary 1.4 . Let $Y = X^{-1}$ and by Schur complement it is easy to show the equivalence between (ii) and (iv). The equivalence between (ii) and (v), (iv) and (vi) are also implied by Schur complement. Since (iii) holds if and only if

$$(G_0(j\omega))^* G_0(j\omega) < \alpha^2 + \frac{1}{\beta}$$

which means $G_0(s) \in RL_\infty$ and $\| G_0(s) \|_\infty < \gamma$, where $G_0(s) = \chi(s-\lambda)+\alpha$ and $\gamma^2 = \alpha^2 + \frac{1}{\beta}$. Therefore the equivalence between (iii) and (vii) can be directly derived by Lemma 2.1 and the equivalence between (vii) and (viii), (viii) and (x) are obvious from Theorem 2.4. It is also obvious that (viii) implies (ix). We shall now show that (ix) implies (iii). Suppose that there exists a matrix $X = X^T$ such that (ix) is satisfied and $A_0 - \alpha\beta bc + \beta bb^T X$ has no eigenvalues on the imaginary axis, where $A_0 = \lambda I + A$. Then

$$M(s) \stackrel{s}{=} \left[\begin{array}{c|c} A_0 & -b \\ \hline b^T X - \alpha c & \dfrac{1}{\beta} \end{array} \right]$$

has no eigenvalues on the imaginary axis since

$$M^{-1}(s) \stackrel{s}{=} \left[\begin{array}{c|c} A_0 - \alpha\beta bc + \beta bb^T X & \beta b \\ \beta(b^T X - \alpha c) & \beta \end{array} \right]$$

has no poles on the imaginary axis. From (ix), we have

$$-X(j\omega I - A_0) - (j\omega I - A_0)^* X - \alpha\beta X bc - \alpha\beta c^T b^T X$$
$$+\beta X bb^T X + (1 + \alpha^2 \beta) c^T c = 0.$$

Multiplying $b^T((j\omega I - A_0)^*)^{-1}$ on the left and $(j\omega I - A_0)^{-1} b$ on the right of the preceding equation, we imply

$$G_0^*(j\omega) G_0(j\omega) = \gamma^2 I - \beta M^*(j\omega) M(j\omega).$$

Since $M(s)$ has no zeros on the imaginary axis, we conclude that $\| G_0(s) \|_\infty < \gamma$ which is equivalent to (iii). □

Remark 5.5. Assume that $\chi(s) = c(A - sI)^{-1} b$ is non-degenerate. Then the corresponding equivalences among (i)-(vi) of Theorem 5.13 for nonstrict inequalities are still true.

Remark 5.6. Consider Example 5.2 given in Section 5.1. Obviously, $\varphi(t, \sigma)$ satisfies the sector condition

$$\sigma\varphi(t, \sigma) \leq \mu = |\beta| + 2|\gamma| + 2|\delta|.$$

Corresponding to Theorem 5.12, in this case $\mu_1 = +\infty\,(\mu^{-1} = 0)$, $\mu_2 = \mu$. And by Remark 5.5, the matrix inequality in Theorem 5.12 is equivalent to

$$\mu^{-1}\mathbf{Re}\chi(iw - \lambda) + |\chi(iw - \lambda)|^2 \leq 0,$$

where $\chi(s) = \frac{1}{s(s+a)}$ for this example. As discussed in [Leonov et al. (1996)], the above frequency-domain inequality takes the form

$$-w^2 + \lambda^2 - a\lambda + \mu \leq 0 \quad w \in \mathbb{R} \cup \{\infty\}.$$

For parameter λ, the assumption of Theorem 5.12 holds if $\lambda \in (0, a)$. Based on the above inequality, the conditions of Theorem 5.12 are satisfied if $a^2 \geq 4\mu$. Therefore, system (5.4) is Lagrange stable if $a^2 \geq 4(|\beta| + 2|\gamma| + 2|\delta|)$.

5.6 Bakaev stability of pendulum-like feedback systems

In this section we consider the first form of pendulum-like system (5.58). For system (5.58), suppose that $\det A = 0$ and $\chi(s) = c(A - sI)^{-1}b$ is non-degenerate, i.e., (A, b) is controllable and (A, c) is observable. $\varphi : \mathbb{R} \times \mathbb{R} \to \mathbb{R}$ is continuous and locally Lipchitz continuous with respect to σ, and satisfies (5.59).

Definition 5.5. System (5.58) is called Bakaev stable if for any solution there exists a $T > 0$ such that $|\sigma(t_1) - \sigma(t_2)| < \Delta$ for all $t_1, t_2 > T$.

Bakaev stability of the system (5.58) means that any solution $\sigma(t)$ of (5.58) in the course of time gets into some kind of "band" of width Δ and never leaves it.

Furthermore we assume that there exists a $\mu > 0$ such that

$$\sigma\varphi(t, \sigma) \leq \mu\sigma^2 \tag{5.72}$$

for all $t \in \mathbb{R}_+, \sigma \in \mathbb{R}$.

The Bakaev stability of system (5.58) has been investigated by method of invariant cones. In order to construct positively invariant cones the following definition and lemma are required [Leonov et al. (1996)].

Definition 5.6. The set $M \subset \mathbb{R}^n$ is called a cone with the top in x_0 if for any $x_1 \in M$ the line $\{x_0 + \lambda(x_1 - x_0), \lambda \in \mathbb{R}\} \subset M$. The dimension of the maximal linear set contained in M is called the dimension of M. For $P = P^T \in \mathbb{R}^{n \times n}$, the set $\{x | x^*Px \leq 0, \det P \neq 0\}$ is called a quadratic cone.

5.6. Bakaev stability of pendulum-like feedback systems

It is clear that the set $\{x|x^*Px \leq 0\}$ is a k-dimensional quadratic cone if P has k negative and $n-k$ positive eigenvalues.

Lemma 5.3. *For given $P = P^T \in \mathbb{R}^{n \times n}$ and $c \in \mathbb{R}^n$, consider k-dimensional quadratic cone $M := \{x|x^TPx \leq 0\}$ and hyperplane $\Pi = \{x|c^Tx = 0\}$ ($c \neq 0$). Let int $M := \{x|x^TPx < 0\}$. Then the relation*

$$intM \cap \{x|c^Tx = 0\} = \emptyset \tag{5.73}$$

is satisfied if and only if $k = 1$, that is, P has one negative eigenvalue and $n - 1$ positive eigenvalues, and $c^TP^{-1}c \leq 0$.

Proof. Firstly, we suppose that P has one negative eigenvalue and $n - 1$ positive eigenvalues, and $c^TP^{-1}c < 0$. Without loss of generality we suppose c is a unit vector. Choosing an orthogonal basis $c, c_1, \cdots, c_{n-1}$ of $\mathbb{R}^n$ and let $Q = [c_1 \ c_2 \ \cdots \ c_{n-1}]$. Then $[c \ Q]$ is an orthogonal matrix. Let $f = P^{-1}(c^TP^{-1}c)^{-1}c$, then $f^Tc = 1$ which means f is not contained in the image space of Q. This implies $[f \ Q]$ is nonsingular and it can be considered as a basis matrix of $\mathbb{R}^n$. Therefore for $\forall x \in \mathbb{R}^n$, we have

$$x = \alpha f + Qa, \quad \alpha \in \mathbb{R}, a \in \mathbb{R}^{n-1}.$$

It is obvious that $c^Tx = \alpha$ and thus

$$x^TPx = x^T(\alpha(c^TP^{-1}c)^{-1}c + PQa) = \frac{(c^Tx)^2}{c^TP^{-1}c} + a^TQ^TPQa. \tag{5.74}$$

Then

$$a^TQ^TPQa = x^TPx - \frac{(c^Tx)^2}{c^TP^{-1}c} = x^TRx \tag{5.75}$$

where

$$R = P - \frac{cc^T}{c^TP^{-1}c} = P\left[I - P^{-1}c\frac{c^T}{c^TP^{-1}c}\right].$$

This implies

$$detR = detPdet\left[1 - \frac{c^TP^{-1}c}{c^TP^{-1}c}\right] = 0$$

and thus there exist an orthogonal matrix $G = [g_1 \ \cdots \ g_n]$ such that

$$G^TRG = diag\{0, \varepsilon_2, \cdots, \varepsilon_n\} = E$$

which derives $x^TRx = x^TGEG^Tx = \sum_{i=2}^{n} \varepsilon_i(g_i^Tx)^2$. By (5.74) and (5.75) it follows that

$$x^TPx = \frac{(c^Tx)^2}{c^TP^{-1}c} + \sum_{i=2}^{n} \varepsilon_i(g_i^Tx)^2 = \beta(c^Tx)^2 + \sum_{i=2}^{n} \varepsilon_i(g_i^Tx)^2 \tag{5.76}$$

where $\beta = \dfrac{1}{c^T P^{-1} c} < 0$. Let $T = [c \quad g_2 \quad \cdots \quad g_n]$ and $T^T x = y$, where $y = [\eta_1 \quad \cdots \quad \eta_n]^T$. Then

$$x^T P x = \beta \eta_1^2 + \sum_{i=2}^n \varepsilon_i \eta_i^2.$$

This shows that $x^T P x$ is transformed into the diagonal form by the transform $y = T^T x$ which implies that T is nonsingular. Since $\beta < 0$ we conclude that $\varepsilon_i > 0$ for $i = 2, \cdots, n$. Let $x \in \Pi$, then $c^T x = 0$. It follows from (5.76) that $x^T P x > 0$ which means x is not contained in the int$M$. Therefore (5.73) holds.

Next, we consider the case of $c^T P^{-1} c = 0$. Suppose that (5.73) is not satisfied. Since int$M$ is open set there exists a vector $h \neq 0$ such that $h^T P^{-1} h < 0$ and int$M \cap \{x | h^T x = 0\} \neq \emptyset$. But this is impossible by above proof.

Finally, Suppose (5.73) holds. We show that $k = 1 = \dim M$, i.e., P has only one negative eigenvalue and $c^T P^{-1} c \leq 0$. Suppose to the contrary, that is, $k > 1$. There must exist a subspace Π_1 with $\dim \Pi_1 \geq 2$ and $\Pi_1 \setminus \{0\} \subset \text{int} M$. Since $\dim \Pi = n - 1$ we imply that $\dim(\Pi_1 \cap \{x | c^T x = 0\}) \geq 1$ which contradicts (5.73). Let $c^T P^{-1} c \neq 0$. Since (5.76) holds for any $x \in \mathbf{R}^n$ we have $x^T P x > 0$ only if $c^T x = 0$ and thus $\varepsilon_i > 0$ for $i = 2, \cdots, n$. By int$M \neq \emptyset$ it follows that $c^T P^{-1} c < 0$. □

Theorem 5.14. *Suppose that $cb \leq 0$ and there exist a positive number $\lambda > 0$ and a number ν, $\nu > \mu$, a matrix $X = X^T$ such that*

(i) the matrix $A + \lambda I$ has $n - 1$ eigenvalues with negative real parts,
(ii) $Xb = \dfrac{1}{2\nu} c^T$, and
(iii) $(\lambda I + A)^T X + X(\lambda I + A) + c^T c \leq 0$.

Then the system (5.58) satisfying (5.72) is Bakaev stable.

Proof. Suppose that there exist a matrix $X = X^T$ such that (ii) and (iii) hold. Let $P = \nu X$. Then we have

$$2Pb = c^T, \quad (\lambda I + A)^T P + P(\lambda I + A) + \nu c^T c \leq 0. \tag{5.77}$$

In terms of Theorem 1.8 in Chapter 1, P has one negative and $n-1$ positive eigenvalues. From (5.77) it is obvious that

$$2x^T P[(A + \lambda I)x + b\xi] + (cx)^T[\nu cx - \xi] \leq 0, \quad x \in \mathbb{R}^n, \xi \in \mathbb{R}. \tag{5.78}$$

5.6. Bakaev stability of pendulum-like feedback systems

By $2Pb = c^T$ we know that $cP^{-1}c^T = 2cb \leq 0$. Then from Lemma 5.3 it follows that

$$\{x|x^T P x \geq 0\} \supset \{ x|cx = 0\}. \tag{5.79}$$

Let $V_j(x) = (x - jd)^T P(x - jd)$, where $j \in \mathbb{Z}$ and d is a nonzero vector with $Ad = 0$ and $cd = \Delta$. By (5.78) the derivative of V_j with respect to t along any solution $x(t)$ of (5.58) can be evaluated as follows,

$$\begin{aligned}\dot{V}_j(x(t)) + 2\lambda V_j(x(t)) &\leq -c(x(t) - jd)[\mu c(x(t) - jd) \\ &\quad -\varphi(t, c(x(t) - jd))] - (\nu - \mu)[c(x(t) - jd)]^2 \\ &\leq -(\nu - \mu)[c(x(t) - jd)]^2.\end{aligned} \tag{5.80}$$

Therefore the sets $\Omega_j = \{x|V_j(x) < 0\}$ are positively invariant.

Let $\xi = 0$ and $x = d$ in (5.78). Then we have $2\lambda d^T P d \leq -\nu\Delta^2$ which implies that $\lim_{j\to\infty} V_j(y) = -\infty$ for any fixed y. It follows that for any $x_0 \in \mathbb{R}^n$ there exists a $N > 0$ such that $V_j(x_0) < 0$ for $|j| > N$. According to $\Omega_j = \{x|V_j(x) < 0\}$ being positively invariant, for any solution of (5.58), we have

$$V_j(x(t, t_0, x_0)) < 0 \quad |j| > N, t > t_0.$$

Then $cx(t, t_0, x_0) \neq j\Delta$ from (5.79).

For $V_j(x(t))$ there are only two cases. One case is that $V_k(x(t)) \geq 0$ for a certain k and for all $t \geq t_0$, where $x(t) = x(t, t_0, x_0)$. The other case is that for any k there exists τ_k such that $V_k(x(\tau_k)) < 0$. For the first case, by (5.80) we imply that

$$V_k(x(t)) - V_k(x_0) \leq -(\nu - \mu) \int_{t_0}^{t} [c(x(\tau) - kd)]^2 d\tau$$

and consequently for any $t \geq t_0$

$$\int_{t_0}^{t} [c(x(\tau) - kd)]^2 d\tau \leq \frac{1}{\nu - \mu} V_k(x_0) \tag{5.81}$$

which means $c(x(t) - kd) \in L^2[0, +\infty)$. Since the conditions of Theorem 5.12 are satisfied here we can conclude that the solution $x(t)$ of (5.58) is bounded on $[t_0, +\infty)$, and thus $\dfrac{d[cx(t)]}{dt}$ is also bounded. This implies that $c(x(t) - kd)$ is uniformly continuous. By (5.81) and Lemma 4.3 in Chapter 4 we conclude that

$$\lim_{t \to +\infty} c(x(t) - kd) = 0. \tag{5.82}$$

For the second case, i.e., for any k there exists τ_k such that $V_k(x(\tau_k)) < 0$. By the set $\{x|V_k(x) < 0\}$ is positively invariant we obtain $V_k(x(t)) < 0$ for

$t \geq \tau_k$. Let $T = \max_{|k| \leq N} \tau_k$. Then from the discussions above it follows that either $\lim_{t \to +\infty} cx(t)$ exists or $V_k(x(t)) < 0$ for $t > T$. In the latter case it follows that $cx(t) \neq k\Delta$ for $t \geq T, |k| \leq N$ by (5.79). Therefore the Bakaev stability is ensured. □

Theorem 5.15. *Suppose that $cb \leq 0$ and there exists a positive number λ and a number ν with $\nu > \mu$ satisfying the following conditions:*
(i) the matrix $A + \lambda I$ has $n - 1$ eigenvalues with negative real parts,
(ii)
$$\pi(\omega) = \text{Re}\chi(i\omega - \lambda) + \nu|\chi(i\omega - \lambda)|^2 \leq 0, \quad \forall \omega \in \mathbb{R}. \tag{5.83}$$
Then system (5.58) satisfying (5.72) is Bakaev stable.

Proof. It is obvious that the frequency condition (ii) is equivalent to the conditions (ii) and (iii) of Theorem 5.14 by Theorem 5.13 and Remark 5.4, where $\mu_1^{-1} = 0$ and $\mu_2 = \nu$. Therefore the result follows by Theorem 5.14. □

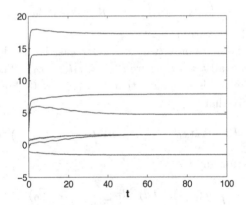

Fig. 5.8   Bakaev stability of (5.4) (solution $\zeta(t)$ with different initial conditions).

Remark 5.7. Reconsider Example 5.2 given in Section 5.1. Obviously, system (5.5) can be written in the form of (5.58) with

$$A = \begin{pmatrix} 0 & 1 \\ 0 & -a \end{pmatrix}, b = \begin{pmatrix} 0 \\ -1 \end{pmatrix}, c = (1, \ 0)$$

with $cb = 0$. Combining Remark 5.6 and Theorem 5.15, we know that system (5.4) is Bakaev stable if $a^2 > 4(|\beta| + 2|\gamma| + 2|\delta|)$, see Fig. 5.8.

5.7 Notes and references

Some results expressed by frequency domain conditions for global properties of pendulum-like feedback systems (dichotomy, Lagrange stability, gradient-like behavior, Bakaev stability) can be found in [Leonov et al. (1996, 1992b); Huang (2003)]. The method of [Leonov et al. (1996)] is closely related to the method in absolute stability of Lur'e systems [Bernstein et al. (1994); Khalil (2002); Narendra and Taylor (1973); de Oliveira et al. (2002a); Park (1997, 2002); Popov (1973); Suykens et al. (1998); Yakubovich (1962)]. In this chapter, we have provided time domain criteria for some global properties of pendulum-like feedback systems in terms of the Lyapunov method, which can easily be tested by LMI method [Boyd et al. (1994); Gahinet et al. (1995)] and used to analyze robustness and design controllers [Wang et al. (2004); Yang et al. (2004)]. Using the celebrated KYP lemma [Lozano et al. (2000); Rantzer (1996); Leonov et al. (1996)], the equivalence between time domain and frequency domain methods can be established. Based on the criterion of dichotomy, some analysis results of the nonexistence of bounded oscillating solutions in Chua's circuit [Chua (1994)] can be found in [Duan et al. (2004b, 2005b)]. Other nonlinear control methods can be found in [Cheng (1988); Feng and Fei (1998); Hong and Cheng (2005); Isidori (1995)].

Chapter 6

Controller Design for a Class of Pendulum-like Systems

In this chapter we consider a class of Lur'e control systems that can also be considered as a class of linear perturbed systems. The particularities are that the linear block contains a single zero eigenvalue and that nonlinearity is periodic. When there is no control, that is, $u = 0$, the systems are pendulum-like but may not be dichotomous, gradient-like or Lagrange stable. We study problems of controller design for this kind of Lur'e control systems to realize Lagrange stability, gradient-like property or dichotomy and to preserve the physical and technical phenomena of the pendulum-like systems at the same time.

6.1 Controller design with dichotomy or gradient-like property

Note that the important properties of the pendulum-like systems with dichotomy are that there are no limit cycles in the systems and oscillation phenomena of the bounded solutions cannot happen. These properties are desired for the pendulum-like systems without dichotomy. From the view of control, we hope to design appropriate controllers for theses systems such that the closed-loop systems not only preserve the pendulum-like property but also have the property of dichotomy. The same desire for the gradient-like behavior is needed for the pendulum-like control systems.

6.1.1 Controller design with dichotomy

Consider a class of autonomous pendulum-like systems given by

$$\dot{x} = Ax + b\varphi(\sigma), \quad \sigma = cx, \quad v = c_*x, \tag{6.1}$$

where $A \in \mathbb{R}^{n \times n}$ is a constant real matrix, $b \in \mathbb{R}^n$, $c^T \in \mathbb{R}^n$ and $c_*^T \in \mathbb{R}^n$ are real vectors. $c \neq c_*$. v is the measurable output which is different from σ. We make the following assumptions.

Assumption 6.1. A has a zero eigenvalue of multiplicity one. And A has no eigenvalues on the imaginary axis other than zero.

Assumption 6.2. $\varphi(\sigma)$ is continuous for all $\sigma \in \mathbb{R}$ and satisfies

$$\varphi(\sigma + \Delta) = \varphi(\sigma), \quad \forall \sigma \in \mathbb{R}$$

where Δ is the period of $\varphi(\sigma)$.

Assumption 6.3. (A, b) is controllable and (A, c) is observable.

Let

$$P_2(s) = c(A - sI)^{-1}b, \quad P_1(s) = c_*(A - sI)^{-1}b.$$

The pendulum-like feedback system (6.1) is shown in Fig. 6.1.

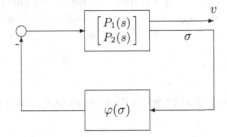

Fig. 6.1 The pendulum-like feedback system (6.1).

The synthesis problem for the pendulum-like systems considered in this section is to design a linear controller $K(s)$ using the measurable output v such that the closed-loop feedback systems shown in Fig. 6.2 is dichotomous or gradient-like.

The following lemma is needed [Leonov et al. (1996)].

Lemma 6.1. *Under Assumptions 6.1, 6.2 and 6.3, if the matrix $A \in \mathbb{R}^{n \times n}$ has $n - 1$ eigenvalues with negative real parts and*

$$\text{Re}\,\{j\omega P_2(j\omega)\} \neq 0, \quad \lim_{\omega \to \infty} \omega^2 \text{Re}\,\{j\omega P_2(j\omega)\} \neq 0 \qquad (6.2)$$

holds for all $\omega \in \mathbb{R}$, then the system (6.1) is dichotomous.

6.1. Controller design with dichotomy or gradient-like property

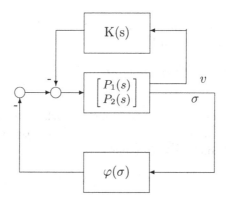

Fig. 6.2 The closed-loop feedback system.

For the system (6.1), since $\det(A) = 0$, $P_2(s)$ can be expressed as $P_2(s) = \dfrac{1}{s} G(s)$ and the condition (6.2) is satisfied if

$$\mathbf{Re}\{G(j\omega)\} \neq 0 \qquad (6.3)$$

holds for all $\omega \in \mathbb{R} \cup \{\infty\}$. It is obvious that (6.3) holds if and only if there exists a scalar $\alpha \in \mathbb{R}, \alpha \neq 0$ such that

$$\alpha G(j\omega) + \alpha G^*(j\omega) > 0, \quad \omega \in \mathbb{R} \cup \{\infty\}. \qquad (6.4)$$

This means the transfer function $\alpha G(s)$ is SPR with $\alpha G(\infty) > 0$ and thus we can obtain the following result.

Lemma 6.2. *Suppose that Assumptions 6.1 and 6.2 hold, $G(s)$ is non-degenerate and zero is not a zero of $G(s)$. If there exists a scalar $\alpha \in \mathbb{R}, \alpha \neq 0$ such that $\alpha G(s)$ is SPR with $\alpha G(\infty) > 0$, then the system (6.1) is dichotomous.*

Remark 6.1. From Corollary 5.3 in Chapter 5, if Assumptions 6.1, 6.2 and 6.3 hold and $\varphi(\sigma)$ has a finite number of isolated zeros on $[0, \Delta)$, then (6.3) ensures that the system (6.1) is still dichotomous.

Under Assumptions 6.1, we can write the transfer matrix $(A - sI)^{-1}b$ as

$$(A - sI)^{-1} b = \frac{1}{s} G_0(s) = \frac{1}{s}(C_0(sI - A_0)^{-1} B_0 + D_0), \qquad (6.5)$$

where $A_0 \in \mathbb{R}^{(n-1)\times(n-1)}, B_0 \in \mathbb{R}^{(n-1)\times 1}, C_0 \in \mathbb{R}^{n\times(n-1)}$ and $D_0 \in \mathbb{R}^{n\times 1}$.

Throughout this section we further assume that

Assumption 6.4. (A_0, B_0) *is controllable, (A_0, C_0) is observable, and zero is not a zero of $G_0(s)$.*

The transfer function of linear part for the closed-loop feedback system from $-\varphi(\sigma)$ to σ shown in Fig. 6.2 is

$$P_{cl}(s) = P_2(s)(1 + K(s)P_1(s))^{-1}$$
$$= c\left[I + (A - sI)^{-1}bK(s)c_*\right]^{-1}(A - sI)^{-1}b. \quad (6.6)$$

In order to make the closed-loop feedback system shown in Fig. 6.2 to be still a pendulum-like feedback system, we assume that the designed controller $K(s)$ contains a differential part, that is,

$$K(s) = sK_1(s), \quad (6.7)$$

where $K_1(s)$ has a state space realization

$$K_1(s) = \left[\begin{array}{c|c} A_k & b_k \\ \hline c_k & 0 \end{array}\right]. \quad (6.8)$$

Thus, the designed controller $K(s)$ can be expressed as

$$K(s) = c_k A_k (sI - A_k)^{-1} b_k + c_k b_k. \quad (6.9)$$

From (6.5), (6.6) and (6.7) the transfer function $P_{cl}(s)$ of linear part for the closed-loop feedback system shown in Fig. 6.2 from $-\varphi(\sigma)$ to σ can be rewritten as

$$P_{cl}(s) = \frac{1}{s} c(I + G_0(s)K_1(s)c_*)^{-1} G_0(s). \quad (6.10)$$

It is obvious that the closed-loop system shown in Fig. 6.2 is still a pendulum-like feedback system if $P_{cl}(s)$ is non-degenerate.

From Assumption 6.4 and (6.10) one implies that if $G_{cl}(s) = c(I + G_0(s)K_1(s)c_*)^{-1} G_0(s)$ is non-degenerate and zero is not a zero of $c(I + G_0(s)K_1(s)c_*)^{-1}$, then $P_{cl}(s) = \frac{1}{s} G_{cl}(s)$ is non-degenerate.

Applying Lemma 6.2, we have

Theorem 6.1. *Suppose that Assumptions 6.1, 6.2, 6.3 and 6.4 hold. If there exist a scalar $\alpha \in \mathbb{R}, \alpha \neq 0$ and a transfer function $K_1(s)$ with state space representation (6.8) such that*

$$G_{cl}(s) = c(I + G_0(s)K_1(s)c_*)^{-1} G_0(s)$$

is non-degenerate, zero is not a zero of $c(I + G_0(s)K_1(s)c_*)^{-1}$ and $\alpha G_{cl}(s)$ is SPR with $\alpha c D_0 > 0$, then the closed-loop feedback system shown in Fig. 6.2 is dichotomous, where $K(s)$ is defined by (6.9).

6.1. Controller design with dichotomy or gradient-like property

According to Theorem 6.1, the problem of designing controller $K(s)$ such that the closed-loop feedback system shown in Fig. 6.2 is dichotomous is transformed into the problem of designing a strictly proper transfer function $K_1(s)$ such that $\alpha G_{cl}(s)$ with $\alpha c D_0 > 0$ is SPR. In the following we will transform the latter problem into the SPR control problem.

In fact, $\alpha G_{cl}(s)$ has a state space realization:

$$\alpha G_{cl} = \left(\begin{array}{cc|c} A_0 & -B_0 c_k & B_0 \\ b_k c_* C_0 & A_k - b_k c_* D_0 c_k & b_k c_* D_0 \\ \hline \alpha c C_0 & -\alpha c D_0 c_k & \alpha c D_0 \end{array} \right), \qquad (6.11)$$

and thus $\alpha G_{cl}(s)$ can be expressed as

$$\alpha G_{cl} = \mathcal{F}_\ell(N, K_0), \qquad (6.12)$$

where $\mathcal{F}_\ell(N, K_0)$ is low linear fractional transformation and N, K_0 have state space realizations:

$$N = \left(\begin{array}{c|cc} A_0 & B_0 & -B_0 \\ \hline \alpha c C_0 & \alpha c D_0 & -\alpha c D_0 \\ c_* C_0 & c_* D_0 & -c_* D_0 \end{array} \right), \qquad (6.13)$$

and

$$K_0 = \left(\begin{array}{c|c} A_k - b_k c_* D_0 c_k & b_k \\ \hline c_k & 0 \end{array} \right), \qquad (6.14)$$

respectively. Therefore, the problem of designing a strictly proper transfer function $K_1(s)$ such that $\alpha G_{cl}(s)$ is SPR with $\alpha c D_0 > 0$ can be transformed into the SPR control problem, that is, for the plant N, design a strictly proper controller K_0 which has state space representation

$$K_0(s) = \left(\begin{array}{c|c} \bar{A}_k & \bar{b}_k \\ \hline \bar{c}_k & 0 \end{array} \right) \qquad (6.15)$$

such that $\alpha G_{cl} = \mathcal{F}_\ell(N, K_0)$ is internally stable and SPR. In this case, from (6.14) and (6.15), the matrices of designed transfer function $K_1(s)$ are expressed as

$$b_k = \bar{b}_k, \quad c_k = \bar{c}_k, \quad A_k = \bar{A}_k + b_k c_* D_0 c_k. \qquad (6.16)$$

The above statements can be summarized as

Theorem 6.2. *Suppose that Assumptions 6.1, 6.2, 6.3 and 6.4 hold. If there exist a scalar $\alpha \in \mathbb{R}, \alpha \neq 0$ and a strictly proper transfer function $K_0(s)$ with state space realization (6.15) such that $\mathcal{F}_\ell(N, K_0)$ is internally stable and SPR with $\alpha c D_0 > 0$, then there exists a transfer function $K_1(s)$*

with state space representation (6.8) such that $\alpha G_{cl}(s)$ is SPR with $\alpha c D_0 > 0$, where A_k, b_k, c_k are defined by (6.16).

Furthermore, if zero is not a zero of $c(I+G_0(s)K_1(s)c_*)^{-1}$ and $G_{cl}(s)$ is non-degenerate, i.e., (A_s, b_s) is controllable and (A_s, c_s) is observable, then the closed-loop feedback system shown in Fig. 6.2 is dichotomous, where $K(s)$ is defined by (6.9) and

$$A_s = \begin{pmatrix} A_0 & -B_0 c_k \\ b_k c_* C_0 & A_k - b_k c_* D_0 c_k \end{pmatrix}, \qquad (6.17)$$

$$b_s = \begin{pmatrix} B_0 \\ b_k c_* D_0 \end{pmatrix}, \quad c_s = (cC_0 \quad -cD_0 c_k). \qquad (6.18)$$

Combining Theorem 3.1 with Theorem 6.2, we have the following LMI conditions for existence of controller $K(s)$ such that the closed-loop feedback system shown in Fig. 6.2 is dichotomous.

Theorem 6.3. *Suppose that Assumptions 6.1, 6.2, 6.3 and 6.4 hold. Then there exist a scalar $\alpha \in \mathbb{R}, \alpha \neq 0$ and a strictly proper transfer function $K_0(s)$ with state space representation (6.15) such that $\mathcal{F}_\ell(N, K_0)$ is internally stable and SPR with $\alpha c D_0 > 0$ if and only if there exist positive definite matrices $W_1 > 0, W_3 > 0$ and matrices W_2, W_4 such that*

(i) $\begin{pmatrix} A_0 W_1 + W_1 A_0^T - B_0 W_2 - W_2^T B_0^T & S_1 \\ S_1^T & -2\alpha c D_0 \end{pmatrix} < 0,$
where $S_1 = \alpha W_1 C_0^T c^T - \alpha W_2^T D_0^T c^T - B_0$;

(ii) $\begin{pmatrix} W_3 A_0 + A_0^T W_3 + W_4 c_* C_0 + C_0^T c_*^T W_4^T & S_2 \\ S_2^T & -2\alpha c D_0 \end{pmatrix} < 0,$
where $S_2 = W_3 B_0 + W_4 c_* D_0 - \alpha C_0^T c^T$;

(iii) $\rho(Y_L X_F) < 1$, where $X_F = W_1^{-1}$ and $Y_L = W_3^{-1}$.

Moreover, if (i)-(iii) hold, then the matrices of state space realization for $K_0(s)$ are expressed as

$$\bar{c}_k = F = W_2 W_1^{-1}, \quad \bar{b}_k = -(I - Y_L X_F)^{-1} L, \quad L = W_3^{-1} W_4,$$

$$\bar{A}_k = A_0 - B_0 F + (I - Y_L X_F)^{-1} L c_* C_0 + \Delta_{FL},$$

where

$$\Delta_{FL} = -(2\alpha c D_0)^{-1}[B_0 + (I - Y_L X_F)^{-1} L c_* D_0][\alpha c C_0 - \alpha c D_0 F - B_0^T X_F]$$
$$-(I - Y_L X_F)^{-1} Y_L F^T [-B_0^T X_F - \frac{1}{2}(\alpha c C_0 - \alpha c D_0 F - B_0^T X_F)]$$
$$+(I - Y_L X_F)^{-1} Y_L R_F(X_F),$$

6.1. Controller design with dichotomy or gradient-like property

$$R_F(X_F) = (A_0 - B_0 F)^T X_F + X_F(A_0 - B_0 F)$$
$$+(2\alpha c D_0)^{-1}(\alpha c C_0 - \alpha c D_0 F - B_0^T X_F)^T(\alpha c C_0 - \alpha c D_0 F - B_0^T X_F)$$

and A_k, b_k, c_k is formulated by (6.16).

Furthermore, if (A_s, b_s) is controllable, (A_s, c_s) is observable and zero is not a zero of $c(I + G_0(s)K_1(s)c_*)^{-1}$, then the closed-loop feedback system shown in Fig. 6.2 is dichotomous, where $K(s)$ is represented by (6.9) and A_s, b_s, c_s are formulated by (6.17) and (6.18).

Remark 6.2. The problems of controller design such that the closed-loop feedback systems are dichotomous or gradient-like are quite interesting when the nonlinearity $\varphi(\sigma)$ is not scalar. In this case, we need to use the results of Chapter 5 for MIMO case instead of Lemma 6.1 and Lemma 6.2 where the frequency-domain inequalities are complicated compared with the case of SISO. The problems of controller design cannot be simply transformed into the SPR control problem but it can be possibly dealt with by using techniques and formulas related to LMI in Chapter 1.

6.1.2 Controller design with gradient-like property

In the following, we make an additional assumption for function $\varphi(\sigma)$ except for Assumption 6.2, that is,

Assumption 6.5. $\int_0^\Delta \varphi(\sigma) d\sigma = 0$, where Δ is the period of $\varphi(\sigma)$.

Assumption 6.5 indicates the nonlinear function $\varphi(\sigma)$ has a zero mean. It has explicit physical meaning in practical systems. For example, for the second pendulum system given by $\ddot{\sigma} + \tau\dot{\sigma} + \delta\sin\sigma = 0$, Assumption 6.5 holds which means the total work done by the elastic force is zero in a period Δ, and thus the average of total energy in a period of the system is decreasing since $\tau > 0$.

The gradient-like property for the system (6.1) is given in [Leonov et al. (1996)] which is summarized as follows.

Lemma 6.3. Suppose that Assumptions 6.1-6.3 and 6.5 hold. If the system (6.1) is dichotomous, then it is also gradient-like.

According to Lemma 6.3 we have

Corollary 6.1. Assume that the conditions of Theorem 6.3 hold and $\varphi(\sigma)$ satisfies Assumption 6.5. Then there exists a controller $K(s)$ such that the closed-loop feedback system shown in Fig. 6.2 is gradient-like. And the controller can be obtained by Theorem 6.3.

Example 6.1. Let us consider the following autonomous pendulum-like system with multiple equilibria:
$$\dot{x} = Ax + b\sin(\sigma), \quad \sigma = cx, \quad v = c_* x, \tag{6.19}$$
where
$$A = \begin{pmatrix} -3 & -2 & 0 \\ 1 & 0 & 0 \\ 0 & 1 & 0 \end{pmatrix}, b = \begin{pmatrix} 1 \\ -7 \\ 1 \end{pmatrix}, c = \begin{pmatrix} 1.4231 & 0.8462 & 2.5 \end{pmatrix}.$$

Assume that the state x_1 is measurable which means $c_* = \begin{pmatrix} 1 & 0 & 0 \end{pmatrix}$. Choosing an initial value $x_0 = (10 \ 2 \ -1)^T$, Fig. 6.3 and Fig. 6.4 show the solution curve of the system (6.19). Obviously, the bounded solution $x(t)$ of (6.19) is not convergent corresponding to this initial value, which indicates the system (6.19) is not dichotomous.

It is easy to test the conditions of Corollary 6.1 are satisfied. Using Theorem 6.3, we can design a controller $K(s)$ as
$$K(s) = \frac{-1.9049s^2 + 5.0468s}{s^2 + 11.7847s + 43.4563}$$
which ensures the closed-loop feedback system shown in Fig. 6.2 is gradient-like.

A state space representation of the closed-loop feedback system from $-\varphi(\sigma)$ to σ is given by
$$\dot{x} = A_{cl} x + b_{cl} \sin(\sigma), \quad \sigma = c_{cl} x \tag{6.20}$$
where
$$A_{cl} = \begin{pmatrix} -12.8798 & -59.1889 & -224.5935 & -86.9125 & 0 \\ 1.0000 & 0 & 0 & 0 & 0 \\ 0 & 1.0000 & 0 & 0 & 0 \\ 0 & 0 & 1.0000 & 0 & 0 \\ 0 & 0 & 0 & 1.0000 & 0 \end{pmatrix},$$
$$b_{cl}^T = \begin{pmatrix} 1 & 0 & 0 & 0 & 0 \end{pmatrix},$$
$$c_{cl} = 1.0 \times 10^3 \cdot \begin{pmatrix} -0.0020 & -0.0306 & -0.2144 & -0.8345 & -1.9555 \end{pmatrix}.$$
It is obvious that (6.20) is also a pendulum-like system with multiple equilibria. Note that if we choose initial value $x_{01} = (10 \ 2 \ -1 \ 1 \ 6)^T$, then the corresponding solution of the system (6.20) is convergent to equilibrium
$$x_e = \begin{pmatrix} 0 & 0 & 0 & 0 & 4.2669 \end{pmatrix}^T,$$
where x_e is an eigenvector of A_{cl} corresponding to zero eigenvalue and $\sigma_e = c_{cl} x_e = -8.3441 \times 10^3$ is a root of $\sin(\sigma) = 0$. The behavior of the corresponding solution for the closed-loop feedback system (6.20) is shown in Fig. 6.5 and Fig. 6.6.

6.2. Controller design with Lagrange stability

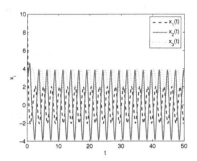

Fig. 6.3 The behavior of $x_i(t)$ for the system (6.19).

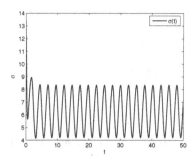

Fig. 6.4 The behavior of $\sigma(t)$ for the system (6.19).

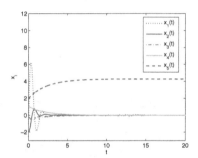

Fig. 6.5 The behavior of $x_i(t)$ for the closed-loop system (6.20).

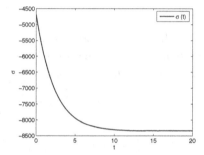

Fig. 6.6 The behavior of $\sigma(t)$ for the closed-loop system (6.20).

6.2 Controller design with Lagrange stability

In this section we consider a control problem for a class of linear perturbed systems which are also called a class of pendulum-like control systems, that is, the linear block of the systems contains zero eigenvalue and nonlinearity is periodic and satisfies sector condition. When there is no control, that is, $u = 0$, the system is pendulum-like but without Lagrange stability. We study controller design for this kind of control system such that the closed-loop system is Lagrange stable and preserve the pendulum-like behavior of the system.

Consider a pendulum-like control system:

$$\dot{x} = Ax + bu + b\varphi(t, \sigma), \quad \sigma = cx \quad (6.21)$$

where $x \in \mathbb{R}^n$ is the system state and $u \in \mathbb{R}$ is control input. $A \in \mathbb{R}^{n \times n}$ is a constant real matrix with Assumption 6.1, $b \in \mathbb{R}^{n \times 1}$ and $c^T \in \mathbb{R}^{n \times 1}$ are real vectors.

The nonlinear function $\varphi(t, \sigma)$ satisfies the following assumptions.

Assumption 6.6. $\varphi : \mathbb{R}_+ \times \mathbb{R} \to \mathbb{R}$ is continuous and locally Lipschitz continuous in the second argument, φ belongs to the sector $M[\mu_1, \mu_2]$, i.e.

$$\mu_1 \leq \frac{\varphi(t, \sigma)}{\sigma} \leq \mu_2. \qquad (6.22)$$

where $\mu_1 < 0, \mu_2 > 0$ are two finite numbers.

Assumption 6.7. There exists a $\Delta > 0$, such that

$$\varphi(t, \sigma + \Delta) = \varphi(t, \sigma), \quad t \in \mathbb{R}_+, \sigma \in \mathbb{R} \qquad (6.23)$$

Assumption 6.8. $P(s) = c(A - sI)^{-1}b$ is non-degenerate, i.e., (A, b) is controllable and (A, c) is observable.

When there is no control, i.e., $u = 0$, the system is a pendulum-like system which can be shown in Fig. 6.7, where $P(s) = -c(sI - A)^{-1}b$ is the transfer function from $-\varphi$ to σ.

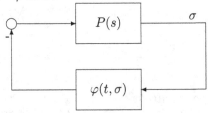

Fig. 6.7 Pendulum-like system (6.21) with $u = 0$.

Since some pendulum-like feedback systems shown in Fig. 6.7 may not be Lagrange stable it is necessary to add an efficient control such that the closed-loop feedback systems are Lagrange stable and preserve the property of pendulum-like systems at the same time. For this purpose we consider the synthesis problem of (6.21), that is, to design a linear controller $K(s)$ such that the closed-loop feedback system shown in Fig. 6.8 is pendulum-like and Lagrange stable.

For the system (6.21), since $\det(A) = 0$, $P(s)$ can be rewritten as

$$P(s) = \frac{1}{s}G(s) = \frac{1}{s}(c_0(sI - A_0)^{-1}b_0 + d), \qquad (6.24)$$

where $A_0 \in \mathbb{R}^{(n-1) \times (n-1)}, b_0 \in \mathbb{R}^{(n-1) \times 1}, c_0 \in \mathbb{R}^{1 \times (n-1)}$. Assumption 6.8 is equivalent to the following assumption:

6.2. Controller design with Lagrange stability

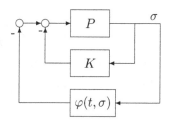

Fig. 6.8   The closed-loop feedback system.

Assumption 6.9. (A_0, b_0) is controllable, (A_0, c_0) is observable, and zero is not a zero of $G(s)$.

In order to design an appropriate controller guaranteeing Lagrange stability of the closed-loop feedback system we first show the following lemma.

Lemma 6.4. *Suppose that Assumptions 6.1, 6.6, 6.7, 6.9 hold, and there exists a number $\lambda > 0$ such that the following conditions for system (6.21) with $u = 0$ are fulfilled*

(i) the matrix $A + \lambda I$ has $n - 1$ eigenvalues with negative real parts;
(ii) $\mathbf{Re}\{H(j\omega)\} \geq 0$ for all $\omega \in \mathbb{R} \cup \{\infty\}$, where

$$H(s) = -\frac{\mu_2}{\mu_1} + \frac{\mu_2}{\mu_1} \cdot \frac{(\mu_2 - \mu_1)\dfrac{1}{s-\lambda}G(s-\lambda)}{1 + \mu_2 \dfrac{1}{s-\lambda}G(s-\lambda)}. \quad (6.25)$$

Then system (6.21) with $u = 0$ is Lagrange stable.

Proof. By Theorem 5.13 and Remark 5.5 in Chapter 5, it is only to show that the frequency-domain inequality

$$\mu_1^{-1}\mu_2^{-1} + (\mu_1^{-1} + \mu_2^{-1})\mathbf{Re}\{P(j\omega - \lambda)\} + |P(j\omega - \lambda)|^2 \leq 0 \quad (6.26)$$

holds for $\forall \omega \in \mathbb{R} \cup \{\infty\}$. By simple manipulation, the frequency-domain inequality (6.26) is equivalent to

$$S^*(j\omega)S(j\omega) \leq 1,$$

where $S(s) = \dfrac{1}{\gamma}P(s-\lambda) + \dfrac{\alpha}{\gamma}$, $\alpha = \dfrac{\mu_1^{-1} + \mu_2^{-1}}{2}$ and $\gamma = \dfrac{\mu_2^{-1} - \mu_1^{-1}}{2}$. Let

$$H(s) = \frac{1 - S(s)}{1 + S(s)}.$$

Then
$$H^*(j\omega) + H(j\omega) = \frac{2(1 - S^*(j\omega)S(j\omega))}{(1 - S(j\omega))(1 - S(j\omega))^*} \geq 0.$$

Thus the frequency-domain inequality (6.26) holds if and only if $\mathbf{Re}\{H(j\omega)\} \geq 0$. From $P(s) = \frac{1}{s}G(s)$, we can write $H(s)$ as the form of (6.25). This completes the proof. □

The transfer function from $-\varphi$ to σ in the closed-loop feedback system shown in Fig. 6.8 is

$$P_{cl}(s) = -(1 + P(s)K(s))^{-1}P(s). \qquad (6.27)$$

Let the designed controller $K(s)$ contain a differential part, that is, $K(s) = sK_1(s)$ where

$$K_1(s) = c_k(sI - A_k)^{-1}b_k \qquad (6.28)$$

and $A_k \in \mathbb{R}^{n_k \times n_k}, b_k \in \mathbb{R}^{n_k \times 1}, c_k \in \mathbb{R}^{1 \times n_k}$. Thus, $K(s)$ can be expressed as

$$K(s) = c_k A_k(sI - A_k)^{-1}b_k + c_k b_k. \qquad (6.29)$$

By (6.27), the transfer function $P_{cl}(s)$ can be rewritten as

$$P_{cl}(s) = -\frac{1}{s}(1 + G(s)K_1(s))^{-1}G(s). \qquad (6.30)$$

It is obvious that the closed-loop feedback system shown in Fig. 6.8 is still a pendulum-like feedback system if $P_{cl}(s)$ is non-degenerate, which is satisfied if

$$G_{cl}(s) = (1 + G(s)K_1(s))^{-1}G(s) \qquad (6.31)$$

is non-degenerate and zero is not a pole of $1 + G(s)K_1(s)$ by Assumption 6.9. On the other hand, $1 + G(s)K_1(s)$ and $G_{cl}(s)$ have state space realizations:

$$1 + G(s)K_1(s) = \left(\begin{array}{cc|c} A_0 & b_0 c_k & 0 \\ 0 & A_k & b_k \\ \hline c_0 & dc_k & 1 \end{array}\right),$$

$$G_{cl}(s) = \left(\begin{array}{c|c} A_g & b_g \\ \hline c_g & -d \end{array}\right),$$

where

$$A_g = \begin{pmatrix} A_0 & -b_0 c_k \\ b_k c_0 & A_k - db_k c_k \end{pmatrix}, \quad b_g = \begin{pmatrix} b_0 \\ db_k \end{pmatrix}, \quad c_g = \begin{pmatrix} -c_0 & dc_k \end{pmatrix}. \qquad (6.32)$$

6.2. Controller design with Lagrange stability

Thus $\det(A_0) \neq 0$ and $\det(A_k) \neq 0$ ensure that zero is not a pole of $1 + G(s)K_1(s)$.

Applying Lemma 6.4 to the closed-loop system (6.30) we have

Theorem 6.4. *Suppose that Assumptions 6.1, 6.6, 6.7, 6.9 hold and $\det(A_0) \neq 0$. If there exists a transfer function $K_1(s)$ expressed by (6.28) and a scalar $\lambda > 0$ such that $\det(A_k) \neq 0$, (A_g, b_g) is controllable and (A_g, c_g) is observable, $\lambda I + A_g$ is Hurwitz stable and $\operatorname{Re}\{H_s(j\omega)\} \geq 0$ for any $\omega \in \mathbb{R}$, then the closed-loop feedback system shown in Fig. 6.8 is pendulum-like and Lagrange stable, where $K(s)$ is defined by (6.29),*

$$H_s(s) = -\frac{\mu_2}{\mu_1} + \frac{\mu_2}{\mu_1} \frac{(\mu_2 - \mu_1)\dfrac{1}{s-\lambda} G_{cl}(s-\lambda)}{1 + \mu_2 \dfrac{1}{s-\lambda} G_{cl}(s-\lambda)} \tag{6.33}$$

and G_{cl} is defined by (6.31).

Theorem 6.5. *Suppose that Assumptions 6.1, 6.6, 6.7, 6.9 hold and $\det(A_0) \neq 0$. If there exist a positive number $\lambda > 0$, a scalar p_2 and a transfer function $K_1(s)$ with the state space representation (6.28) such that*

(i) $\det(A_k) \neq 0$, (A_g, b_g) is controllable and (A_g, c_g) is observable,

(ii) $d^2 p_2^2 - \dfrac{2\mu_2}{\mu_1}[2\lambda - (\mu_1 + \mu_2)d]p_2 + \mu^2 > 0$,

(iii) the transfer function matrix

$$R(s) = C_r(sI - \lambda I - A_g)^{-1} B_r + D_r \tag{6.34}$$

with $D_r + D_r^T > 0$ is SPR and $\lambda I + A_g$ is stable,

then the closed-loop feedback system shown in Fig. 6.8 is pendulum-like and Lagrange stable, where

$$B_r = b_g \begin{pmatrix} \mu_2 & 1 \end{pmatrix}, \quad C_r = \begin{pmatrix} p_2 \\ 0 \end{pmatrix} c_g,$$

$$D_r = -\begin{pmatrix} (\lambda + \mu_2 d)p_2 & \dfrac{1}{2}(\mu + p_2 d) \\ \dfrac{1}{2}(\mu + p_2 d) & \dfrac{\mu_2}{\mu_1} \end{pmatrix}, \quad \mu = \dfrac{\mu_2}{\mu_1}(\mu_2 - \mu_1) \tag{6.35}$$

and A_g, b_g, c_g are defined by (6.32).

Proof. From Lemma 3.8, (iii) holds if and only if there exists a positive definite matrix $P_1 > 0$ such that

$$\begin{pmatrix} P_1(\lambda I + A_g) + (\lambda I + A_g)^T P_1 & C_r^T - P_1 B_r \\ C_r - B_r^T P_1 & -(D_r + D_r^T) \end{pmatrix} < 0 \tag{6.36}$$

which is equivalent to

$$\begin{pmatrix} PA_q + A_q^T P & c_q^T - Pb_q \\ c_q - b_q^T P & \dfrac{2\mu_2}{\mu_1} \end{pmatrix} < 0, \qquad (6.37)$$

where $P = \mathrm{diag}(P_1, P_2)$ and $(A_q, b_q, c_q, -\dfrac{\mu_2}{\mu_1})$ is a state space realization of $H_s(s)$ in (6.33). By Lemma 6.4 and Corollary 1.4 one implies that $\mathbf{Re}\{H_s(j\omega)\} > 0$ for all $\omega \in \mathbb{R} \cup \{\infty\}$, and thus the result follows from Theorem 6.4. $\square$

In what follows, we provide strategies for designing a strictly proper transfer function $K_1(s)$ such that the conditions of Theorem 6.5 are satisfied.

In fact, from (6.32), (6.34) and (6.35), $R(s)$ defined by (6.34) can be expressed as $R(s) = \mathcal{F}_\ell(P_s, K_s)$, where $\mathcal{F}_\ell(P_s, K_s)$ is a lower linear fractional transformation, P_s and K_s have the following state space realizations:

$$P_s = \left(\begin{array}{c|cc} A_p & B_1 & B_2 \\ \hline C_1 & D_{11} & D_{12} \\ C_2 & D_{21} & 0 \end{array} \right), \quad K_s = \left(\begin{array}{c|c} \lambda I + A_k - db_k c_k & b_k \\ \hline c_k & 0 \end{array} \right), \qquad (6.38)$$

where

$$A_p = \lambda I + A_0, \quad B_1 = b_0 \begin{pmatrix} \mu_2 & 1 \end{pmatrix}, \quad D_{21} = d \begin{pmatrix} \mu_2 & 1 \end{pmatrix},$$

$$B_2 = -b_0, \quad C_1 = \begin{pmatrix} -p_2 \\ 0 \end{pmatrix} c_0, \quad D_{12} = d \begin{pmatrix} p_2 \\ 0 \end{pmatrix},$$

$$C_2 = c_0, \quad D_{11} = - \begin{pmatrix} (\lambda + \mu_2 d) p_2 & \dfrac{1}{2}(\mu + p_2 d) \\ \dfrac{1}{2}(\mu + p_2 d) & \dfrac{\mu_2}{\mu_1} \end{pmatrix}.$$

Therefore, the problem of designing a transfer function $K_1(s)$ such that $R(s)$ is SPR with $D_r + D_r^T > 0$ and $\lambda I + A_g$ is stable, is equivalent to finding K_s such that $\mathcal{F}_\ell(P_s, K_s)$ is internally stable and SPR with $D_r + D_r^T > 0$. By Theorem 3.1, the following results follow immediately.

Theorem 6.6. *Suppose that Assumptions 6.1, 6.6, 6.7, 6.9 hold and $\det(A_0) \neq 0$. Then for a given scalar p_2 and a positive number λ satisfying (ii) of Theorem 6.5 there exist matrices A_k, b_k and c_k such that $R(s) = \mathcal{F}_\ell(P_s, K_s)$ is SPR with $D_r + D_r^T > 0$ and $\lambda I + A_g$ is stable if and only if the following LMIs*

$$\begin{pmatrix} A_p W_1 + W_1 A_p^T + B_2 W_2 + W_2^T B_2^T & W_1 C_1^T + W_2^T D_{12}^T - B_1 \\ C_1 W_1 + D_{12} W_2 - B_1^T & -(D_{11} + D_{11}^T) \end{pmatrix} < 0, \qquad (6.39)$$

6.2. Controller design with Lagrange stability

$$\begin{pmatrix} W_3 A_p + A_p^T W_3 + W_4 C_2 + C_2^T W_4^T & W_3 B_1 + W_4 D_{21} - C_1^T \\ B_1^T W_3 + D_{21}^T W_4^T - C_1 & -(D_{11} + D_{11}^T) \end{pmatrix} < 0$$
(6.40)

have solutions W_1, W_2, W_3, W_4 with $W_1 > 0, W_3 > 0$, and the spectral radius $\rho(YX) < 1$, where $X = W_1^{-1}, Y = W_3^{-1}$.

Moreover, when these conditions are satisfied the matrices A_k, b_k, c_k can be expressed as

$$b_k = -(I - YX)^{-1} L, \quad c_k = F,$$
$$A_k = A_p + B_2 F + (I - YX)^{-1} LC_2 + \Delta - \lambda I + db_k c_k$$
(6.41)

where $F = W_2 W_1^{-1}, L = W_3^{-1} W_4$ and

$$\Delta = -(B_1 + (I - YX)^{-1} LD_{21})(D_{11} + D_{11}^T)^{-1}(C_1 - B_1^T X + D_{12} F)$$
$$-(I - YX)^{-1} Y F^T (B_2^T X + D_{12}^T (D_{11} + D_{11}^T)^{-1}(C_1 - B_1^T X + D_{12} F))$$
$$+(I - YX)^{-1} Y R_F,$$
$$R_F = (A_p + B_2 F)^T X + X(A_p + B_2 F)$$
$$+(C_1 + D_{12} F - B_1^T X)^T (D_{11} + D_{11}^T)^{-1}(C_1 + D_{12} F - B_1^T X).$$

In addition, if $det(A_k) \neq 0$, (A_g, b_g) is controllable, (A_g, b_g) is observable, then the closed-loop system shown in Fig. 6.8 is pendulum-like and Lagrange stable, where A_g, b_g, c_g are defined by (6.32).

Combining Theorem 6.5 with Theorem 6.6, the following procedure of controller design is provided.

Procedure 6.1:

(i) Choosing a scalar p_2 and a positive number $\lambda > 0$ satisfying

$$d^2 p_2^2 - \frac{2\mu_2}{\mu_1}[2\lambda - (\mu_1 + \mu_2)d]p_2 + \mu^2 > 0.$$

(ii) Solve LMIs (6.39), (6.40) and test $\rho(YX) < 1$.
(iii) Calculate A_k, b_k, c_k by (6.41) and verify $det(A_k) \neq 0$, (A_g, b_g) is controllable and (A_g, b_g) is observable. Then a controller $K(s)$ is obtained.

By Lemma 3.8, it is easy to imply the following corollary.

Corollary 6.2. *Assume that there exist scalars γ, η, p_2 and a positive number λ such that the condition (ii) of Theorem 6.5 holds and $G_1(s)$, $G_2(s)$ are all SPR with $D_{11} + D_{11}^T > 0$, where*

$$G_1(s) = (C_1 - \gamma D_{12} B_2^T)(sI - A_p + \gamma B_2 B_2^T)^{-1} B_1 + D_{11},$$
$$G_2(s) = C_1(sI - A_p + \eta C_2^T C_2)^{-1}(B_1 - \eta C_2^T D_{21}) + D_{11}.$$

Then there exist matrices $W_1 > 0, W_2 = -\gamma B_2^T W_1, W_3 > 0, W_4 = -\eta W_3 C_2^T$ such that (6.39) and (6.40) hold. Furthermore, if $\rho(YX) < 1$, then the results of Theorem 6.6 hold, where $X = W_1^{-1}$ and $Y = W_3^{-1}$.

Example 6.2. Consider a pendulum-like system given by

$$\begin{cases} \begin{pmatrix} \dot{x}_1 \\ \dot{x}_2 \end{pmatrix} = \begin{pmatrix} 0 & 1 \\ 0 & -\alpha \end{pmatrix} \begin{pmatrix} x_1 \\ x_2 \end{pmatrix} + \begin{pmatrix} 0 \\ -1 \end{pmatrix} u + \begin{pmatrix} 0 \\ -1 \end{pmatrix} \varphi(t,\sigma), \\ \sigma = \begin{pmatrix} 1 & 0 \end{pmatrix} \begin{pmatrix} x_1 \\ x_2 \end{pmatrix}, \end{cases} \quad (6.42)$$

where $\varphi(t,\sigma) = \beta \cos t \sin\sigma - \gamma \sin 2\sigma - \delta \cos 2t \sin 2\sigma$ and $\alpha > 0, \beta, \gamma, \delta$ are real numbers. It is obvious that $\varphi(t,\sigma)$ satisfies the conditions (6.22) and (6.23), where $-\mu_1 = \mu_2 = |\beta| + 2|\gamma| + 2|\delta|$. When $u = 0$, that is, there is no control in the system we can get

$$\ddot{\eta} + \alpha\dot{\eta} + \beta \cos t \cos\eta + \gamma \sin 2\eta + \delta \cos 2t \sin 2\eta = 0 \quad (6.43)$$

where $\eta = \sigma - \dfrac{\pi}{2}$. Note that above system (6.43) is a pendulum-like system which describes the work of synchronous single-phase motor with pulsing vibrating moment [Leonov et al. (1996); Morary (1970)]. The boundedness of solutions for above system is basic and important.

Let $\alpha = 0.001, \beta = 1, \gamma = -0.2, \delta = 1$. Choosing an initial value $x_0 = (30, -10)^T$, from Fig. 6.9, we find that the solution of system (6.42) with $u = 0$ is not bounded.

From (6.42) we know that $A_0 = -0.001, b_0 = 1, c_0 = 1, d = 0$. Choosing $\lambda = 1, p_2 = -3$, by solving LMIs (6.39) and (6.40) we find that $A_k = -90.7176$, $b_k = 64.5385$, $c_k = 9.0076$. In terms of Procedure 6.1 we obtain a linear controller $K(s)$ as

$$K(s) = c_k A_k (sI - A_k)^{-1} b_k + c_k b_k = \frac{581.3402s}{s + 90.7176}$$

which ensures that the closed-loop feedback system shown in Fig. 6.8 is Lagrange stable. The form of state space for the closed-loop feedback system can be expressed as

$$\dot{x} = A_{cl} x + b_{cl} \varphi(t,\sigma), \quad \sigma = c_{cl} x, \quad (6.44)$$

where

$$A_{cl} = \begin{pmatrix} -90.7186 & -581.4309 & 0 \\ 1.0000 & 0 & 0 \\ 0 & 1.0000 & 0 \end{pmatrix}, \quad b_{cl} = \begin{pmatrix} 1 \\ 0 \\ 0 \end{pmatrix},$$

$$c_{cl} = \begin{pmatrix} 0 & -1.0000 & -90.7176 \end{pmatrix}.$$

Fig. 6.10 describes the states of the closed-loop nonlinear feedback system with an initial value $x_0 = (30, -10, 10)^T$ which is bounded. It is easy to verify that the closed-loop system is also a pendulum-like system with respect to $\Gamma = \{j\bar{d}, j \in Z\}$, where $\bar{d}^T = [0 \quad 0 \quad -0.0693]$.

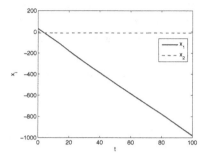 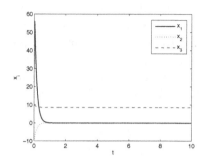

Fig. 6.9 The states of the pendulum-like system (6.42) with $u = 0$.

Fig. 6.10 The states of the closed-loop feedback system (6.44).

6.3 Notes and references

The boundedness of solutions for systems is a basic and natural requirement. Lagrange stability is a necessary condition for every solution of a system to converge to an equilibrium. This convergence can be achieved if the system is Lagrange stable and dichotomous. In the framework of absolute stability the boundedness of trajectories was established by Arcak et.al which the linear block must be relative degree one or two and minimum phase, and the nonlinearity is required to grow faster than linear, that is, to be stiffening [Arcak et al. (2002)]. In [Arcak and Teel (2002b)] the boundedness of the systems with relative degree one in feedback with a stiffening nonlinearity was shown with a positive real condition on the zero dynamics, which relaxed the earlier results in [Arcak et al. (2002)].

The frequency-domain method established in the fifties and sixties of last century is an important tool in system analysis not only for the stability of systems with a single stationary point but also for the global properties of solutions such as dichotomy, gradient-like property and Lagrange stability for the nonlinear pendulum-like systems [Popov (1962); Khalil (2002); Anderson (1967); Leonov et al. (2001); Rasvan (1998); Leonov et al. (1996); Huang (2003)]. The frequency-domain and real-domain inequality conditions guaranteeing the dichotomy, gradient-like property and Lagrange stability for a class of pendulum-like system given in [Leonov et al. (1996)] and Chapter 5 become the basis of controller design for nonlinear systems with multiple equilibria. This chapter is based on the results in [Wang et al. (2004); Wang et al. (2006a)] where the conditions for existences of controllers for the pendulum-like systems guaranteeing dichotomy, gradient-like property or Lagrange stability of the closed-loop systems and preserving the pendulum-like behavior of systems were given. The results of SPR control problem are applied to design controllers with some performance. Yang and Huang [Yang and Huang (2003)] gave different results for stabilizing a pendulum-like system in the Lagrange sense which are within the framework of H_∞ control.

Chapter 7

Controller Designs for Systems with Input Nonlinearities

This chapter studies control problems for a class of systems with input nonlinearities. The existence conditions and design methods of dynamical output feedback controllers that guarantee the closed-loop systems to be Lagrange stable, Bakaev stable or dichotomous are given.

7.1 Lagrange stabilizing for systems with input nonlinearities

Let us consider a class of nonlinear control systems given by:

$$\begin{cases} \dot{x} = Ax + b\varphi(t, u), \\ y = cx, \end{cases} \quad (7.1)$$

where x is the system state, u is the control input and y is the system output. $A \in \mathbb{R}^{n \times n}$ is a constant real matrix satisfying $\det(A) = 0$, $b \in \mathbb{R}^{n \times 1}$ and $c \in \mathbb{R}^{1 \times n}$ are real vectors and (A, b) is controllable. $\varphi : \mathbb{R}_+ \times \mathbb{R} \to \mathbb{R}$ is continuous and locally Lipschitz continuous in the second argument and satisfies

$$\varphi(t, \sigma + \Delta) = \varphi(t, \sigma), \quad t \in \mathbb{R}_+, \, \sigma \in \mathbb{R} \quad (7.2)$$

and

$$\mu_1 \leq \frac{\varphi(t, \sigma)}{\sigma} \leq \mu_2 \quad (7.3)$$

where μ_1, μ_2 are two finite numbers with $\mu_1 < 0$ and $\mu_2 > 0$.

Consider a dynamical output feedback controller $K(s)$ with the state space representation

$$\begin{cases} \dot{x}_k = A_k x_k + b_k y, \\ u = c_k x_k + d_k y, \end{cases} \quad (7.4)$$

where $A_k \in \mathbb{R}^{n_k \times n_k}, b_k \in \mathbb{R}^{n_k \times 1}, c_k \in \mathbb{R}^{1 \times n_k}, d_k \in \mathbb{R}$. The transfer function of the controller $K(s)$ is given by

$$K(s) = c_k(sI - A_k)^{-1}b_k + d_k. \tag{7.5}$$

The closed-loop system defined by (7.1) and (7.4) is written as

$$\begin{cases} \begin{pmatrix} \dot{x} \\ \dot{x}_k \end{pmatrix} = \begin{pmatrix} A & 0 \\ b_k c & A_k \end{pmatrix} \begin{pmatrix} x \\ x_k \end{pmatrix} + \begin{pmatrix} b \\ 0 \end{pmatrix} \varphi(t, u), \\ u = (d_k c \quad c_k) \begin{pmatrix} x \\ x_k \end{pmatrix} \end{cases} \tag{7.6}$$

which is shown in Fig. 7.1. The transfer function of the linear part for the closed-loop system (7.6) from φ to u is expressed as $T_{cl}(s) = K(s)G(s)$, where $G(s) = c(sI - A)^{-1}b$ and $K(s)$ has state space representation (7.5).

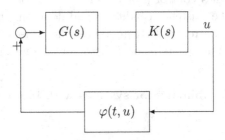

Fig. 7.1 The closed-loop feedback system.

Applying Theorem 5.12 to the closed-loop system (7.6), we have

Lemma 7.1. *Suppose that there exist a real number $\lambda > 0$, a symmetric matrix $P = P^T$ and a transfer function $K(s)$ with the state space representation (7.5) such that the following conditions are fulfilled:*

(i) the matrix $A + \lambda I$ has n-1 eigenvalues with negative real part;
(ii) (A_s, b_s) is controllable and (A_s, c_s) is observable, where

$$A_s = \begin{pmatrix} A & 0 \\ b_k c & A_k \end{pmatrix}, \quad b_s = \begin{pmatrix} b \\ 0 \end{pmatrix}, \quad c_s = (d_k c \quad c_k); \tag{7.7}$$

(iii) $A_k + \lambda I$ is Hurwitz stable;
(iv)

$$\begin{pmatrix} 2\lambda P + PA_s + A_s^T P + c_s^T c_s & Pb_s - \alpha c_s^T \\ b_s^T - \alpha c_s & \mu_1^{-1}\mu_2^{-1} \end{pmatrix} < 0, \tag{7.8}$$

where $\alpha = \dfrac{\mu_1^{-1} + \mu_2^{-1}}{2}.$

7.1. Lagrange stabilizing for systems with input nonlinearities

Then the closed-loop system (7.6) is Lagrange stable.

In the following sufficient conditions for the existence of dynamical output feedback controllers such that the closed-loop system (7.6) be Lagrange stable are given.

Theorem 7.1. *Assume that there exist a positive number $\lambda > 0$, a symmetric matrix $P_1 = P_1^T$ and a nonsingular symmetric matrix $Y_1 = Y_1^T$ such that*

(i) the matrix $A + \lambda I$ has n-1 eigenvalues with negative real parts;

(ii)

$$\begin{pmatrix} Y_1(\lambda I + A) + (\lambda I + A)^T Y_1 & Y_1 b \\ b^T Y_1 & \mu_1^{-1}\mu_2^{-1} - \alpha^2 \end{pmatrix} < 0, \quad (7.9)$$

where α is given as in Lemma 7.1;

(iii)

$$\begin{pmatrix} c^{T\perp} & 0 \\ 0 & 1 \end{pmatrix} \begin{pmatrix} P_1(\lambda I + A) + (\lambda I + A)^T P_1 & P_1 b \\ b^T P_1 & \mu_1^{-1}\mu_2^{-1} \end{pmatrix} \begin{pmatrix} c^{T\perp T} & 0 \\ 0 & 1 \end{pmatrix} < 0; \quad (7.10)$$

(iv)

$$P_1 - Y_1 \geq 0. \quad (7.11)$$

Then there exist matrices A_k, b_k, c_k, d_k such that the conditions (iii) and (iv) of Lemma 7.1 hold.

Furthermore, if the condition (ii) of Lemma 7.1 holds, then the closed-loop system (7.6) is Lagrange stable.

Proof. By the condition (iv), there exists a matrix P_2 satisfying $P_1 - Y_1 = P_2 P_2^T$. Let $P = \begin{pmatrix} P_1 & P_2 \\ P_2^T & I \end{pmatrix}$. We show that there exist control matrices A_k, b_k, c_k and d_k such that the inequality (7.8) in Lemma 7.1 holds for above P. For this purpose we prove that there exist matrices A_k, b_k, c_k and d_k such that (7.8) holds if and only if (7.9) and (7.10) hold. Multiplying $\text{diag}(P^{-1}, 1)$ on the left and the right in (7.8) and in terms of Schur complement, above inequality can be written as its equivalent form

$$\begin{pmatrix} 2\lambda Q + A_s Q + Q A_s^T & b_s - \alpha Q c_s^T & Q c_s^T \\ b_s^T - \alpha c_s Q & \mu_1^{-1}\mu_2^{-1} & 0 \\ c_s Q & 0 & -1 \end{pmatrix} < 0, \quad (7.12)$$

where $Q = P^{-1}$. Rewriting (7.12) as
$$\Gamma \bar{K} \Lambda + (\Gamma \bar{K} \Lambda)^T + \Theta < 0, \qquad (7.13)$$
where
$$\Theta = \begin{pmatrix} 2\lambda Q + A_0 Q + Q A_0^T & b_s & 0 \\ b_s^T & \mu_1^{-1}\mu_2^{-1} & 0 \\ 0 & 0 & -1 \end{pmatrix}, \quad \Gamma = \begin{pmatrix} \begin{pmatrix} 0 & 0 \\ 0 & I \end{pmatrix} \\ -\alpha \begin{pmatrix} I & 0 \end{pmatrix} \\ \begin{pmatrix} I & 0 \end{pmatrix} \end{pmatrix},$$
$$\Lambda = \left(\begin{pmatrix} c & 0 \\ 0 & I \end{pmatrix} Q \quad 0 \quad 0 \right), \quad \bar{K} = \begin{pmatrix} d_k & c_k \\ b_k & A_k \end{pmatrix}, \quad A_0 = \begin{pmatrix} A & 0 \\ 0 & 0 \end{pmatrix}.$$

According to Lemma 1.1, (7.13) holds if and only if the following two inequalities
$$\Gamma^\perp \Theta \Gamma^{\perp T} < 0 \qquad (7.14)$$
and
$$\Lambda^{T\perp} \Theta \Lambda^{T\perp T} < 0 \qquad (7.15)$$
hold. Let
$$Q = \begin{pmatrix} Q_1 & Q_2 \\ Q_2^T & Q_3 \end{pmatrix}, \quad P = Q^{-1} = \begin{pmatrix} P_1 & P_2 \\ P_2^T & P_3 \end{pmatrix}.$$
Then $Q_1 = Y_1^{-1}$ follows by $Q_1 = (P_1 - P_2 P_2^T)^{-1}$. Choosing
$$\Gamma^\perp = \begin{pmatrix} (I & 0) & 0 & 0 \\ 0 & 1 & \alpha \end{pmatrix}, \quad \Lambda^{T\perp} = \begin{pmatrix} \begin{pmatrix} c^T & 0 \\ 0 & I \end{pmatrix}^\perp Q^{-1} & 0 & 0 \\ 0 & I & 0 \\ 0 & 0 & I \end{pmatrix}.$$

Then by algebraic manipulation, one implies the equivalences between (7.14) and (7.9), (7.15) and (7.10) which guarantee the existences of A_k, b_k, c_k and d_k. In this case, A_k, b_k, c_k, d_k can be obtained by solving LMI (7.13) for variable $\bar{K}$, where $Q = P^{-1}$ (In fact, we can obtain the parameterized form of $\bar{K}$ from Lemma 1.1).

Next, from (7.8), we have
$$P(\lambda I + A_s) + (\lambda I + A_s)^T P + c_s^T c_s < 0$$
which yields
$$\lambda I + A_k + (\lambda I + A_k)^T + c_k^T c_k < 0, \qquad (7.16)$$
and thus $\lambda I + A_k$ is Hurwitz stable, i.e., the condition (iii) of Lemma 7.1 holds. In terms of Lemma 7.1, if (v) also holds, then the closed-loop system (7.6) is Lagrange stable. This completes the proof. □

7.1. Lagrange stabilizing for systems with input nonlinearities

From the proof of Theorem 7.1, we propose the following procedure for designing dynamical output feedback controllers such that the closed-loop systems are Lagrange stable.

Algorithm 7.1.

(i) Choose an appropriate scalar $\lambda > 0$, solve LMIs (7.9), (7.10) and (7.11) for the symmetric matrices P_1 and Y_1.
(ii) Choose a matrix P_2 such that $P_2 P_2^T = P_1 - Y_1$.
(iii) Let $P = \begin{pmatrix} P_1 & P_2 \\ P_2^T & I \end{pmatrix}$ and $Q = P^{-1}$. Solve LMI (7.13) for variable $\bar{K}$ and thus A_k, b_k, c_k and d_k are obtained. (In fact, we can obtain the parameterized form of $\bar{K} = \begin{pmatrix} d_k & c_k \\ b_k & A_k \end{pmatrix}$ from Lemma 1.1.)
(iv) Verify (A_s, b_s) is controllable and (A_s, c_s) is observable.

Corollary 7.1. *Suppose that there exist scalars $\delta \in R$ and $\lambda > 0$ such that*

(i) the matrix $A + \lambda I$ has n-1 eigenvalues with negative real parts;
(ii) Algebraic Riccati Inequality:

$$P_1(\lambda I + A) + (\lambda I + A)^T P_1 + \tau P_1 bb^T P_1 - \delta c^T c < 0 \quad (7.17)$$

has a positive definite solution $P_1 > 0$, where $\tau = -\mu_1 \mu_2 > 0$.

Then the results of Theorem 7.1 are still true.

Proof. Firstly, the LMI (7.10) in Theorem 7.1 is rewritten as

$$c^{T\perp}(P_1(\lambda I + A) + (\lambda I + A)^T P_1 + \tau P_1 bb^T P_1) c^{T\perp T} < 0, \quad (7.18)$$

where $\tau = -\mu_1 \mu_2 > 0$. By Lemma 1.3, (7.18) holds if and only if there exists $\delta \in \mathbb{R}$ such that (7.17) holds.

Next, assumption (i) implies that there exists a nonsingular matrix $Q_1 = Q_1^T$ such that

$$(\lambda I + A)Q_1 + Q_1(\lambda I + A)^T < 0$$

which implies that there exists a positive number $\eta > 0$ large enough such that

$$(\lambda I + A)(\eta Q_1) + (\eta Q_1)(\lambda I + A)^T + \gamma bb^T < 0$$

and

$$P_1 - (\eta Q_1)^{-1} \geq 0$$

for $P_1 > 0$, where $\dfrac{1}{\gamma} = \alpha^2 - \mu_1^{-1}\mu_2^{-1} > 0$. Let $Y_1 = (\eta Q_1)^{-1}$. By schur complement, one implies that the conditions (ii) and (iv) of Theorem 7.1 are satisfied. Consequently, the conditions of Theorem 7.1 are all fulfilled and the results follow. □

Remark 7.1. If $\mu_1 = -\infty$ or $\mu_2 = +\infty$, the problem of controller design with the closed-loop Lagrange stable can be investigated along the line of Bakaev stabilizing in next section.

Remark 7.2. Note that the control variable u in the system (7.1) is contained in the time-varying nonlinear function φ and the system is linear when $u = 0$ in this section. It is different from the system (6.21) in Section 6.2 where it depends on the control variable u linearly but the system is a nonlinear pendulum-like system when $u = 0$. The objective in both cases is to design a linear dynamical output feedback controller K such that the closed-loop system is Lagrange stable and preserves the properties of the pendulum-like systems. The transfer function $P_{cl}(s)$ of the closed-loop system in Section 6.2 is in feedback form, i.e., $P_{cl} = -(1 + P(s)K(s))^{-1}P(s)$, here it is in product form, i.e., $P_{cl}(s) = K(s)P(s)$.

Example 7.1. Consider a nonlinear control system:

$$\dot{x} = \begin{pmatrix} -5 & 0 \\ 1 & 0 \end{pmatrix} x + \begin{pmatrix} 1 \\ -1 \end{pmatrix} \sin u, \quad y = (1 \quad 1)x, \tag{7.19}$$

where x is the system state, u is the control input and y is the system output. $\varphi(t, \sigma) = \sin \sigma$ satisfies

$$-1 \leq \dfrac{\varphi(t, \sigma)}{\sigma} \leq 1, \quad \varphi(t, \sigma + 2\pi) = \varphi(t, \sigma).$$

For above system we can apply Theorem 7.1 to design a dynamical output feedback controller such that the closed-loop feedback system is Lagrange stable. Choosing $\lambda = 1$, by Algorithm 7.1, a dynamical output feedback controller (7.4) with

$$A_k = \begin{pmatrix} -8.0198 & -0.9237 \\ -0.9237 & -3.6639 \end{pmatrix}, \quad b_k = \begin{pmatrix} 0.2589 \\ 0.9904 \end{pmatrix},$$

$$c_k = (0.2589 \quad 0.9904), \quad d_k = -0.3833$$

is easily obtained such that the closed-loop feedback system (7.6) is Lagrange stable, that is, all the solutions of the closed-loop feedback system are bounded.

7.2 Bakaev stabilizing for systems with input nonlinearities

In this section, we consider problems of controller design for the system (7.1) via state feedback and output feedback such that the closed-loop systems are Bakaev stable.

Consider the nonlinear control system (7.1) with $\det(A) = 0$. $\varphi : \mathbb{R}_+ \times \mathbb{R} \to \mathbb{R}$ is continuous and locally Lipchitz continuous in the second argument and satisfies (7.2) and

$$\sigma\varphi(t,\sigma) \leq \mu\sigma^2, \quad t \in \mathbb{R}_+, \sigma \in \mathbb{R}. \tag{7.20}$$

First, we consider Bakaev stabilizing problem for the system (7.1) with state feedback, i.e., to design a state feedback $u = Kx$ such that the closed-loop feedback system:

$$\begin{cases} \dot{x} = Ax + b\varphi(t, u), \\ u = Kx \end{cases} \tag{7.21}$$

is Bakaev stable.

Theorem 7.2. *Suppose that (A, b) is controllable and there exist a positive number $\lambda > 0$, a real number ν with $\nu > \mu$ and a symmetric matrix $X = X^T$ such that*

(i) the matrix $A + \lambda I$ has $n - 1$ eigenvalues with negative real parts;

(ii) $\begin{pmatrix} (\lambda I + A)^T X + X(\lambda I + A) & 2\nu Xb \\ 2\nu b^T X & -I \end{pmatrix} \leq 0;$

(iii) $b^T X b \leq 0$.

Then there exists a control matrix $K = 2\nu b^T X$. Furthermore, if (A, K) is observable, then the closed-loop feedback system (7.21) is Bakaev stable.

Proof. Let $K = 2\nu b^T X$. Then

$$Xb = \frac{1}{2\nu}K^T, \quad Kb \leq 0$$

and

$$(\lambda I + A)^T X + X(\lambda I + A) + K^T K \leq 0$$

hold by conditions (ii) and (iii). In terms of Theorem 5.14, the closed-loop feedback system (7.21) is Bakaev stable. □

Next, we discuss the problem of designing a dynamical output feedback controller $K(s)$ with state space representation (7.4) such that the closed-loop feedback system (7.6) is Bakaev stable.

Applying Theorem 5.14 to the closed-loop system (7.6), we have

Lemma 7.2. *Suppose that there exist a scalar $\lambda > 0$, a real number ν with $\nu > \mu$ and a matrix $P = P^T$ such that*

(i) the matrix $A + \lambda I$ has $n - 1$ eigenvalues with negative real parts;
(ii) (A_s, b_s) is controllable and (A_s, c_s) is observable;
(iii) $A_k + \lambda I$ is Hurwitz stable;
(iv) $Pb_s = \dfrac{1}{2\nu} c_s^T$;
(v) $(\lambda I + A_s)^T P + P(\lambda I + A_s) + c_s^T c_s < 0$;
(vi) $c_s b_s \leq 0$;

where A_s, b_s, c_s are defined by (7.7) in Lemma 7.1. Then the closed-loop feedback system (7.6) is Bakaev stable.

Based on Lemma 7.2, conditions for existence of a dynamical output feedback controller are given in the following theorem.

Theorem 7.3. *Suppose that $cb \neq 0$ and there exist scalars $\lambda > 0, d_k \in \mathbb{R}, \delta \in \mathbb{R}$, a real number ν with $\nu > \mu$, a symmetric matrix $X = X^T$ and a symmetric nonsingular matrix $Y_1 = Y_1^T$ such that*

(i) the matrix $A + \lambda I$ has $n - 1$ eigenvalues with negative real parts;
(ii) $d_k cb \leq 0$;
(iii)

$$\begin{pmatrix} Y_1(\lambda I + A) + (\lambda I + A)^T Y_1 & Y_1 b \\ b^T Y_1 & -\beta^2 \end{pmatrix} < 0; \qquad (7.22)$$

(iv)

$$\begin{pmatrix} N + b^{\perp T} X b^{\perp}(\lambda I + A) + (\lambda I + A)^T b^{\perp T} X b^{\perp} - \delta c^T c & d_k c^T \\ d_k c & -I \end{pmatrix} < 0, \qquad (7.23)$$

where

$$N = \beta d_k c^T (cb)^{-1} c(\lambda I + A) + \beta d_k (\lambda I + A)^T c^T (cb)^{-1} c;$$

(v)

$$\beta d_k c^T (cb)^{-1} c + b^{\perp T} X b^{\perp} - Y_1 \geq 0, \qquad (7.24)$$

7.2. Bakaev stabilizing for systems with input nonlinearities

where $\beta = \dfrac{1}{2\nu}$. Then there exist matrices A_k, b_k, c_k and d_k such that (iii)-(vi) of Lemma 7.2 hold.

Furthermore, if the condition (ii) of Lemma 7.2 holds, then the closed-loop feedback system (7.6) is Bakaev stable.

Proof. Let
$$P_1 = \beta d_k c^T (cb)^{-1} c + b^{\perp T} X b^{\perp}.$$

Then
$$P_1 b = \beta d_k c^T. \tag{7.25}$$

The condition (v) implies that $P_1 - Y_1 \geq 0$. Let
$$P_2 P_2^T = P_1 - Y_1, \quad c_k = \frac{1}{\beta} b^T P_2, \tag{7.26}$$

and $P = \begin{pmatrix} P_1 & P_2 \\ P_2^T & I \end{pmatrix}$. Then it follows from (7.25) and (7.26) that

$$P b_s = \frac{1}{2\nu} c_s^T, \tag{7.27}$$

i.e., the condition (iv) of Lemma 7.2 holds. The condition (vi) of Lemma 7.2 is obviously true by the assumption (ii).

In what follows we show that there exist control matrices A_k, b_k such that the condition (v) of Lemma 7.2 holds. By (7.27), the condition (v) of Lemma 7.2 is rewritten as

$$\Theta + P \begin{pmatrix} 0 \\ I \end{pmatrix} \begin{pmatrix} b_k & A_k \end{pmatrix} \begin{pmatrix} c & 0 \\ 0 & I \end{pmatrix} + \left(P \begin{pmatrix} 0 \\ I \end{pmatrix} \begin{pmatrix} b_k & A_k \end{pmatrix} \begin{pmatrix} c & 0 \\ 0 & I \end{pmatrix} \right)^T < 0, \tag{7.28}$$

where

$$\Theta = \begin{pmatrix} A + \lambda I & 0 \\ 0 & \lambda I \end{pmatrix}^T P + P \begin{pmatrix} A + \lambda I & 0 \\ 0 & \lambda I \end{pmatrix} + 4\nu^2 P \begin{pmatrix} bb^T & 0 \\ 0 & 0 \end{pmatrix} P.$$

Let $Q = P^{-1} = \begin{pmatrix} Q_1 & Q_2 \\ Q_2^T & Q_3 \end{pmatrix}$. Then $Q_1^{-1} = Y_1$. And (7.28) is rewritten as

$$\Omega + U K_0 V + (U K_0 V)^T < 0, \tag{7.29}$$

where

$$U = \begin{pmatrix} 0 \\ I \end{pmatrix}, \quad V = \begin{pmatrix} c & 0 \\ 0 & I \end{pmatrix} Q, \quad K_0 = \begin{pmatrix} b_k & A_k \end{pmatrix},$$

$$\Omega = Q \begin{pmatrix} A + \lambda I & 0 \\ 0 & \lambda I \end{pmatrix}^T + \begin{pmatrix} A + \lambda I & 0 \\ 0 & \lambda I \end{pmatrix} Q + 4\nu^2 \begin{pmatrix} bb^T & 0 \\ 0 & 0 \end{pmatrix}.$$

In terms of Lemma 1.1, it is not difficult to show that (7.29) holds if and only if (7.22) and

$$c^{T\perp}(P_1(\lambda I + A) + (\lambda I + A)^T P_1 + 4\nu^2 P_1 bb^T P_1)c^{T\perp T} < 0 \qquad (7.30)$$

hold. Combining Lemma 1.3 with (7.25), (7.30) holds if and only if

$$P_1(\lambda I + A) + (\lambda I + A)^T P_1 + (d_k^2 - \delta)c^T c < 0$$

holds, which is equivalent to (7.23). Consequently, the conditions (iii) and (iv) guarantee that there exist control matrices A_k, b_k such that the condition (v) of Lemma 7.2 holds. In this case, the matrices A_k, b_k can be obtained by solving (7.29). Furthermore, from (7.28), we have

$$\lambda I + A_k + (\lambda I + A_k)^T + c_k^T c_k < 0$$

which implies that $\lambda I + A_k$ is Hurwitz stable, and thus the condition (iii) of Lemma 7.2 holds.

From the above statements, the conditions (iii)-(vi) of Lemma 7.2 are satisfied and thus the conclusions of Theorem 7.3 follow by Lemma 7.2. □

From the proof of Theorem 7.3, we propose the following procedure of designing dynamical output feedback controllers such that the closed-loop systems are Bakaev stable.

Algorithm 7.2.

(i) Solving (ii)-(v) of Theorem 7.3 for variables $d_k \in R$, $X = X^T$ and nonsingular matrix $Y_1 = Y_1^T$.
(ii) Let $P_1 = \alpha d_k c^T (cb)^{-1} c + b^{\perp T} X b^{\perp}$.
(iii) Choosing a matrix P_2 such that $P_2 P_2^T = P_1 - Y_1$.
(iv) Let $P = \begin{bmatrix} P_1 & P_2 \\ P_2^T & P_3 \end{bmatrix}$, $Q = P^{-1}$, solving LMI (7.29) for variable $K_0 = (b_k \ A_k)$ (In fact, we can obtain the parameterized form of K_0 by Lemma 1.1.).
(v) Computing $c_k = \dfrac{1}{\alpha} b^T P_2$.
(vi) Verify (A_s, b_s) is controllable and (A_s, c_s) is observable.

7.3 Control for systems with input nonlinearities guaranteeing dichotomy

Let us consider a class of autonomous nonlinear control system:
$$\begin{cases} \dot{x} = Ax + b\varphi(u), \\ y = cx, \end{cases} \quad (7.31)$$

where x is the system state, u is the control input and y is the system output. $A \in \mathbb{R}^{n \times n}$ is a constant real matrix satisfying $\det(A) = 0$, $b \in \mathbb{R}^n$ and $c^T \in \mathbb{R}^n$ are real vectors. $\varphi(\sigma)$ is continuous for all $\sigma \in \mathbb{R}$ and $\varphi(\sigma + \Delta) = \varphi(\sigma)$. The transfer function of linear part for the system (7.31) is $P(s) = c(A - sI)^{-1}b$.

Writing $P(s)$ as
$$P(s) = \frac{1}{s}G(s) = \frac{1}{s}(c_0(sI - A_0)^{-1}b_0 + d).$$

In this section, we make the following assumptions:

Assumption 7.1. (A_0, b_0) is controllable and (A_0, c_0) is observable.

Assumption 7.2. $d \neq 0$ and $\det(A_0 - d^{-1}b_0 c_0) \neq 0$.

Assumption 7.3. A_0 is Hurwitz stable.

Assumptions 7.1 and 7.2 ensure that $P(s)$ is non-degenerate.

The aim of this section is to find a dynamical output feedback $K(s)$ with the state space representation
$$\begin{cases} \dot{x}_k = A_k x_k + b_k y, \\ u = c_k x_k + d_k y \end{cases} \quad (7.32)$$

such that the closed-loop system
$$\begin{cases} \begin{pmatrix} \dot{x} \\ \dot{x}_k \end{pmatrix} = \begin{pmatrix} A & 0 \\ b_k c & A_k \end{pmatrix} \begin{pmatrix} x \\ x_k \end{pmatrix} + \begin{pmatrix} b \\ 0 \end{pmatrix} \varphi(u), \\ u = (d_k c \quad c_k) \begin{pmatrix} x \\ x_k \end{pmatrix} \end{cases} \quad (7.33)$$

is dichotomous, where $A_k \in \mathbb{R}^{n_k \times n_k}, b_k \in \mathbb{R}^{n_k \times 1}, c_k \in \mathbb{R}^{1 \times n_k}, d_k \in \mathbb{R}, d_k \neq 0$. The transfer function of controller $K(s)$ is given by
$$K(s) = c_k(sI - A_k)^{-1}b_k + d_k \quad (7.34)$$

and the transfer function of the linear part for closed-loop system (7.33) from the input φ to the output u can be written as
$$T_{cl}(s) = K(s)P(s) = \frac{1}{s}K(s)G(s) \quad (7.35)$$

Applying Lemma 6.2 to the closed-loop system (7.33), we have

Lemma 7.3. *Suppose that Assumptions 7.1, 7.2 and 7.3 hold. If there exist a scalar $\alpha \in \mathbb{R}, \alpha \neq 0$ and matrices A_k, b_k, c_k, d_k such that $T_{cl}(s)$ is non-degenerate, $\alpha d d_k > 0$ and $\alpha K(s)G(s)$ is SPR, then the closed-loop system (7.33) is dichotomous.*

Under Assumptions 7.1 and 7.2, from (7.35) one implies that if

$$\det(A_k - d_k^{-1}b_k c_k) \neq 0, \quad \Lambda(A) \cap \Lambda(A_k - d_k^{-1}b_k c_k) = \emptyset \tag{7.36}$$

and

$$\Lambda(A_k) \cap \Lambda(A_0 - d^{-1}b_0 c_0) = \emptyset \tag{7.37}$$

hold, then $T_{cl}(s)$ is non-degenerate, where $\Lambda(M)$ denotes the set of all the eigenvalues of M.

The following theorem gives conditions for the existence of a dynamical output feedback controller such that the closed-loop system (7.33) is dichotomous.

Theorem 7.4. *Suppose that Assumptions 7.1, 7.2 and 7.3 hold. Then there exist control matrices A_k, b_k, c_k, d_k such that $\alpha d d_k > 0$ and $\alpha K(s)G(s)$ is SPR if and only if $A_0 - d^{-1}b_0 c_0$ is a stable matrix. Furthermore, if $A_0 - d^{-1}b_0 c_0$ is stable and (7.36) and (7.37) hold, then the closed-loop system (7.33) is dichotomous.*

Proof. Note that the transfer function $K(s)G(s)$ has a state space realization

$$K(s)G(s) = \left(\begin{array}{c|c} A_s & b_s \\ \hline c_s & d_k d \end{array} \right) \tag{7.38}$$

where

$$A_s = \begin{pmatrix} A_0 & 0 \\ b_k c_0 & A_k \end{pmatrix}, \quad b_s = \begin{pmatrix} b_0 \\ b_k d \end{pmatrix}, \quad c_s = (d_k c_0 \ \ c_k).$$

Therefore by the positive real lemma, $\alpha K(s)G(s)$ is SPR with $\alpha d d_k > 0$ if and only if there exists a positive definite matrix $P > 0$ such that

$$\begin{pmatrix} PA_s + A_s^T P & Pb_s - \alpha c_s^T \\ b_s^T P - \alpha c_s & -2\alpha d_k d \end{pmatrix} < 0. \tag{7.39}$$

Multiplying $\mathrm{diag}(P^{-1}, 1)$ on the left and right of above inequality and by Schur complement, (7.39) is written as its equivalent form

$$\begin{pmatrix} A_s Q + Q A_s^T & b_s - \alpha Q c_s^T \\ b_s^T - \alpha c_s Q & -2\alpha d_k d \end{pmatrix} < 0, \tag{7.40}$$

7.3. Control for systems with input nonlinearities guaranteeing dichotomy

where $Q = P^{-1}$. Rewriting (7.40) as
$$\Gamma \bar{K} \Lambda + (\Gamma \bar{K} \Lambda)^T + \Theta < 0 \tag{7.41}$$
where
$$\Theta = \left(\begin{pmatrix} A_0 & 0 \\ 0 & 0 \end{pmatrix} Q + Q \begin{pmatrix} A_0 & 0 \\ 0 & 0 \end{pmatrix}^T \begin{pmatrix} b_0 \\ 0 \end{pmatrix} \right), \quad \Gamma = \begin{pmatrix} \begin{pmatrix} 0 & 0 \\ 0 & I \end{pmatrix} \\ -\alpha (I \quad 0) \end{pmatrix},$$
$$\Lambda = \left(\begin{pmatrix} c_0 & 0 \\ 0 & I \end{pmatrix} Q \quad \begin{pmatrix} d \\ 0 \end{pmatrix} \right), \quad \bar{K} = \begin{pmatrix} d_k & c_k \\ b_k & A_k \end{pmatrix}.$$

By Lemma 1.1, there exist matrices A_k, b_k, c_k and d_k such that (7.41) holds if and only if the following two inequalities hold, i.e.,

$$\begin{pmatrix} \begin{pmatrix} 0 & 0 \\ 0 & I \end{pmatrix} \\ -\alpha(I \quad 0) \end{pmatrix}^{\perp} \begin{pmatrix} A_1 Q + Q A_1^T & \begin{pmatrix} b_0 \\ 0 \end{pmatrix} \\ (b_0^T \quad 0) & 0 \end{pmatrix} \begin{pmatrix} \begin{pmatrix} 0 & 0 \\ 0 & I \end{pmatrix} \\ -\alpha(I \quad 0) \end{pmatrix}^{\perp T} < 0, \tag{7.42}$$

$$\begin{pmatrix} Q \begin{pmatrix} c_0^T & 0 \\ 0 & I \end{pmatrix} \\ (d \quad 0) \end{pmatrix}^{\perp} \begin{pmatrix} A_1 Q + Q A_1^T & \begin{pmatrix} b_0 \\ 0 \end{pmatrix} \\ (b_0^T \quad 0) & 0 \end{pmatrix} \begin{pmatrix} Q \begin{pmatrix} c_0^T & 0 \\ 0 & I \end{pmatrix} \\ (d \quad 0) \end{pmatrix}^{\perp T} < 0, \tag{7.43}$$

where $A_1 = \begin{pmatrix} A_0 & 0 \\ 0 & 0 \end{pmatrix}$. Let

$$Q = \begin{pmatrix} Q_1 & Q_2 \\ Q_2^T & Q_3 \end{pmatrix}, \quad P = Q^{-1} = \begin{pmatrix} P_1 & P_2 \\ P_2^T & P_3 \end{pmatrix}.$$

Choosing

$$\begin{pmatrix} \begin{pmatrix} 0 & 0 \\ 0 & I \end{pmatrix} \\ -\alpha(I \quad 0) \end{pmatrix}^{\perp} = \begin{pmatrix} (I \quad 0) & 0 \end{pmatrix},$$

$$\begin{pmatrix} Q \begin{pmatrix} c_0^T & 0 \\ 0 & I \end{pmatrix} \\ (d \quad 0) \end{pmatrix}^{\perp} = \begin{pmatrix} (I \quad 0) Q^{-1} & -d^{-1} c_0^T \end{pmatrix}.$$

Then (7.42) and (7.43) can be written as
$$A_0 Q_1 + Q_1 A_0^T < 0 \tag{7.44}$$
and
$$P_1 (A_0 - d^{-1} b_0 c_0) + (A_0 - d^{-1} b_0 c_0)^T P_1 < 0 \tag{7.45}$$

respectively. On the other hand, $P = Q^{-1} > 0$ derives that
$$P_1 - Q_1^{-1} \geq 0. \tag{7.46}$$
From Assumption 7.3, one implies that (7.44), (7.45) and (7.46) hold if and only if $A_0 - d^{-1}b_0c_0$ is stable. Consequently, the results follow from Lemma 7.3. □

Combining Lemma 1.1 and the proof of Theorem 7.4, we give the following algorithm for designing a dynamical output feedback controller such that the closed-loop systems (7.33) is dichotomous.

Algorithm 7.3.

(i) Solving LMIs (7.44), (7.45) and (7.46) for matrix variables $P_1 > 0$ and $Q_1 > 0$.
(ii) Choosing matrices $P_3 > 0$ and P_2 satisfying $P_2 P_2^T = P_1 - Q_1^{-1}$. Let
$$P = \begin{pmatrix} P_1 & P_2 \\ P_2^T & P_3 \end{pmatrix} \text{ and } Q = P^{-1}.$$
(iii) Solving LMI (7.41) for variable $\bar{K}$ and A_k, b_k, c_k and d_k are obtained.
(iv) Testing (7.36) and (7.37).

7.4 Notes and references

Within the framework of absolute stability, Haddad and Kapila [Haddad and Kapila (1997)] and Kapila [Kapila et al. (2001)] discussed the problem of controller design for linear time-invariant systems subject to plant input and plant output time-varying nonlinearities. The obtained results for systems subject to input nonlinearies with a single stationary point are different from that of this chapter.

Chapter 8

Analysis and Control for Uncertain Feedback Nonlinear Systems

This chapter discusses analysis and control problems for two types of uncertain feedback nonlinear systems where nonlinear feedback functions are piece-wise continuously differentiable and with bounded derivatives and periodic nonlinearities, respectively. Section 8.1 studies the property of dichotomy for nonlinear systems with norm bounded uncertainties corresponding to the first form of the nonlinear feedback functions. The pendulum-like systems in engineering have a specific structure that enables a more detailed analysis. As usual, there also exist uncertainties or perturbations in such systems. Section 8.2 to Section 8.7 consider some global properties of the pendulum-like systems with uncertainties and periodic nonlinearities. Section 8.2 studies the dichotomy of the pendulum-like systems with addictive, multiplicative and H_∞ uncertainties. Section 8.3 gives approaches of controller design guaranteeing dichotomy for the corresponding uncertain pendulum-like systems. Section 8.4 and Section 8.5 discuss Lagrange stability and a gradient-like property for the same uncertain pendulum-like systems, respectively. The control design problems for gradient-like property are investigated in Section 8.6. The gradient-like property for the pendulum-like systems with norm bounded uncertainties is studied in Section 8.7 and notes and references conclude this Chapter in Section 8.8.

8.1 Dichotomy of systems with norm bounded uncertainties

In this section the property of dichotomy for nonlinear systems with norm-bounded uncertainties is first studied. A state feedback controller is designed guaranteeing the dichotomy of the closed-loop systems and the corresponding results are applied to Chua's circuit for chaos control.

8.1.1 Robust analysis for dichotomy

Consider an uncertain nonlinear feedback system given by

$$\begin{cases} \dot{x} = (A + \Delta A)x + (b + \Delta b)\varphi(y), \\ \dot{y} = cx + \rho\varphi(y), \end{cases} \quad (8.1)$$

where $A \in \mathbb{R}^{n \times n}$ is a real matrix, $b \in \mathbb{R}^n$ and $c^T \in \mathbb{R}^n$ are real vectors and ρ is a real number. $x \in \mathbb{R}^n, y \in \mathbb{R}$. We suppose that the nonlinear function $\varphi : \mathbb{R} \to \mathbb{R}$ is piece-wise continuously differentiable on $\mathbb{R}$ and

$$-\infty < \mu_1 \leq \frac{d\varphi}{d\sigma} \leq \mu_2 < +\infty \quad (8.2)$$

holds for all $\sigma \in \mathbb{R}$ where $\varphi'(\sigma)$ exists. Assume that parameter uncertainties ΔA and Δb can be expressed as the form of norm uncertainties.

$$\Delta A = H_1 F E_1, \quad \Delta b = H_2 F E_2 \quad (8.3)$$

where $H_1 \in \mathbb{R}^{n \times m}, E_1 \in \mathbb{R}^{p \times n}, H_2 \in \mathbb{R}^{n \times m}, E_2 \in \mathbb{R}^{p \times 1}$ and $F \in \mathbb{R}^{m \times p}$ indicates uncertainty which satisfies $F^T F \leq I$. When $\Delta A = 0$ and $\Delta b = 0$ the transfer function of its linear part from the input φ to the output $-\dot{y}$ is expressed as

$$P(s) = c(A - sI)^{-1}b - \rho. \quad (8.4)$$

When there are no uncertainties, i.e., $\Delta A = 0$ and $\Delta b = 0$, the frequency condition of dichotomy for the system (8.1) is given in [Leonov et al. (1996)] which can be described as

Lemma 8.1. *Suppose that* (A, b) *is controllable,* (A, c) *is observable and* $P(0) \neq 0$. *Suppose also that there exist numbers* $\gamma, \varepsilon > 0$ *and* $\tau \geq 0$ *such that the following frequency-domain inequality is true:*

$$\text{Re}\left\{\gamma P(i\omega) + \tau[\mu_1 P(i\omega) + i\omega]^*[\mu_2 P(i\omega) + i\omega]\right\} - \varepsilon|P(i\omega)|^2 \geq 0, \quad \forall \omega \in \mathbb{R}. \quad (8.5)$$

If the matrix A *is Hurwitz stable or* $\varphi(y)$ *has a finite number of isolated zeros then the system (8.1) with* $\Delta A = 0$ *and* $\Delta b = 0$ *is dichotomous.*

Remark 8.1. It has been shown that in [Duan et al. (2004b)] the result of Lemma 8.1 is still true if $P(0) = 0$.

By Corollary 1.4, Lemma 8.1 can be written as the following lemma with time-domain form.

Lemma 8.2. *Suppose that* $\det(j\omega I - A) \neq 0$ *for all* $\omega \in \mathbb{R}$ *and there exist numbers* $\gamma, \varepsilon > 0, \tau \geq 0$ *and a matrix* $P = P^T$ *such that*

8.1. Dichotomy of systems with norm bounded uncertainties

(i)
$$\begin{pmatrix} A^T P + PA - \alpha c^T c & Pb - E \\ b^T P - E^T & h \end{pmatrix} < 0, \qquad (8.6)$$

where
$$\alpha = \tau \mu_1 \mu_2 - \varepsilon, \quad E = \alpha \rho c^T - \frac{1}{2}\gamma c^T + \frac{\tau(\mu_1 + \mu_2)}{2} A^T c^T, \qquad (8.7)$$
$$h = -\alpha \rho^2 + \gamma \rho - \tau(\mu_1 + \mu_2)cb;$$

(ii) (A, b) is controllable, (A, c) is observable;
(iii) A is a stable matrix or $\varphi(y)$ has a finite number of isolated zeros.

Then the system (8.1) with $\Delta A = 0$ and $\Delta b = 0$ is dichotomous.

Remark 8.2. The result of Lemmas 8.1 and 8.2 are still true for $\tau = 0$ if φ does not have the property (11.5).

The following corollary gives in H_∞ norm condition for the dichotomy of the system (8.1) with $\Delta A = 0$ and $\Delta b = 0$ which can be easily shown by the bounded real lemma (Theorem 2.7).

Corollary 8.1. Suppose that $det(j\omega I - A) \neq 0$ holds for all $\omega \in \mathbb{R}$ and there exist real numbers $\gamma, \varepsilon > 0, \tau \geq 0$ such that

(i) the conditions (ii) and (iii) of Lemma 8.2 hold;
(ii) $EE^T - \gamma_0 \alpha c^T c \geq 0$ and $\gamma_0 = -h > 0$;
(iii) $G_0(s) = c_0(sI - A_0)^{-1} b \in \mathcal{RH}_\infty$ and $||G_0||_\infty < \gamma_0$, where α, h, E, A_0, c_0 are defined by (8.7) and

$$A_0 = A - \frac{1}{\gamma_0} b E^T, \quad c_0^T c_0 = EE^T - \gamma_0 \alpha c^T c.$$

Then the system (8.1) with $\Delta A = 0$ and $\Delta b = 0$ is dichotomous.

In order to derive the property of dichotomy for nonlinear systems with norm bounded uncertainty. The following lemma is needed [Wang et al. (1992)].

Lemma 8.3. For matrices M, N, Ω and a symmetric matrix Φ with appropriate dimensions, the inequality

$$\Phi + M\Omega N + (M\Omega N)^T < 0$$

holds for $\Omega^T \Omega \leq I$ if and only if there exists a positive number $\nu > 0$ such that

$$\Phi + \nu M M^T + \nu^{-1} N^T N < 0.$$

Theorem 8.1. *Suppose that* $det(j\omega I - A - \Delta A) \neq 0$ *for all* $\omega \in \mathbb{R}$ *and there exist real numbers* $\gamma, \varepsilon > 0, \tau \geq 0, \delta > 0$ *and matrices* $P = P^T$ *such that*

(i)

$$\begin{pmatrix} PA + A^T P - \alpha c^T c + \dfrac{1}{\delta} E_1^T E_1 & PL - g_1 - g_2 A^T c^T & PG \\ L^T P - g_1^T - g_2 cA & J & 0 \\ G^T P & 0 & -\dfrac{1}{\delta} \end{pmatrix} < 0, \tag{8.8}$$

where

$$\alpha = \tau\mu_1\mu_2 - \varepsilon, \quad g_1 = \alpha\rho c^T - \frac{1}{2}\gamma c^T, \quad g_2 = \frac{1}{2}\tau(\mu_1 + \mu_2),$$
$$g_3 = -\alpha\rho^2 + \gamma\rho, \quad J = g_3 - 2g_2 cb + \frac{1}{\delta}E_2^T E_2 + \delta g_2^2 cGG^T c^T, \tag{8.9}$$
$$L = b - \delta g_2 H_1 H_1^T c^T - \delta g_2 H_2 H_2^T c^T$$

and

$$GG^T = H_1 H_1^T + H_2 H_2^T; \tag{8.10}$$

(ii) (A, b) *is controllable and* (A, c) *is observable;*
(iii)

$$\max\{m^{\frac{1}{2}}||H_1||_1||E_1||_1, \ m^{\frac{1}{2}}||H_2||_1||E_2||_1\} < \frac{1}{||Q^{-1}(A,b)||_1}$$

and

$$p||H_1||_\infty ||E_1||_\infty < \frac{1}{||H^{-1}(A,c)||_1},$$

where $Q(A, b)$ *and* $H(A, c)$ *are defined by (1.5) and (1.6) in Chapter 1;*

(iv) $\varphi(y)$ *has a finite number of isolated zeros, or* A *is a Hurwitz stable matrix and* $||E_1(sI - A)^{-1}H_1||_\infty < 1$.

Then the uncertain nonlinear system (8.1) is dichotomous.

Proof. Applying Lemma 8.2 to the uncertain nonlinear system (8.1), the corresponding inequality (8.6) can be written as

$$\begin{pmatrix} PA_\Delta + A_\Delta^T P - \alpha c^T c & P(b + \Delta b) - g_1 - g_2 A_\Delta^T c^T \\ (b + \Delta b)^T P - g_1^T - g_2 cA_\Delta & g_3 - 2g_2 c(b + \Delta b) \end{pmatrix} < 0 \tag{8.11}$$

8.1. Dichotomy of systems with norm bounded uncertainties

where $A_\Delta = A + \Delta A$ and g_1, g_2 and g_3 are defined by (8.9). Putting $\Delta A = H_1 F E_1$ and $\Delta b = H_2 F E_2$ into (8.11) leads to

$$\Omega + \begin{pmatrix} PH_1 & PH_2 \\ -g_2 c H_1 & -g_2 c H_2 \end{pmatrix} \begin{pmatrix} F & 0 \\ 0 & F \end{pmatrix} \begin{pmatrix} E_1 & 0 \\ 0 & E_2 \end{pmatrix}$$
$$+ \begin{pmatrix} E_1 & 0 \\ 0 & E_2 \end{pmatrix}^T \begin{pmatrix} F & 0 \\ 0 & F \end{pmatrix}^T \begin{pmatrix} PH_1 & PH_2 \\ -g_2 c H_1 & -g_2 c H_2 \end{pmatrix}^T < 0 \quad (8.12)$$

where

$$\Omega = \begin{pmatrix} PA + A^T P - \alpha c^T c & Pb - g_1 - g_2 A^T c^T \\ b^T P - g_1^T - g_2 c A & g_3 - 2 g_2 c b \end{pmatrix}.$$

From Lemma 8.3 and $F^T F \leq I$, (8.12) is equivalent to

$$\begin{pmatrix} PA + A^T P - \alpha c^T c + \frac{1}{\delta} E_1^T E_1 + \delta P G G^T P & PL - g_1 - g_2 A^T c^T \\ L^T P - g_1^T - g_2 c A & J \end{pmatrix} < 0 \quad (8.13)$$

for some positive number δ. Using the Schur complement lemma the above inequality (8.13) holds if and only if (8.8) holds, where L, J and G are defined by (8.9) and (8.10) respectively.

In the following we show that $(A + \Delta A, b + \Delta b)$ is controllable and $(A + \Delta A, c)$ is observable. For $\forall x \in \mathbb{R}^n, y \in \mathbb{R}^n$, by Cauchy-Schwarz inequality $(x^T y)^2 \leq x^T x \cdot y^T y$ we have:

$$(|y_1| + |y_2| + \cdots + |y_n|)^2 \leq n(y_1^2 + y_2^2 + \cdots + y_n^2) \quad (8.14)$$

where $y_i \in \mathbb{R}, i = 1, 2, \cdots, n$. For the uncertain matrix $F = (f_{ij}) \in \mathbb{R}^{m \times p}$ satisfying $F^T F \leq I$, we imply

$$f_{1j}^2 + f_{2j}^2 + \cdots + f_{mj}^2 \leq 1$$

for $j = 1, 2, \cdots, p$. Combining (8.14) we have

$$|f_{1j}| + |f_{2j}| + \cdots + |f_{mj}| \leq m^{\frac{1}{2}}$$

for $j = 1, 2, \cdots, p$, which implies $||F||_1 \leq m^{\frac{1}{2}}$. Therefore

$$||\Delta A||_1 \leq m^{\frac{1}{2}} ||H_1||_1 ||E_1||_1, \quad ||\Delta b||_1 \leq m^{\frac{1}{2}} ||H_2||_1 ||E_2||_1.$$

By conditions (ii), (iii) and Theorem 1.5 one implies that $(A + \Delta A, b + \Delta b)$ is controllable. Similarly by $F^T F \leq I$, for $i = 1, 2, \cdots, m$ we have $f_{i1}^2 + f_{i2}^2 + \cdots + f_{ip}^2 \leq \operatorname{tr}(F^T F) \leq p$. Cauchy-Schwarz inequality derives $|f_{i1}| + \cdots + |f_{ip}| \leq p$ for $i = 1, 2, \cdots, m$, which implies $||F||_\infty \leq p$, and thus

$$||\Delta A||_\infty \leq p ||H_1||_\infty ||E_1||_\infty.$$

From conditions (ii), (iii) and Theorem 1.6 it follows that $(A + \Delta A, c)$ is observable.

According to above discussion and combining with Lemma 8.2 the result follows. This completes the proof. □

Remark 8.3. Suppose that $\frac{1}{\delta}E_1^T E_1 - \alpha cc^T \geq 0$ and A is a stable matrix. It is obvious that $P > 0$ by (8.8). Multiplying diag(P^{-1}, I, I) in the left and in the right respectively on both side of inequality (8.8) and let $Q = P^{-1}$, it follows that

$$\begin{pmatrix} AQ + QA^T - \alpha Qcc^T Q + \frac{1}{\delta}QE_1^T E_1 Q \ L + Q(-g_1 - g_2 A^T c) & G \\ L^T + (-g_1 - g_2 A^T c)^T Q & J & 0 \\ G^T & 0 & -\frac{1}{\delta} \end{pmatrix} < 0$$

which is obviously equivalent to

$$\begin{pmatrix} AQ + QA^T + \delta GG^T & L + Q(-g_1 - g_2 A^T c) & QS \\ L^T + (-g_1 - g_2 A^T c)^T Q & J & 0 \\ S^T Q & 0 & -I \end{pmatrix} < 0 \qquad (8.15)$$

where $SS^T = \frac{1}{\delta}E_1^T E_1 - \alpha cc^T$. The inequality (8.15) will be used to discuss design problems subsequently.

Remark 8.4. From [Khargonekar et al. (1990)], it can be deduced that condition (iv) that A is a stable matrix and $\|E_1(sI - A)^{-1}H_1\|_\infty < 1$ holds if and only if $A + \Delta A$ is quadratically stable for $\Delta A = H_1 F E_1$ where F satisfies $F^T F \leq I$. In this case the assumption that $\det(j\omega I - A - \Delta A) \neq 0$ for all $\omega \in R$ can be deleted.

Remark 8.5. If there are no uncertainties in (8.1), that is, E_1, E_2, H_1, H_2 can be considered as zero matrix, then the result of Theorem 8.1 is the same as the result of Lemma 8.2.

8.1.2 Robust control for systems with dichotomy

Consider a nonlinear uncertain control system:

$$\begin{cases} \dot{x} = (A + \Delta A)x + (b + \Delta b)\varphi(\sigma) + b_1 u, \\ \dot{\sigma} = cx + \rho\varphi(\sigma), \end{cases} \qquad (8.16)$$

where $A, \Delta A, b, \Delta b, c, \rho$ and $\varphi(\sigma)$ are defined in subsection 8.1.1. $u \in \mathbb{R}$ is the control variable.

In what follows, we consider to design a state feedback $u = Kx$ such that the closed-loop nonlinear feedback system

$$\begin{cases} \dot{x} = (A + b_1 K + \Delta A)x + (b + \Delta b)\varphi(\sigma), \\ \dot{\sigma} = cx + \rho\varphi(\sigma) \end{cases} \qquad (8.17)$$

8.1. Dichotomy of systems with norm bounded uncertainties   169

is dichotomous. Applying Theorem 8.1 and Remark 8.3 to the closed-loop system (8.17) and let $KQ = Y$, we have

Theorem 8.2. *Assume that there exist real numbers $\gamma, \varepsilon > 0, \tau \geq 0, \delta > 0$ and matrices $Q = Q^T > 0$ and Y such that*

(i) $\dfrac{1}{\delta} E_1^T E_1 - \alpha cc^T \geq 0;$

(ii)
$$\begin{pmatrix} AQ + QA^T + b_1 Y + Y^T b_1^T + \delta GG^T & QR - g_2 Y^T b_1^T c + L & QS \\ R^T Q - g_2 c^T b_1 Y + L^T & J & 0 \\ S^T Q & 0 & -I \end{pmatrix} < 0$$
(8.18)

where $SS^T = \dfrac{1}{\delta} E_1^T E_1 - \alpha cc^T$, $R = -g_2 A^T c - g_1$ *and* $\alpha, g_1, g_2, g_3, J, L, G$ *are defined by (8.9) and (8.10);*

(iii)
$$\begin{pmatrix} AQ + QA^T + b_1 Y + Y^T b_1^T + H_1^T H_1 & QE_1^T \\ E_1 Q & -I \end{pmatrix} < 0.$$
(8.19)

Then there exists a control matrix $K = YQ^{-1}$.

Furthermore, if the conditions (ii) and (iii) of Theorem 8.1 hold for A instead of $A + b_1 K$, then the closed-loop nonlinear system (8.17) is dichotomous

Note that the condition (iii) is equivalent to $\|E_1(sI - A - b_1 K)^{-1} H_1\|_\infty < 1$ which guarantees $A + \Delta A + b_1 K$ is quadratically stable. In the following, an example is considered.

Example 8.1. Consider Chua's circuit system [Chua et al. (1986); Chua (1994); Madan (1993); Leonov et al. (1996)] described by:

$$\begin{cases} \dot{v}_1 = \dfrac{1}{C_1} \left[\dfrac{v_2 - v_1}{R} - g(v_1) \right], \\ \dot{v}_2 = \dfrac{1}{C_2} \left[\dfrac{v_1 - v_2}{R} + i_3 \right], \\ \dot{i}_3 = \dfrac{1}{L} [-v_2 - R_0 i_3], \end{cases}$$
(8.20)

where v_1 and v_2 are the voltages across the capacitors C_1 and C_2 respectively, i_3 is the current through the inductor L, R, R_0 are resistors, and $g(v_1)$ is the current through the nonlinear resistor which is defined as

$$g(v_1) = G_b v_1 + \dfrac{1}{2}(G_a - G_b)[|v_1 + B_p| - |v_1 - B_p|].$$
(8.21)

As we known Chua's circuit exhibits a wide variety of nonlinear phenomena such as bifurcations and chaos. In order to control chaos we inject external current signals i_u and i_v in the RCL-node. It is shown in Fig. 8.1 which the state equations describing the circuit are as follows

$$\begin{cases} \dfrac{dv_1}{dt} = \dfrac{1}{C_1}\left[\dfrac{v_2 - v_1}{R} - g(v_1) + i_v\right], \\ \dfrac{dv_2}{dt} = \dfrac{1}{C_2}\left[\dfrac{v_1 - v_2}{R} + i_3 + i_u\right], \\ \dfrac{di_3}{dt} = \dfrac{1}{L}[-v_2 - R_0 i_3]. \end{cases} \quad (8.22)$$

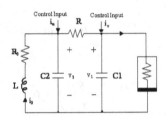 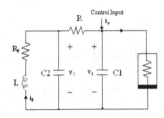

Fig. 8.1 Chua's circuit injected control signals i_v and i_u.

Fig. 8.2 Chua's circuit injected control signal i_v.

Fig. 8.2 corresponds to the case $i_u = 0$, that is, only an external current signal i_v is injected in the RLC-node.

As in [Leonov et al. (1996); Wang et al. (2006c,d)], using a transform of variables (8.22) is converted into the following form

$$\begin{cases} \dot{\bar{x}} = \overline{A}\bar{x} + bf(\sigma) + b_1 u, \\ \dot{\sigma} = c\bar{x} + \rho f(\sigma), \end{cases} \quad (8.23)$$

where

$$\overline{A} = \begin{pmatrix} -k\mu & k\mu & 0 \\ k & -k & k \\ 0 & -k\beta & -k\eta \end{pmatrix}, \quad b = \begin{pmatrix} -k\mu \\ 0 \\ 0 \end{pmatrix}, \quad \bar{x} = \begin{pmatrix} x \\ y \\ z \end{pmatrix},$$

$$c = \begin{pmatrix} -k\mu & k\mu & 0 \end{pmatrix}, \quad \rho = -k\mu, \quad \sigma = x, \quad k = \pm 1.$$

The nonlinear function $f(x)$ is defined as

$$f(x) = m_1 x + \dfrac{1}{2}(m_0 - m_1)[|x+1| - |x-1|]. \quad (8.24)$$

It is obvious that $\mu_1 = \min\{m_0, m_1\} \leq f'(x) \leq \mu_2 = \max\{m_0, m_1\}$. u in (8.23) is the injected control variable and b_1 in (8.23) is defined by

$b_1 = (1 \ 1 \ 0)^T$ for the case of Fig. 8.1 and $b_1 = (1 \ 0 \ 0)^T$ for the case of Fig. 8.2.

Choosing $\mu = 9$, $\eta = 0$, $k = 1$, $m_0 = -1.142$, $m_1 = -0.7142$, and β is respectively chosen as 11.63,11.93,12.35,12.85,12.95,13.01,13.35,13.65,13.95, 14.35, 14.16,14.28,14.38,15.01,15.35,15.65, 16.28,17.28, 18.28, 18.68, 19.28, 19.63. By numerical simulations, it is shown that the system (8.23) with $u = 0$ appears chaos attractors or periodic solutions for above groups of parameters. e.g., see Fig. 8.3, Fig. 8.4, Fig. 8.5 and Fig. 8.6. The initial values are all chosen as $\overline{x}(0) = (0.1 \ 0.1 \ -0.4)^T$.

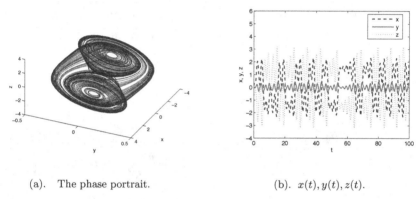

(a). The phase portrait.   (b). $x(t), y(t), z(t)$.

Fig. 8.3 The dynamics behavior of (8.23) without control for $\beta = 12.28$.

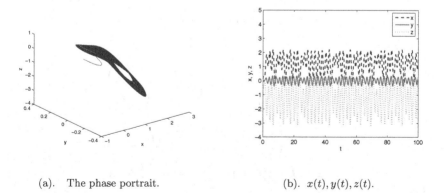

(a). The phase portrait.   (b). $x(t), y(t), z(t)$.

Fig. 8.4 The dynamics behavior of (8.23) without control for $\beta = 15.01$.

In order to apply Theorem 8.2 we write system (8.23) in the form of

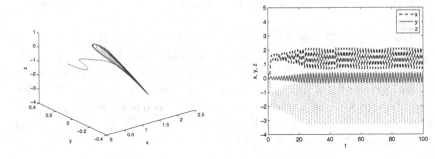

(a). The phase portrait.   (b). $x(t), y(t), z(t)$.

Fig. 8.5 The dynamics behavior of (8.23) without control for $\beta = 19.63$.

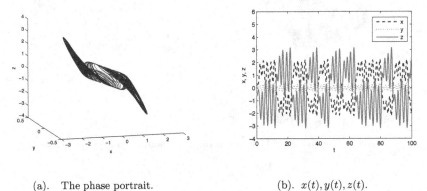

(a). The phase portrait.   (b). $x(t), y(t), z(t)$.

Fig. 8.6 The dynamics behavior of (8.23) without control for $\beta = 14.28$.

(8.17), where

$$A = \begin{pmatrix} -k\mu & k\mu & 0 \\ k & -k & k \\ 0 & -k\beta_0 & -k\eta \end{pmatrix}, \quad b = \begin{pmatrix} -k\mu \\ 0 \\ 0 \end{pmatrix},$$

$$\Delta A = \begin{pmatrix} 0 \\ 0 \\ m \end{pmatrix} \left(\frac{-k\beta + k\beta_0}{m} \right) \begin{pmatrix} 0 & 1 & 0 \end{pmatrix},$$

$$c = \begin{pmatrix} -k\mu & k\mu & 0 \end{pmatrix}, \quad \rho = -k\mu, \quad \Delta b = 0.$$

Let

$$H_1 = \begin{pmatrix} 0 \\ 0 \\ m \end{pmatrix}, E_1 = \begin{pmatrix} 0 & 1 & 0 \end{pmatrix}, F = \frac{-k\beta + k\beta_0}{m},$$

8.1. Dichotomy of systems with norm bounded uncertainties

where β_0 is a given real number. The variable F denotes uncertainty which is defined by the uncertain parameter β. $F^T F \leq I$ holds if and only if $|\beta - \beta_0| \leq m$. Obviously, the larger the positive number m is, the larger is the range of parameter β. For $\mu = 9, \eta = 0, k = 1, m_0 = -1.142, m_1 = -0.7142, b_1 = [1, \ 1 \ \ 0]^T$ and $\beta_0 = 14.28$, by Theorem 8.2, LMIs (8.18) and (8.19) are feasible and the condition (i) of Theorem 8.2 holds for $m = 12.82$, and control matrix

$$K = (K_1 \ \ K_2 \ \ K_3) = 10^5 \cdot (-0.1120 \ \ 0.0064 \ \ 1.9875). \quad (8.25)$$

Note that (ii) and (iii) of Theorem 8.1 for A instead of $A + b_1 K$ are only sufficient conditions guaranteeing the controllability and observability of the uncertain closed-loop systems. If we require above conditions are satisfied the obtained parameter $m \ll 12.82$. In order to reduce conservativeness in this example, we consider the controllability and observability of system (8.23) directly. It is not difficult to test that $(\overline{A} + b_1 K, b)$ is controllable and $(\overline{A} + b_1 K, c)$ is observable if and only if

$$\begin{aligned} \mu\beta(1 + K_3) &\neq 0, \\ \beta &\neq (K_2 - K_1)\left((K_2 - K_1)(1 + K_3) - \mu - 1 - K_1 - K_2\right) \end{aligned} \quad (8.26)$$

hold.

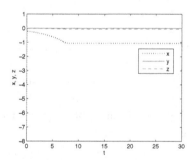
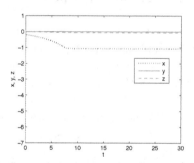

Fig. 8.7 $x(t), y(t), z(t)$ of the system (8.23) with control $u = K\overline{x}$ for $\beta = 12.28$.

Fig. 8.8 $x(t), y(t), z(t)$ of the system (8.23) with control $u = K\overline{x}$ for $\beta = 15.01$.

It is obvious that the state feedback control matrix K defined in (8.25) obtained by solving LMIs (8.18) and (8.19) satisfies the conditions (8.26), and thus by injecting an identical control $u = K\overline{x}$ the closed-loop systems are dichotomous for $|\beta - \beta_0| \leq m$ with $\beta_0 = 14.28, m = 12.82$, i.e., $1.46 \leq \beta \leq 27.10$, where K is given in (8.25). Figs. 8.7, 8.8, 8.9 and 8.10 show dynamics $\overline{x}(t)$ of the Chua's system (8.23) with control $u = K\overline{x}$ which

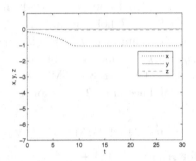
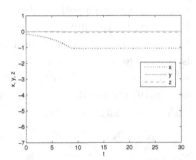

Fig. 8.9 $x(t), y(t), z(t)$ of the system (8.23) with control $u = K\bar{x}$ for $\beta = 19.63$.

Fig. 8.10 $x(t), y(t), z(t)$ of the system (8.23) with control $u = K\bar{x}$ for $\beta = 14.28$.

correspond to Figs. 8.3, 8.4, 8.5 and 8.6, respectively. The initial values are still chosen as $\bar{x}(0) = (0.1 \quad 0.1 \quad -0.4)^T$. Obviously, the solutions are convergent.

In the same way, we can consider the case that $b_1 = (1 \quad 0 \quad 0)^T$.

8.2 Dichotomy of pendulum-like systems with uncertainties

In this section, we discuss the dichotomy of the uncertain feedback system shown in Fig. 8.11, where the SISO transfer function $P_\Delta(s)$ denotes an uncertain system which can be additive uncertainty, or multiplicative uncertainty or H_∞ uncertainty. The uncertainties may be caused by parameter changes or by neglected dynamics, or by a host of other unspecified effects. $\varphi(\sigma)$ is the part of nonlinear feedback. The uncertainties of the systems and the function $\varphi(\sigma)$ satisfy the following assumptions:

Assumption 8.1. $P_\Delta(s)$ is non-degenerate.

Assumption 8.2. $P_\Delta(s)$ has a unique zero pole and $P_\Delta(\infty) = 0$.

Assumption 8.3. $\varphi(\sigma) : \mathbb{R} \to \mathbb{R}$ is continuous and satisfies

$$\varphi(\sigma + T) = \varphi(\sigma), \varphi(\sigma) \not\equiv 0, \forall \sigma \in \mathbb{R} \tag{8.27}$$

where T is the period of $\varphi(\sigma)$.

It is obvious that the uncertain feedback system shown in Fig. 8.11 is a pendulum-like system under Assumptions 8.1, 8.2 and 8.3. From Assump-

8.2. Dichotomy of pendulum-like systems with uncertainties

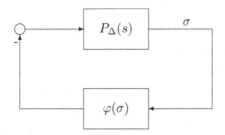

Fig. 8.11 The uncertain feedback system for dichotomy.

tion 8.2, $P_\Delta(s)$ can be expressed as

$$P_\Delta(s) = \frac{1}{s} G_\Delta(s). \tag{8.28}$$

Obviously, Assumptions 8.1 and 8.2 hold if and only if the following conditions hold:

Assumption 8.4. $G_\Delta(s)$ is non-degenerate, and zero is not a zero of $G_\Delta(s)$.

Applying Lemma 6.2 to the uncertain system shown in Fig. 8.11, we have

Lemma 8.4. *Suppose that Assumptions 8.3 and 8.4 hold, and $G_\Delta(s) \in \mathcal{RH}_\infty$. If there exists a scalar $\alpha \in \mathbb{R}, \alpha \neq 0$ such that $\alpha G_\Delta(s)$ is SPR with $\alpha G_\Delta(\infty) > 0$, i.e.,*

$$\alpha G_\Delta(i\omega) + \alpha G_\Delta^T(-i\omega) > 0, \quad \omega \in \mathbb{R} \cup \{\infty\},$$

then the uncertain feedback system shown in Fig. 8.11 is dichotomous.

We first consider the feedback systems with additive uncertainties.

Case 1. Additive uncertainty

Consider additive uncertainty:

$$P_\Delta(s) = \frac{1}{s} G_\Delta(s), \quad G_\Delta(s) = G(s) + W_1(s)\Delta(s)W_2(s) \tag{8.29}$$

where $\Delta(s)$ is uncertainty, $W_1(s)$ and $W_2(s)$ are weighted functions that characterize the spatial and frequency structure of the uncertainty. $P(s) = \frac{1}{s} G(s)$ is the transfer function of nominal model.

Theorem 8.3. *Suppose that*

(i) Assumptions 8.3 and 8.4 hold;
(ii) $G(s), W_1(s), W_2(s), \Delta(s)$ are stable transfer functions and $||\Delta||_\infty < 1$;
(iii) $W_1(j\omega) \neq 0$ or $W_2(j\omega) \neq 0$ for $\omega \in \mathbb{R} \cup \{\infty\}$;
(iv) there exists a scalar $\alpha \in \mathbb{R}, \alpha \neq 0$ satisfying
$$1 + \alpha G(\infty) \neq 0, 1 + \alpha W_2(\infty)(1 + \alpha G(\infty))^{-1} W_1(\infty)\Delta(\infty) \neq 0;$$
(v) $T_G(s) \in \mathcal{RL}_\infty$ and $||T_G(s)||_\infty \leq 1$, where
$$T_G = \begin{pmatrix} (\alpha G - 1)(\alpha G + 1)^{-1} & 2\alpha(\alpha G + 1)^{-1} W_1 \\ W_2(\alpha G + 1)^{-1} & -\alpha W_2(\alpha G + 1)^{-1} W_1 \end{pmatrix}. \quad (8.30)$$

Then the uncertain feedback system with additive uncertainty shown in Fig. 8.11 is dichotomous.

Proof. Let
$$S_\Delta = \mathcal{F}_\ell(T_G, \Delta) \quad (8.31)$$
where T_G is defined by (8.30). The condition (iv) guarantees that $\mathcal{F}_\ell(T_G, \Delta)$ is well-posed. By the conditions (iii), (v) and Lemma 2.2 one implies that $||S_\Delta||_\infty < 1$. Let
$$T_\Delta = (1 - S_\Delta)^{-1}(1 + S_\Delta). \quad (8.32)$$
Then
$$1 - S_\Delta^*(j\omega) S_\Delta(j\omega) = \frac{2(T_\Delta^*(j\omega) + T_\Delta(j\omega))}{(T_\Delta^*(j\omega) + 1)(T_\Delta(j\omega) + 1)}.$$
Thus by $||S_\Delta||_\infty < 1$, $T_\Delta^*(j\omega) + T_\Delta(j\omega) > 0$ holds for $\omega \in \mathbb{R} \cup \{\infty\}$. On another aspect, from (8.31) and the property of star product, we have
$$T_\Delta = 1 + 2(1 - S_\Delta)^{-1} S_\Delta = \mathcal{F}_\ell\left(\begin{pmatrix} 1 & 2 \\ 1 & 1 \end{pmatrix}, S_\Delta\right)$$
$$= \mathcal{F}_\ell\left(\begin{pmatrix} 1 & 2 \\ 1 & 1 \end{pmatrix}, \mathcal{F}_\ell(T_G, \Delta)\right) = \mathcal{F}_\ell\left(\begin{pmatrix} \alpha G & 2\alpha W_1 \\ \frac{1}{2}W_2 & 0 \end{pmatrix}, \Delta\right)$$
$$= \alpha G + \alpha W_1 \Delta W_2.$$

Combining with (ii) implies that T_Δ is a stable transfer function. Consequently, in terms of Lemma 8.4, the uncertain feedback system with additive uncertainty shown in Fig. 8.11 is dichotomous. □

Remark 8.6. When there is no uncertainty in the systems, i.e., $\Delta = 0$, we have $S_\Delta = (\alpha G - 1)(\alpha G + 1)^{-1}$ by (8.31). According to Theorem 8.3, the result of Lemma 6.2 is true if the condition that there exists a scalar $\alpha \in \mathbb{R}$, $\alpha \neq 0$ such that $\alpha G(s)$ is SPR with $\alpha G(\infty) > 0$ is substituted by $G(s) \in \mathcal{RH}_\infty$ and $||S(s)||_\infty < 1$.

8.2. Dichotomy of pendulum-like systems with uncertainties

Remark 8.7. Using Lemma 6.3, it is easy to imply that the uncertain feedback system with additive uncertainty shown in Fig. 8.11 is gradient-like if $\varphi(\sigma)$ has zero mean:

$$\int_0^T \varphi(\sigma)d\sigma = 0 \tag{8.33}$$

and the conditions of Theorem 8.3 hold.

Case 2. Multiplicative uncertainty

Consider the case of multiplicative uncertainty:

$$P_\Delta(s) = \frac{1}{s}G_\Delta(s), \quad G_\Delta(s) = (1 + W_1(s)\Delta(s)W_2(s))G(s).$$

Along the line of Theorem 8.3 we can show that

Theorem 8.4. *Suppose that*

(i) Assumptions 8.3 and 8.4 hold;
(ii) $G(s), W_1(s), W_2(s), \Delta(s)$ are stable transfer functions and $\|\Delta(s)\|_\infty < 1$;
(iii) $W_1(j\omega) \neq 0$ or $W_2(j\omega)G(j\omega) \neq 0$ for $\omega \in \mathbb{R} \cup \{\infty\}$;
(iv) there exists a scalar $\alpha \neq 0$ satisfying

$$1 + \alpha G(\infty) \neq 0, \quad 1 + \alpha W_2(\infty)G(\infty)(1 + \alpha G(\infty))^{-1}W_1(\infty)\Delta(\infty) \neq 0;$$

(v) $T_G(s) \in \mathcal{RL}_\infty$ and $\|T_G(s)\|_\infty \leq 1$, where

$$T_G = \begin{pmatrix} (\alpha G - 1)(\alpha G + 1)^{-1} & 2\alpha(\alpha G + 1)^{-1}W_1 \\ W_2 G(\alpha G + 1)^{-1} & -\alpha W_2 G(\alpha G + 1)^{-1}W_1 \end{pmatrix}.$$

Then the uncertain feedback system with multiplicative uncertainty shown in Fig. 8.11 is dichotomous.

Remark 8.8. If $\varphi(\sigma)$ satisfies (8.33) and the conditions of Theorem 8.4 are fulfilled, then the uncertain feedback system with multiplicative uncertainty shown in Fig. 8.11 is gradient-like.

Case 3. H_∞ uncertainty

Consider the case of H_∞ uncertainty:

$$P_\Delta = \frac{1}{s}G_\Delta(s), \quad G_\Delta(s) = \mathcal{F}_\ell(G, \Delta) = G_{11} + G_{12}\Delta(1 - G_{22}\Delta)^{-1}G_{21}$$

where Δ denotes the uncertainties of the system.

Theorem 8.5. *Suppose that*

(i) Assumptions 8.3 and 8.4 hold;

(ii) $G = \begin{pmatrix} G_{11} & G_{12} \\ G_{21} & G_{22} \end{pmatrix} \in \mathcal{RH}_\infty, \|G_{22}\|_\infty \leq 1$ and $\Delta \in \mathcal{RH}_\infty, \|\Delta\|_\infty < 1$;

(iii) $G_{12}(j\omega) \neq 0$ or $G_{21}(j\omega) \neq 0$ for $\omega \in \mathbb{R} \cup \{\infty\}$;

(iv) there exists a scalar $\alpha \neq 0$ satisfying

$$1 + \alpha G_{11}(\infty) \neq 0, \text{ and}$$
$$1 + \left(\alpha G_{21}(\infty)(1 + \alpha G_{11}(\infty))^{-1} G_{12}(\infty) - G_{22}(\infty)\right) \Delta(\infty) \neq 0;$$

(v) $T_G(s) \in \mathcal{RL}_\infty$ and $\|T_G(s)\|_\infty \leq 1$, where

$$T_G = \begin{pmatrix} (\alpha G_{11} - 1)(\alpha G_{11} + 1)^{-1} & 2\alpha(\alpha G_{11} + 1)^{-1} G_{12} \\ G_{21}(\alpha G_{11} + 1)^{-1} & -\alpha G_{21}(\alpha G_{11} + 1)^{-1} G_{12} + G_{22} \end{pmatrix}.$$

Then the uncertain feedback system with H_∞ uncertainty shown in Fig. 8.11 is dichotomous.

Proof. By Theorem 2.10, the assumption (ii) guarantees that $G_\Delta \in \mathcal{RH}_\infty$. Let

$$S_\Delta = (\alpha G_\Delta - 1)(\alpha G_\Delta + 1)^{-1}. \tag{8.34}$$

Then S_Δ can be expressed as $S_\Delta = \mathcal{F}_\ell(T_G, \Delta)$. Combining Lemma 2.2 with the condition (v) leads to $\|S_\Delta\| < 1$ which implies that αG_Δ is SPR with $\alpha G_\Delta(\infty) > 0$. In terms of Lemma 3.9 and Lemma 8.4 the result follows. □

Remark 8.9. If $\varphi(\sigma)$ satisfies (8.33) and the conditions of Theorem 8.5 are fulfilled, then the uncertain feedback system with H_∞ uncertainty shown in Fig. 8.11 is gradient-like.

Remark 8.10. When there is no uncertainty in the systems, that is, $\Delta = 0$, the results of Theorems 8.3, 8.4 and 8.5 are identical and degenerate into the results of Remark 8.6 which is equivalent to Lemma 6.2.

Example 8.2. Consider the uncertain feedback system shown in Fig. 8.11 where $\varphi(\sigma)$ satisfies Assumption 8.3 and

$$P_\Delta(s) = \frac{1}{s}(1 + \Delta(s)).$$

Assume that $\Delta(s) \in RH_\infty, \|\Delta(s)\|_\infty < 1, \Delta(\infty) = 0$ and $(1 + \Delta(s))^{-1}$ has no zero pole. The system can be viewed as uncertain system with multiplicative uncertainty, where $G(s) = 1, W_1(s) = W_2(s) = 1$. It is easy to verify that there exists a scalar $\alpha = 0.5$ such that $\|T_G(S)\|_\infty = 1$ and

8.3. Controller design with dichotomy for uncertain pendulum-like systems

thus the conditions of Theorem 8.4 hold. The uncertain feedback system shown in Fig. 8.11 is dichotomous.

In special, Let $\Delta(s) = \dfrac{\delta}{1+\eta s}$, where δ and η are uncertain scalar parameters with $|\delta| < 1, \delta \neq 0, \eta \geq 0$. In this case, the uncertain feedback system shown in Fig. 8.11 is dichotomous.

8.3 Controller design with dichotomy for uncertain pendulum-like systems

In this section, we consider problems of designing a linear controller K for SISO uncertain system P_Δ such that the uncertain closed-loop feedback system shown in Fig. 8.12 is a pendulum-like system and dichotomous.

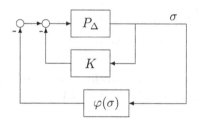

Fig. 8.12 The uncertain closed-loop feedback system.

Through this section we assume that the nonlinear function $\varphi(\sigma)$ satisfies Assumption 8.3. Without loss of generality we only consider the case of additive uncertainty defined by (8.29). In order to make the closed-loop feedback system shown in Fig. 8.12 to be still a pendulum-like feedback system, we assume that the designed controller $K(s)$ contains a differential part, that is,

$$K(s) = sK_1(s) \tag{8.35}$$

where $K_1(s)$ has a state space realization

$$K_1(s) = \left(\begin{array}{c|c} A_k & b_k \\ \hline c_k & 0 \end{array}\right). \tag{8.36}$$

Thus, the designed controller $K(s)$ can be expressed as

$$K(s) = c_k A_k (sI - A_k)^{-1} b_k + c_k b_k. \tag{8.37}$$

The transfer function of the linear part for the closed-loop feedback system from $-\varphi(\sigma)$ to σ shown in Fig. 8.12 is

$$P_{cl} = (1 + P_\Delta K)^{-1} P_\Delta = \frac{1}{s}(1 + G_\Delta K_1)^{-1} G_\Delta, \qquad (8.38)$$

where $G_\Delta(s)$ is defined by (8.29).
Denotes

$$G_{cl} = (1 + G_\Delta K_1)^{-1} G_\Delta. \qquad (8.39)$$

We have the following lemma.

Lemma 8.5. *Suppose that Assumption 8.3 holds and there exists a scalar $\alpha \in \mathbb{R}, \alpha \neq 0$ such that*

(i) the conditions (ii), (iii) of Theorem 8.3 hold and $1 + \alpha G(\infty) \neq 0$;
(ii) G_{cl} defined by (8.39) is non-degenerate, zero is not a zero of G_{cl};
(iii) $T_K(s) \in \mathcal{RH}_\infty$ and $||T_K(s)||_\infty < 1$.

Then the closed-loop feedback system shown in Fig. 8.12 is dichotomous, where

$$T_K = \begin{pmatrix} -1 + 2\alpha N_K & 2\alpha(1 + GK_1 + \alpha G)^{-1} W_1 \\ W_2 N_K & -W_2 K_1 (1 + GK_1)^{-1} W_1 - \alpha W_2 N_K (1 + GK_1)^{-1} W_1 \end{pmatrix} \qquad (8.40)$$

and

$$N_K = G(1 + K_1 G + \alpha G)^{-1}.$$

Proof. According to Lemma 8.4 we need to show that $\alpha G_{cl}(s) \in \mathcal{RH}_\infty$ and

$$\alpha G_{cl}(j\omega) + \alpha G_{cl}^T(j\omega) > 0, \ \forall \omega \in \mathbb{R} \cup \{\infty\} \qquad (8.41)$$

holds, which means $\alpha G_{cl}(s)$ is SPR with $\alpha G_\Delta(\infty) > 0$. Let

$$S_{cl}(s) = (\alpha G_{cl}(s) - 1)(\alpha G_{cl}(s) + 1)^{-1}. \qquad (8.42)$$

It is required to prove that $S_{cl}(s) \in \mathcal{RH}_\infty$ and $||S_{cl}(s)||_\infty < 1$. In the following we first show that $S_{cl} = \mathcal{F}_\ell(T_K, \Delta)$. From (8.29) and (8.39) G_{cl} can be expressed as

$$G_{cl} = \mathcal{F}_\ell(M, \Delta), \qquad (8.43)$$

where

$$M = \begin{pmatrix} G(1 + K_1 G)^{-1} & (1 + GK_1)^{-1} W_1 \\ W_2(1 + K_1 G)^{-1} & -W_2 K_1 (1 + GK_1)^{-1} W_1 \end{pmatrix}. \qquad (8.44)$$

8.3. Controller design with dichotomy for uncertain pendulum-like systems

By (8.42), S_{cl} is written as

$$S_{cl} = \mathcal{F}_\ell \left(\begin{pmatrix} -1 & 2\alpha \\ 1 & -\alpha \end{pmatrix}, G_{cl} \right). \tag{8.45}$$

In terms of (8.43), substituting $\mathcal{F}_\ell(M, \Delta)$ for G_{cl} in (8.45) gives

$$S_{cl} = \mathcal{F}_\ell(T_K, \Delta) \tag{8.46}$$

where $T_K = S\left(\begin{pmatrix} -1 & 2\alpha \\ 1 & -\alpha \end{pmatrix}, M \right)$, $S(\cdot)$ denotes the star product. It is easy to express T_K as (8.40) by the definition of star product in Chapter 2. From (8.46) and conditions (iii) as well as $\Delta \in \mathcal{RH}_\infty$, combining Theorem 2.10 with Lemma 2.2, one implies that $S_{cl} \in \mathcal{RH}_\infty$ and $\|S_{cl}\|_\infty < 1$, and thus αG_{cl} is SPR with $\alpha G_\Delta(\infty) > 0$. This completes the proof. □

By algebraic calculation the transfer function T_K defined in (8.40) can be expressed as a form of low linear fractional transformation, i.e., $T_K = \mathcal{F}_\ell(N, K_1)$, where

$$N = \left(\begin{pmatrix} 1 - 2(1+\alpha G)^{-1} & 2\alpha(1+\alpha G)^{-1}W_1 \\ W_2(1+\alpha G)^{-1} & -\alpha W_2(1+\alpha G)^{-1}W_1 \\ (G(1+\alpha G)^{-1} & (1+\alpha G)^{-1}W_1) \end{pmatrix} \begin{pmatrix} -2\alpha(1+\alpha G)^{-1}G \\ -W_2(1+\alpha G)^{-1} \\ -(1+\alpha G)^{-1}G \end{pmatrix} \right). \tag{8.47}$$

Therefore Lemma 8.5 can be rewritten as follows.

Lemma 8.6. *Suppose that Assumption 8.3 holds and there exist a scalar $\alpha \in \mathbb{R}$, $\alpha \neq 0$ such that*

(i) the conditions (ii) and (iii) of Theorem 8.3 hold;
(ii) $1 + \alpha G(\infty) \neq 0$;
(iii) there exist a strictly proper transfer function $K_1(s)$ with the state space realization (8.36) such that $\mathcal{F}_\ell(N, K_1)$ is internally stable and $\|\mathcal{F}_\ell(N, K_1)\|_\infty < 1$, where N is defined by (8.47);
(iv) $(1 + G_\Delta K_1)^{-1} G_\Delta$ is non-degenerate and zero is not a zero of $(1 + G_\Delta K_1)^{-1} G_\Delta$.

Then the closed-loop feedback system shown in Fig. 8.12 is dichotomous, where $K(s) = sK_1(s)$.

When there are no uncertainties, that is, $\Delta = 0$, we have the following corollary.

Corollary 8.2. *Assume that there exist a scalar $\alpha \in \mathbb{R}$, $\alpha \neq 0$ such that*

(i) Assumption 8.3 holds;

(ii) there exist a strictly proper transfer function $K_1(s)$ with the state space realization (8.36) such that $\mathcal{F}_\ell(\overline{N}, K_1) = \alpha(1+GK_1)^{-1}G$ is SPR with $\alpha G(\infty) > 0$, where

$$\overline{N} = \begin{pmatrix} \alpha G & -\alpha G \\ G & -G \end{pmatrix};$$

(iii) $(1 + GK_1)^{-1}G$ is non-degenerate and zero is not a zero of $(1 + GK_1)^{-1}G$.

Then the closed-loop feedback system shown in Fig. 8.12 with $P_\Delta = P = \frac{1}{s}G(s)$ is dichotomous, where $K(s) = sK_1(s)$.

Proof. In the case of $\Delta = 0$, N defined by (8.47) becomes

$$N = \begin{pmatrix} 1 - 2(1+\alpha G)^{-1} & -2\alpha(1+\alpha G)^{-1}G \\ G(1+\alpha G)^{-1} & -(1+\alpha G)^{-1}G \end{pmatrix}.$$

It is easy to test

$$(I - \mathcal{F}_\ell(N, K_1))^{-1}(I + \mathcal{F}_\ell(N, K_1)) = \mathcal{F}_\ell(\overline{N}, K_1)$$

holds. In terms of Lemma 3.9, $\mathcal{F}_\ell(N, K_1)$ is SBR if and only if $\mathcal{F}_\ell(\overline{N}, K_1)$ is SPR with $\alpha G(\infty) > 0$. Combining with Lemma 8.6 implies the result. □

The result of Lemma 8.6 shows that the problem of designing a controller $K(s)$ with the property of dichotomy for closed-loop feedback system shown in Fig. 8.12 is transformed into H_∞ control problem, which can be realized by LMI approach in Section 2.7.

Assume that G, W_1 and W_2 have the following state space realizations:

$$G = \left(\begin{array}{c|c} A & b \\ \hline c & d \end{array}\right), \quad W_1 = \left(\begin{array}{c|c} A_{W_1} & B_{W_1} \\ \hline C_{W_1} & D_{W_1} \end{array}\right), \quad W_2 = \left(\begin{array}{c|c} A_{W_2} & B_{W_2} \\ \hline C_{W_2} & D_{W_2} \end{array}\right).$$

Then by (8.47), N has a state space realization

$$N = \left(\begin{array}{c|cc} A_N & B_1 & B_2 \\ \hline C_1 & D_{11} & D_{12} \\ C_2 & D_{21} & -d \end{array}\right),$$

where

$$A_N = \begin{pmatrix} A_{W_2} & \alpha(1+\alpha d)^{-1}B_{W_2}c & \alpha(1+\alpha d)^{-1}B_{W_2}C_{W_1} \\ 0 & A - \alpha(1+\alpha d)^{-1}bc & -\alpha(1+\alpha d)^{-1}bC_{W_1} \\ 0 & 0 & A_{W_1} \end{pmatrix},$$

8.3. Controller design with dichotomy for uncertain pendulum-like systems

$$B_1 = \begin{pmatrix} (1+\alpha d)^{-1}B_{W_2} & \alpha(1+\alpha d)^{-1}B_{W_2}D_{W_1} \\ (1+\alpha d)^{-1}b & -\alpha(1+\alpha d)^{-1}bD_{W_1} \\ 0 & -B_{W_1} \end{pmatrix},$$

$$B_2 = \begin{pmatrix} (1+\alpha d)^{-1}B_{W_2} \\ -(1+\alpha d)^{-1}b \\ 0 \end{pmatrix},$$

$$C_1 = \begin{pmatrix} 0 & 2\alpha(1+\alpha d)^{-1}c & 2\alpha(1+\alpha d)^{-1}C_{W_1} \\ -C_{W_2} & -\alpha(1+\alpha d)^{-1}D_{W_2}c & -\alpha(1+\alpha d)^{-1}D_{W_2}C_{W_1} \end{pmatrix},$$

$$C_2 = \begin{pmatrix} 0 & (1+\alpha d)^{-1}c & (1+\alpha d)^{-1}C_{W_1} \end{pmatrix}, \quad D_{22} = -d(1+\alpha d)^{-1},$$

$$D_{11} = \begin{pmatrix} 1 - 2(1+\alpha d)^{-1} & 2\alpha(1+\alpha d)^{-1}D_{W_1} \\ (1+\alpha d)^{-1}D_{W_2} & -\alpha(1+\alpha d)^{-1}D_{W_2}D_{W_1} \end{pmatrix},$$

$$D_{12} = \begin{pmatrix} -2\alpha(1+\alpha d)^{-1}d \\ -(1+\alpha d)^{-1}D_{W_2} \end{pmatrix}, \quad D_{21} = \begin{pmatrix} d(1+\alpha d)^{-1} & (1+\alpha d)^{-1}D_{W_1} \end{pmatrix}.$$

Applying Theorem 3.3, we have

Theorem 8.6. *Suppose that Assumption 8.3 holds and there exist a scalar $\alpha \in \mathbb{R}$, $\alpha \neq 0$ such that*

(i) the conditions (ii) and (iii) of Theorem 8.3 hold;
(ii) $D_{11} + D_{11}^T > 0$;
(iii) $1 + \alpha d \neq 0$.

Then there exist a strictly proper transfer function K_1 with the state space realization (8.36) such that $\mathcal{F}_\ell(N, K_1)$ is internally stable and $\|\mathcal{F}_\ell(N, K_1)\|_\infty < 1$ if and only if there exist matrices $P_1 > 0, P_3 > 0$ and P_2, P_4 satisfying

$$\begin{pmatrix} A_N P_1 + P_1 A_N^T + B_2 P_2 + P_2^T B_2^T & P_1 C_1^T + P_2^T D_{12}^T - B_1 \\ C_1 P_1 + D_{12} P_2 - B_1^T & -(D_{11} + D_{11}^T) \end{pmatrix} < 0,$$

$$\begin{pmatrix} P_3 A_N + A_N^T P_3 + P_4 C_2 + C_2^T P_4^T & P_3 B_1 + P_4 D_{21} - C_1^T \\ B_1^T P_3 + D_{21}^T P_4^T - C_1 & -(D_{11} + D_{11}^T) \end{pmatrix} < 0$$

and the spectral radius $\rho(YX) < 1$, where $X = P_1^{-1}$ and $Y = P_3^{-1}$. Moreover, when these conditions are satisfied a strictly proper controller K_1 can be formulated by Theorem 3.3.

Furthermore, if $(1+G_\Delta K_1)^{-1}G_\Delta$ is non-degenerate and zero is not a zero of $(1+G_\Delta K_1)^{-1}G_\Delta$, then the closed-loop feedback system shown in Fig. 8.12 is dichotomous, where $K(s) = sK_1(s)$.

Remark 8.11. For system (6.1), when $c = c_*$, that is, the output σ is equal to the measurable output v, it becomes
$$\dot{x} = Ax + b\varphi(\sigma), \quad \sigma = cx.$$
In this case, the result of Corollary 8.2 is the same as that of Theorem 6.1.

Remark 8.12. Similar method can also be applied to the uncertain systems with multiplicative or H_∞ uncertainties.

8.4 Lagrange stability for uncertain pendulum-like systems

Consider an uncertain feedback system shown in Fig. 8.13, where the SISO transfer function $P_\Delta(s)$ denotes a system with additive uncertainty, or multiplicative uncertainty or H_∞ uncertainty. $\varphi(t,\sigma)$ is the part of feedback. The function $\varphi(t,\sigma)$ satisfy the following assumptions:

Assumption 8.5. $\varphi(t,\sigma) : \mathbb{R}_+ \times \mathbb{R} \to \mathbb{R}$ is continuous and locally Lipchitz continuous in the second argument and there exists a $T > 0$, such that
$$\varphi(t, \sigma + T) = \varphi(t, \sigma), \quad t \in \mathbb{R}_+, \sigma \in \mathbb{R} \quad (8.48)$$
Furthermore we assume that φ belongs to the sector $M[\mu_1, \mu_2]$, i.e.
$$\mu_1 \leq \frac{\varphi(t,\sigma)}{\sigma} \leq \mu_2, \quad (8.49)$$
where μ_1 is either a certain negative number or $-\infty$ and μ_2 ia either a certain positive number or $+\infty$. In the case $\mu_1 = -\infty$ (resp. $\mu_2 = +\infty$) we assume that $\mu_1^{-1} = 0$ (resp. $\mu_2^{-1} = 0$).

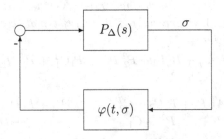

Fig. 8.13 The uncertain feedback system for Lagrange stability.

The uncertain feedback system shown in Fig. 8.13 is obviously a pendulum-like system under Assumptions 8.1, 8.2 and 8.5. By Theorem 5.13, we have

8.4. Lagrange stability for uncertain pendulum-like systems

Lemma 8.7. *Suppose that Assumptions 8.1, 8.2 and 8.5 hold and there exists a positive number $\lambda > 0$ such that*

(i) $P_\Delta(s)$ has a zero pole of multiplicity one and $P_\Delta(s - \lambda)$ has precisely $n - 1$ poles with negative real parts, and
(ii)

$$\mu_1^{-1}\mu_2^{-1} + (\mu_1^{-1} + \mu_2^{-1})\mathbf{Re}\{P_\Delta(i\omega - \lambda)\} + |P_\Delta(i\omega - \lambda)|^2 \leq 0 \quad (8.50)$$

for all $\omega \in \mathbb{R} \cup \{\infty\}$.

Then the uncertain feedback system shown in Fig. 8.13 is Lagrange stable.

Since (8.50) holds if and only if $(G_\Delta(j\omega))^* G_\Delta(j\omega) \leq \gamma^2$ which means $\| G_\Delta(s) \|_\infty \leq \gamma$, where

$$G_\Delta(s) = P_\Delta(s - \lambda) + \beta, \quad \gamma = \frac{\mu_2^{-1} - \mu_1^{-1}}{2}, \quad \beta = \frac{\mu_1^{-1} + \mu_2^{-1}}{2}. \quad (8.51)$$

Therefore the result of Lemma 8.7 can be rewritten as:

Lemma 8.8. *Suppose that Assumptions 8.1, 8.2 and 8.5 hold, and there exists a scalar $\lambda > 0$ such that $P_\Delta(s)$ has a zero pole of multiplicity one and $P_\Delta(s - \lambda)$ has precisely $n - 1$ poles with negative real parts. If $\|G_\Delta(s)\|_\infty \leq \gamma$, where $G_\Delta(s)$ and γ are defined by (8.51), then the uncertain feedback system shown in Fig. 8.13 is Lagrange stable.*

In the following, conditions of Lagrange stability for the uncertain feedback systems with additive uncertainty and multiplicative uncertainty shown in Fig. 8.13 are given.

Firstly, we consider the case of additive uncertainty, i.e.,

$$P_\Delta(s) = P(s) + W_1(s)\Delta(s)W_2(s).$$

Theorem 8.7. *Suppose that Assumptions 8.1, 8.2, 8.5 hold and*

(i) $P_\Delta(s)$ has a zero pole of multiplicity one and there exists a scalar $\lambda > 0$ such that $P_\Delta(s - \lambda)$ has precisely $n - 1$ poles with negative real parts;
(ii) $W_1(j\omega - \lambda) \neq 0$ or $W_2(j\omega - \lambda) \neq 0$ for $\omega \in \mathbb{R} \cup \{\infty\}$;
(iii) $\Delta(s - \lambda) \in \mathcal{RL}_\infty$ and $\|\Delta(s - \lambda)\|_\infty < 1$;
(iv) $T_G(s - \lambda) \in \mathcal{RL}_\infty$ and $\|T_G(s - \lambda)\|_\infty \leq 1$, where

$$T_G(s - \lambda) = \begin{pmatrix} \frac{1}{\gamma}P(s - \lambda) + \frac{\beta}{\gamma} & \frac{1}{\gamma}W_1(s - \lambda) \\ W_2(s - \lambda) & 0 \end{pmatrix}, \quad (8.52)$$

β and γ are defined by (8.51).

Then the uncertain feedback system with additive uncertainty shown in Fig. 8.13 is Lagrange stable.

Proof. $\frac{1}{\gamma}P_\Delta(s-\lambda) + \frac{\beta}{\gamma}$ can be written as

$$\frac{1}{\gamma}P_\Delta(s-\lambda) + \frac{\beta}{\gamma} = \mathcal{F}_\ell(T_G(s-\lambda), \Delta(s-\lambda))$$

where $T_G(s-\lambda)$ is defined by (8.52). Combining Lemma 8.8 with Lemma 2.2 the result follows. □

For the uncertain feedback system shown in Fig. 8.13 where the uncertainty is described by multiplicative uncertainty, i.e.,

$$P_\Delta(s) = (1 + W_1(s)\Delta(s)W_2(s))P(s)$$

we have

Theorem 8.8. *Suppose that Assumptions 8.1, 8.2, 8.5 hold and*

(i) $P_\Delta(s)$ *has a zero pole of multiplicity one and there exists a scalar* $\lambda > 0$ *such that* $P_\Delta(s-\lambda)$ *has precisely* $n-1$ *poles with negative real parts;*
(ii) $W_1(j\omega - \lambda) \neq 0$ *or* $W_2(j\omega - \lambda)P(j\omega - \lambda) \neq 0$ *for* $\omega \in \mathbb{R} \cup \{\infty\}$;
(iii) $\Delta(s-\lambda) \in \mathcal{RL}_\infty$ *and* $\|\Delta(s-\lambda)\|_\infty < 1$;
(iv) $T_G(s-\lambda) \in \mathcal{RL}_\infty$ *and* $\|T_G(s-\lambda)\|_\infty \leq 1$, *where*

$$T_G(s-\lambda) = \begin{pmatrix} \frac{1}{\gamma}P(s-\lambda) + \frac{\beta}{\gamma} & \frac{1}{\gamma}W_1(s-\lambda) \\ W_2(s-\lambda)P(s-\lambda) & 0 \end{pmatrix},$$

β *and* γ *are defined by (8.51).*

Then the uncertain feedback system with multiplicative uncertainty shown in Fig. 8.13 is Lagrange stable.

Remark 8.13. The result of Lagrange stability for uncertain feedback system with H_∞ uncertainty can also be given in the same method as Theorem 8.7 and Theorem 8.8.

Remark 8.14. When there are no uncertainties in the systems, that is, $\Delta = 0$, the condition $\|T_G(s-\lambda)\|_\infty \leq 1$ in Theorem 8.7 or Theorem 8.8 becomes $\|\frac{1}{\gamma}P(s-\lambda) + \frac{\beta}{\gamma}\|_\infty \leq 1$ and thus the result of Theorem 8.7 is identical to that of Theorem 8.8 which is the same as the result of Lemma 8.8.

8.5 Gradient-like property for pendulum-like systems with uncertainties

Consider an uncertain feedback system defined by

$$\dot{x} = A_\Delta x + B_\Delta \varphi(\sigma), \quad \dot{\sigma} = C_\Delta x + R_\Delta \varphi(\sigma) \quad (8.53)$$

where $A_\Delta, B_\Delta, C_\Delta$ and R_Δ are uncertain real matrices of orders $n \times n, n \times m, m \times n$ and $m \times m$ respectively. Let $P_\Delta(s) = C_\Delta(A_\Delta - sI)^{-1}B_\Delta - R_\Delta$ be the transfer function of the linear part of system (8.53) from the input $-\varphi$ to the output $\dot{\sigma}$. The uncertain feedback system (8.53) is shown in Fig. 8.14.

This section gives some conditions that the uncertain feedback system shown in 8.14 is gradient-like. Here the uncertainty $P_\Delta(s)$ includes three cases, that is, additive uncertainty, multiplicative uncertainty and H_∞ uncertainty.

Let $P_\Delta(s)$ and the function $\varphi(\sigma)$ satisfy the following assumptions:

Assumption 8.6. $P_\Delta(s)$ is non-degenerate and $P_\Delta(0) \neq 0$.

Assumption 8.7. $\varphi(\sigma) : \mathbb{R}^m \to \mathbb{R}^m$ is a vector-valued function having the components $\varphi_i(\sigma) = \varphi_i(\sigma_i)$ with $\sigma = (\sigma_1, \sigma_2, \cdots, \sigma_m)$. Assume that every component $\varphi_i : \mathbb{R} \to \mathbb{R}$ is T_i-periodic, satisfies a local Lipschitz condition and possesses a finite number of zeros on $[0, T_i)$.

Assumptions 8.6, 8.7 ensure that the uncertain feedback system shown in Fig. 8.14 is a pendulum-like system. Let us describe the equilibria set of system (8.53). If (x_{eq}, σ_{eq}) is an equilibrium, then $(C_\Delta A_\Delta^{-1} B_\Delta - R_\Delta)\varphi(\sigma_{eq}) = 0$. By $P_\Delta(0) \neq 0$, the equilibria set of system (8.53) is described as

$$\{(x_{eq}, \sigma_{eq}) : x_{eq} = 0, (\sigma_{eq})_i = \hat{\sigma}_i \quad \text{with} \quad \varphi_i(\hat{\sigma}_i) = 0, i = 1, 2, \cdots, m\}$$

Applying Corollary 5.5 and Corollary 5.6 to the uncertain feedback system shown in Fig. 8.14, we have

Lemma 8.9. *Suppose that Assumptions 8.6, 8.7 hold and $P_\Delta(s) \in \mathcal{RH}_\infty$. If there exist diagonal $m \times m$ real matrices $\kappa = \text{diag}(\kappa_1, \cdots, \kappa_m)$, $\delta = \text{diag}(\delta_1, \cdots, \delta_m)$ and $\epsilon = \text{diag}(\epsilon_1, \cdots, \epsilon_m)$ with $\delta > 0$ and $\varepsilon > 0$ satisfying the following conditions:*

(i) $\text{Re}\{\kappa P_\Delta(j\omega)\} - P_\Delta^*(j\omega)\varepsilon P(j\omega) - \delta \geq 0$ *for all* $\omega \in \mathbb{R}$, *and*

(ii) $2\sqrt{\varepsilon_i \delta_i} \int_0^{T_i} |\varphi_i(\tau)| d\tau > |\kappa_i| \left| \int_0^{T_i} \varphi_i(\tau) d\tau \right|$ *for* $i = 1, 2, \cdots, m$,

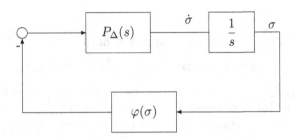

Fig. 8.14 The uncertain feedback system for gradient-like property.

then the uncertain feedback system shown in 8.14 is gradient-like.

By simple manipulation the condition (i) of Lemma 8.9 can be rewritten as $G^*(j\omega)G(j\omega) \leq I$, where

$$G(s) = \left(\alpha P(s) - \frac{1}{2}\alpha^{-1}\kappa\right)\gamma \qquad (8.54)$$

and

$$\gamma^{-2} = \frac{1}{4}\kappa\alpha^{-2}\kappa - \delta, \quad \alpha = diag(\varepsilon_1^{\frac{1}{2}}, \cdots, \varepsilon_m^{\frac{1}{2}}). \qquad (8.55)$$

Thus Lemma 8.9 can be re-stated as follows.

Lemma 8.10. *Suppose that Assumptions 8.6, 8.7 hold and $P_\Delta(s) \in \mathcal{RH}_\infty$. If there exist diagonal matrices $\kappa = diag(\kappa_1, \cdots, \kappa_m)$, $\delta = diag(\delta_1, \cdots, \delta_m)$ and $\alpha = diag(\alpha_1, \cdots, \alpha_m)$ with $\delta > 0$ and $\alpha > 0$ satisfying*

(i) $\frac{1}{4}\kappa\alpha^{-2}\kappa^T - \delta > 0$,

(ii) $2\alpha_i\sqrt{\delta_i}\int_0^{T_i}|\varphi_i(\tau)|d\tau > |\kappa_i|\left|\int_0^{T_i}\varphi_i(\tau)d\tau\right|$, $i = 1, 2, \cdots, m$, and

(iii) $\|G(s)\|_\infty \leq 1$, where $G(s)$ is defined by (8.54),

then the uncertain feedback system shown in 8.14 is gradient-like.

In the following, criteria for testing the gradient-like property of the uncertain feedback system are given for three kinds of uncertainties based on Lemma 8.10.

Firstly, we assume that the model uncertainty in Fig. 8.14 is represented by an additional perturbation:

$$P_\Delta(s) = P(s) + W_1(s)\Delta(s)W_2(s) \qquad (8.56)$$

8.5. Gradient-like property for pendulum-like systems with uncertainties 189

where $W_1(s)$ and $W_2(s)$ are transfer matrices that characterize the spatial and frequency structure of the uncertainty. $P(s)$ is the transfer matrix of nominal model.

Theorem 8.9. *Suppose that Assumptions 8.6, 8.7 hold and there exist diagonal matrices $\kappa = diag\ (\kappa_1, \cdots, \kappa_m)$, $\delta = diag\ (\delta_1, \cdots, \delta_m)$ and $\alpha = diag\ (\alpha_1, \cdots, \alpha_m)$ with $\delta > 0$, $\alpha > 0$ such that*

(i) *the conditions (i) and (ii) of Lemma 8.10 hold;*
(ii) $P(s), W_1(s), W_2(s), \Delta(s)$ *are stable transfer matrices and $||\Delta||_\infty < 1$;*
(iii) $W_1(j\omega)$ *has full column rank for all $\omega \in \mathbb{R} \cup \{\infty\}$, or $W_2(j\omega)$ has full row rank for all $\omega \in \mathbb{R} \cup \{\infty\}$;*
(iv) $T_G(s) \in \mathcal{RL}_\infty$ *and $||T_G(s)||_\infty \leq 1$, where*

$$T_G = \begin{pmatrix} \alpha P\gamma - \frac{1}{2}\alpha^{-1}\kappa\gamma & \alpha W_1 \\ W_2\gamma & 0 \end{pmatrix} \tag{8.57}$$

and γ is defined by (8.55).

Then the uncertain feedback system with additive uncertainty shown in Fig. 8.14 is gradient-like.

Proof. Let

$$G_\Delta = \left(\alpha P_\Delta - \frac{1}{2}\alpha^{-1}\kappa\right)\gamma.$$

Then G_Δ can be written as

$$G_\Delta = \mathcal{F}_\ell(T_G, \Delta)$$

where T_G is defined by (8.57). It follows from (ii)-(iv) and Lemma 2.2 that $||G_\Delta||_\infty < 1$. Consequently, by Lemma 8.10 the uncertain feedback system with additive uncertainty shown in Fig. 8.14 is gradient-like. □

Next, we consider the second case, that is, the model uncertainty in Fig. 8.14 is represented by multiplicative perturbation

$$P_\Delta(s) = (1 + W_1(s)\Delta(s)W_2(s))P(s).$$

Analogous to the proof of Theorem 8.9 we have

Theorem 8.10. *Suppose that Assumptions 8.6, 8.7 hold and there exist diagonal matrices $\kappa = diag\ (\kappa_1, \cdots, \kappa_m)$, $\delta = diag\ (\delta_1, \cdots, \delta_m)$ and $\alpha = diag\ (\alpha_1, \cdots, \alpha_m)$ with $\delta > 0$, $\alpha > 0$ such that*

(i) the conditions (i) and (ii) of Lemma 8.10 hold;
(ii) $G(s), W_1(s), W_2(s), \Delta(s)$ are stable transfer matrices and $||\Delta(s)||_\infty < 1$;
(iii) $W_1(j\omega)$ has full column rank for all $\omega \in \mathbb{R} \cup \{\infty\}$. or $W_2(j\omega)P(j\omega)$ has full row rank for all $\omega \in \mathbb{R} \cup \{\infty\}$;
(iv) $T_G(s) \in \mathcal{RL}_\infty$ and $||T_G(s)||_\infty \leq 1$, where

$$T_G = \begin{pmatrix} \alpha P\gamma - \frac{1}{2}\alpha^{-1}\kappa\gamma & \alpha W_1 \\ W_2 P\gamma & 0 \end{pmatrix}$$

and γ is defined by (8.55).

Then the uncertain feedback system with multiplicative uncertainty shown in Fig. 8.14 is gradient-like.

Finally, we consider the uncertain feedback system shown in Fig.8.15, where the transfer matrix P_Δ of linear part from $-\varphi(\sigma)$ to $\dot\sigma$ for the system is described by H_∞ uncertainty

$$P_\Delta = \mathcal{F}_\ell(P, \Delta) = P_{11} + P_{12}\Delta(I - P_{22}\Delta)^{-1}P_{21}$$

where $P = \begin{pmatrix} P_{11} & P_{12} \\ P_{21} & P_{22} \end{pmatrix}$ is a nominal model and Δ denotes uncertainties.

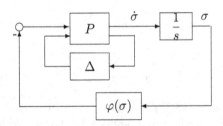

Fig. 8.15 The closed-loop feedback system for H_∞ uncertainty.

Theorem 8.11. *Suppose that Assumptions 8.6, 8.7 hold and there exist diagonal matrices $\kappa = \text{diag}(\kappa_1, \cdots, \kappa_m)$, $\delta = \text{diag}(\delta_1, \cdots, \delta_m)$ and $\alpha = \text{diag}(\alpha_1, \cdots, \alpha_m)$ with $\delta > 0$, $\alpha > 0$ such that*

(i) the conditions (i) and (ii) of Lemma 8.10 hold;

(ii) $P(s) = \begin{pmatrix} P_{11} & P_{12} \\ P_{21} & P_{22} \end{pmatrix} \in \mathcal{RH}_\infty$, $||P_{22}||_\infty \leq 1$, $\Delta(s) \in \mathcal{RH}_\infty$ and $||\Delta||_\infty < 1$;

8.6. Control of uncertain systems guaranteeing gradient-like property

(iii) $P_{12}(j\omega)$ *has full column rank for all* $\omega \in \mathbb{R} \cup \{\infty\}$, *or* $P_{21}(j\omega)$ *has full row rank for all* $\omega \in \mathbb{R} \cup \{\infty\}$;

(iv) $T_G(s) \in \mathcal{RL}_\infty$ *and* $||T_G(s)||_\infty \leq 1$, *where*

$$T_G = \begin{pmatrix} \alpha P_{11}\gamma - \dfrac{1}{2}\alpha^{-1}\kappa\gamma & \alpha P_{12} \\ P_{21}\gamma & P_{22} \end{pmatrix}$$

and γ *is defined by (8.55).*

Then the uncertain feedback system with H_∞ *uncertainty shown in Fig. 8.15 is gradient-like.*

Proof. Firstly, we note that the condition (ii) implies $P_\Delta \in \mathcal{RH}_\infty$. Next, G_Δ can be written as

$$G_\Delta = (\alpha P_\Delta - \dfrac{1}{2}\alpha^{-1}\kappa)\gamma = \mathcal{F}_\ell(T_G, \Delta)$$

which leads to $||G_\Delta||_\infty < 1$ by Lemma 2.2. Combining with Lemma 8.9 the result follows. □

Remark 8.15. When there are no uncertainties in the systems, that is, $\Delta = 0$, the results of Theorems 8.9, 8.10 and 8.11 are identical and degenerate into the result of Lemma 8.10.

8.6 Control of uncertain systems guaranteeing gradient-like property

In this section we consider controller design for the uncertain systems with gradient-like property. To this end, we first discuss the gradient-like property for closed-loop systems with unstructured uncertainty as shown in Fig. 8.16, where K is a given linear controller and P_Δ is subject to additional uncertainty.

Firstly, consider the case of additive uncertainty shown in Fig. 8.16, i.e.,

$$P_\Delta(s) = P(s) + W_1(s)\Delta(s)W_2(s).$$

In this case, $P_{cl}(s) = (I + P_\Delta K)^{-1}P_\Delta$ is the transfer function from $-\varphi(\sigma)$ to $\dot{\sigma}$.

Lemma 8.11. *Suppose that Assumption 8.7 holds and there exist diagonal* $m \times m$ *matrices* $\kappa = \text{diag}(\kappa_1, \cdots, \kappa_m)$, $\delta = \text{diag}(\delta_1, \cdots, \delta_m)$ *and* $\alpha = \text{diag}(\alpha_1, \cdots, \alpha_m)$ *with* $\delta > 0$, $\alpha > 0$ *such that*

192   *Analysis and Control for Uncertain Feedback Nonlinear Systems*

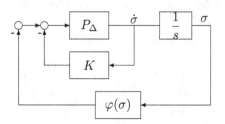

Fig. 8.16   The uncertain closed-loop feedback system.

(i) the conditions (i) and (ii) of Lemma 8.10 hold;
(ii) $P_{cl}(s) = (I + P_\Delta K)^{-1} P_\Delta$ is non-degenerate and $\det(P_{cl}(0)) \neq 0$;
(iii) $\Delta(s) \in \mathcal{RH}_\infty$ and $\|\Delta(s)\|_\infty < 1$;
(iv) $W_1(s) \in \mathcal{RH}_\infty, W_2(s) \in \mathcal{RH}_\infty$ and $W_1(j\omega)$ has full column rank for all $\omega \in \mathbb{R} \cup \{\infty\}$, or $W_2(j\omega)$ has full row rank for all $\omega \in \mathbb{R} \cup \{\infty\}$;
(v) $N(s) \in \mathcal{RH}_\infty$ and $\|N(s)\|_\infty < 1$, where

$$N = \begin{pmatrix} \alpha(I+PK)^{-1}P\gamma - \frac{1}{2}\alpha^{-1}\kappa\gamma & \alpha(I+PK)^{-1}W_1 \\ W_2(I+KP)^{-1}\gamma & -W_2K(I+PK)^{-1}W_1 \end{pmatrix} \quad (8.58)$$

and γ is described by (8.55).

Then the uncertain closed-loop feedback system with additive uncertainty shown in Fig. 8.16 is gradient-like.

Proof.   By

$$P_\Delta = P + W_1 \Delta W_2 = \mathcal{F}_\ell\left(\begin{pmatrix} P & W_1 \\ W_2 & 0 \end{pmatrix}, \Delta\right)$$

one implies that

$$P_{cl} = P_\Delta(I + KP_\Delta)^{-1} = \mathcal{F}_\ell\left(\begin{pmatrix} 0 & I \\ I & -K \end{pmatrix}, P_\Delta\right)$$
$$= \mathcal{F}_\ell\left(\begin{pmatrix} 0 & I \\ I & -K \end{pmatrix}, \mathcal{F}_\ell\left(\begin{pmatrix} P & W_1 \\ W_2 & 0 \end{pmatrix}, \Delta\right)\right) = \mathcal{F}_\ell(T, \Delta),$$

where

$$T = \begin{pmatrix} P(I+KP)^{-1} & (I+PK)^{-1}W_1 \\ W_2(I+KP)^{-1} & -W_2K(I+PK)^{-1}W_1 \end{pmatrix}.$$

Therefore, $G_{cl} = \alpha P_{cl}\gamma - \frac{1}{2}\alpha^{-1}\kappa\gamma$ can be represented as

$$G_{cl} = \mathcal{F}_\ell(N, \Delta), \quad (8.59)$$

8.6. Control of uncertain systems guaranteeing gradient-like property

where N is defined by (8.58). From (iv), (v) and Lemma 2.2 it follows that $||G_{cl}||_\infty < 1$. In order to show that the uncertain closed-loop feedback system with additive uncertainty shown in Fig. 8.16 is gradient-like, we only need to show that $G_{cl}(s) \in \mathcal{RH}_\infty$ which implies that $P_{cl}(s) \in \mathcal{RH}_\infty$. In fact, condition (v) implies that $|| - W_2 K(I+PK)^{-1}W_1||_\infty < 1$ and thus the conclusion that $G_{cl}(s) \in \mathcal{RH}_\infty$ follows by (iii), (8.59) and Theorem 2.10. This completes the proof. □

In what follows, we consider the uncertain closed-loop feedback system shown in Fig. 8.16 where the uncertainty is described by multiplicative uncertainty, i.e.,

$$P_\Delta(s) = (1 + W_1(s)\Delta(s)W_2(s))P(s).$$

Similar to Lemma 8.11, we have the the following result.

Lemma 8.12. *Suppose that Assumption 8.7 holds and there exist diagonal $m \times m$ matrices $\kappa = diag\ (\kappa_1, \cdots, \kappa_m)$, $\delta = diag\ (\delta_1, \cdots, \delta_m)$ and $\alpha = diag\ (\alpha_1, \cdots, \alpha_m)$ with $\delta > 0$, $\alpha > 0$ such that*

(i) the conditions (i) and (ii) of Lemma 8.10 hold;
(ii) $P_{cl}(s) = (I + P_\Delta K)^{-1}P_\Delta$ is non-degenerate and $det(P_{cl})(0) \neq 0$;
(iii) $\Delta(s) \in \mathcal{RH}_\infty$ and $||\Delta(s)||_\infty < 1$;
(iv) $W_1(s) \in \mathcal{RH}_\infty, W_2(s) \in \mathcal{RH}_\infty$ and $W_1(j\omega)$ has full column rank for all $\omega \in \mathbb{R} \cup \{\infty\}$, or $W_2(j\omega)P(j\omega)$ has full row rank for all $\omega \in \mathbb{R} \cup \{\infty\}$;
(v) $N(s) \in \mathcal{RH}_\infty$ and $||N(s)||_\infty < 1$, where

$$N = \begin{pmatrix} \alpha(I+PK)^{-1}P\gamma - \frac{1}{2}\alpha^{-1}\kappa\gamma & \alpha(I+PK)^{-1}W_1 \\ W_2 P(I+KP)^{-1}\gamma & -W_2 PK(I+PK)^{-1}W_1 \end{pmatrix}$$

and γ is defined by (8.55).

Then the uncertain closed-loop feedback system with multiplicative uncertainty shown in Fig. 8.16 is gradient-like.

Remark 8.16. The gradient-like property for uncertain closed-loop feedback system with H_∞ uncertainty shown in Fig. 8.16 can also be given in the same form as Lemma 8.11 and Lemma 8.12.

In the following, we consider problems of designing a linear controller K such that the uncertain closed-loop feedback system shown in Fig. 8.16 is gradient-like.

Consider the case of additional uncertainty:

$$P_\Delta(s) = P(s) + W_1(s)\Delta(s)W_2(s).$$

By the analysis of gradient-like property for the closed-loop feedback system above, we have

Theorem 8.12. *Suppose that there exist diagonal matrices* $\kappa = diag$ $(\kappa_1, \cdots, \kappa_m)$, $\delta = diag(\delta_1, \cdots, \delta_m)$ *and* $\alpha = diag(\alpha_1, \cdots, \alpha_m)$ *with* $\delta > 0$, $\alpha > 0$ *such that the conditions (i), (iii) and (iv) of Lemma 8.11 are satisfied. If there exists a linear controller* K *such that* $P_{cl}(s) = (I + P_\Delta K)^{-1} P_\Delta$ *is non-degenerate,* $\det(P_{cl}(0)) \neq 0$, $\mathcal{F}_\ell(T_N, K)$ *is internally stable and* $\|\mathcal{F}_\ell(T_N, K)\|_\infty < 1$, *then the uncertain closed-loop feedback system shown in Fig. 8.16 is gradient-like, where*

$$T_N = \left(\begin{pmatrix} \alpha P\gamma - \frac{1}{2}\alpha^{-1}\kappa\gamma & \alpha W_1 \\ W_2\gamma & 0 \\ (P\gamma & W_1) \end{pmatrix} \begin{pmatrix} -\alpha P \\ -W_2 \\ -P \end{pmatrix}\right) \quad (8.60)$$

and γ *is defined by (8.55).*

Proof. From (8.58) in Lemma 8.11, the transfer matrix N can be expressed as

$$N = \begin{pmatrix} \alpha P\gamma - \frac{1}{2}\alpha^{-1}\kappa\gamma & \alpha W_1 \\ W_2\gamma & 0 \end{pmatrix} - \begin{pmatrix} \alpha P \\ W_2 \end{pmatrix} K(I + PK)^{-1} \begin{pmatrix} P\gamma & W_1 \end{pmatrix}$$
$$= \mathcal{F}_\ell(T_N, K),$$

(8.61)

where T_N is defined by (8.60). Thus in terms of Lemma 8.11, the problem of design controller K is equivalent to that of finding K such that $\mathcal{F}_\ell(T_N, K)$ is internally stable and $\|\mathcal{F}_\ell(T_N, K)\|_\infty < 1$. $\square$

According to Theorem 8.12 the problem of controller design is transformed into H_∞ control problem. Therefore, the procedure of design controller K can be realized by LMI approach. Assume that P, W_1 and W_2 have the following state space realizations:

$$P = \left(\begin{array}{c|c} A & B \\ \hline C & D \end{array}\right), \quad W_1 = \left(\begin{array}{c|c} A_{W_1} & B_{W_1} \\ \hline C_{W_1} & D_{W_1} \end{array}\right), \quad W_2 = \left(\begin{array}{c|c} A_{W_2} & B_{W_2} \\ \hline C_{W_2} & D_{W_2} \end{array}\right).$$

Then by (8.60), T_N has a state space realization

$$T_N = \left(\begin{array}{c|cc} A_N & B_1 & B_2 \\ \hline C_1 & D_{11} & D_{12} \\ C_2 & D_{21} & -D \end{array}\right), \quad (8.62)$$

8.6. Control of uncertain systems guaranteeing gradient-like property

where

$$A_N = \begin{pmatrix} A & 0 & 0 \\ 0 & A_{W_1} & 0 \\ 0 & 0 & A_{W_2} \end{pmatrix}, \quad B_1 = \begin{pmatrix} B\gamma & 0 \\ 0 & B_{W_1} \\ -B_{W_2}\gamma & 0 \end{pmatrix}, \quad B_2 = \begin{pmatrix} -B \\ 0 \\ B_{W_2} \end{pmatrix},$$

$$C_1 = \begin{pmatrix} \alpha C & \alpha C_{W_1} & 0 \\ 0 & 0 & -C_{W_2} \end{pmatrix}, \quad D_{11} = \begin{pmatrix} \alpha D\gamma - \frac{1}{2}\alpha^{-1}\kappa\gamma & \alpha D_{W_1} \\ D_{W_2}\gamma & 0 \end{pmatrix},$$

$$D_{12} = \begin{pmatrix} -\alpha D \\ -D_{W_2} \end{pmatrix}, \quad C_2 = \begin{pmatrix} C & C_{W_1} & 0 \end{pmatrix}, \quad D_{21} = \begin{pmatrix} D\gamma & D_{W_1} \end{pmatrix}.$$

Let

$$\overline{T}_N = \left(\begin{array}{c|cc} A_N & B_1 & B_2 \\ \hline C_1 & D_{11} & D_{12} \\ C_2 & D_{21} & 0 \end{array} \right), \quad \overline{K} = K(I+DK)^{-1}. \tag{8.63}$$

Then

$$\mathcal{F}_\ell(T_N, K) = \mathcal{F}_\ell(\overline{T}_N, \overline{K})$$

and thus there exists a controller K such that $\mathcal{F}_\ell(T_N, K)$ is internally stable and $\|\mathcal{F}_\ell(T_N, K)\|_\infty < 1$ if and only if there exists $\overline{K}$ such that $\mathcal{F}_\ell(\overline{T}_N, \overline{K})$ is internally stable and $\|\mathcal{F}_\ell(\overline{T}_N, \overline{K})\|_\infty < 1$. The conditions for existence of $\bar{K}$ can be given by applying approach in Section 2.7 (see also [Iwasaki and Skelton (1994)]) which is summarized as follows.

Theorem 8.13. *Suppose that there exist diagonal matrices $\kappa = \mathrm{diag}\,(\kappa_1, \cdots, \kappa_m)$, $\delta = \mathrm{diag}\,(\delta_1, \cdots, \delta_m)$ and $\alpha = \mathrm{diag}\,(\alpha_1, \cdots, \alpha_m)$ with $\delta > 0$, $\alpha > 0$ such that the conditions (i), (iii) and (iv) of Lemma 8.11 are satisfied. Then for $\overline{T}_N$ defined by (8.63), there exists $\overline{K}$ such that $\mathcal{F}_\ell(\overline{T}_N, \overline{K})$ is internally stable and $\|\mathcal{F}_\ell(\overline{T}_N, \overline{K})\|_\infty < 1$ if and only if $L_D \neq \emptyset$, where*

$$L_D = \left\{ (X,Y) : X \in L_B, Y \in L_C, \begin{pmatrix} X & I \\ I & Y \end{pmatrix} \geq 0 \right\} \tag{8.64}$$

and

$$L_B = \left\{ X : X > 0, \begin{pmatrix} B_2 \\ D_{12} \end{pmatrix}^\perp \right. \\ \left. \times \begin{pmatrix} A_N X + X A_N^T + B_1 B_1^T & X C_1^T + B_1 D_{11}^T \\ C_1 X + D_{11} B_1^T & D_{11} D_{11}^T - I \end{pmatrix} \begin{pmatrix} B_2 \\ D_{12} \end{pmatrix}^{\perp T} < 0 \right\}, \tag{8.65}$$

$$L_C = \left\{ Y : Y > 0, \begin{pmatrix} C_2^T \\ D_{21}^T \end{pmatrix}^\perp \right.$$

$$\left. \times \begin{pmatrix} YA_N + A_N^T Y + C_1^T C_1 & YB_1 + C_1^T D_{11} \\ B_1^T Y + D_{11}^T C_1 & D_{11}^T D_{11} - I \end{pmatrix} \begin{pmatrix} C_2^T \\ D_{21}^T \end{pmatrix}^{\perp T} < 0 \right\}.$$

(8.66)

Furthermore, if $L_D \neq \emptyset$, then transfer function $\overline{K}$ can be obtained by the method given in Section 2.7 and be parameterized (see [Iwasaki and Skelton (1994)] in detail), and thus the controller K can also be obtained by $K = (I - \overline{K}D)^{-1}\overline{K}$.

Moreover, if $P_{cl}(s) = (I + P_\Delta K)^{-1} P_\Delta$ is non-degenerate and $\det(P_{cl}(0)) \neq 0$, then the uncertain closed-loop feedback system shown in Fig. 8.16 is gradient-like.

Remark 8.17. Similar to the case of additional uncertainty the related results for multiplicative uncertainty can also be discussed.

The results of controller design for the pendulum-like system without uncertainty can be obtained by Theorem 8.12 and Theorem 8.13 as a corollary. In this case, $\Delta(s) = 0, P_\Delta(s) = P(s)$, the corresponding result of Theorem 8.13 is still true but T_N defined by (8.60) in Theorem 8.12 should be substituted by

$$T_N = \begin{pmatrix} \alpha P\gamma - \frac{1}{2}\alpha^{-1}\kappa\gamma & -\alpha P \\ P\gamma & -P \end{pmatrix}, \qquad (8.67)$$

and T_N has a state space realization defined as (8.62) but the matrices there should be

$$A_N = A, \quad B_1 = B\gamma, \quad B_2 = -B, \quad C_1 = \alpha C, \qquad (8.68)$$

$$C_2 = C, \quad D_{11} = \alpha D\gamma - \frac{1}{2}\alpha^{-1}\kappa\gamma, \quad D_{12} = -\alpha D, \quad D_{21} = D\gamma. \qquad (8.69)$$

As a corollary of Theorem 8.13, we have

Corollary 8.3. *Suppose that there exist diagonal matrices* $\kappa = \text{diag}(\kappa_1, \cdots, \kappa_m)$, $\delta = \text{diag}(\delta_1, \cdots, \delta_m)$ *and* $\alpha = \text{diag}(\alpha_1, \cdots, \alpha_m)$ *with* $\delta > 0$, $\alpha > 0$ *such that the conditions (i) and (ii) of Lemma 8.10 are satisfied. Then for* $\overline{T}_N$ *defined by (8.63), there exists* $\overline{K}$ *such that* $\mathcal{F}_\ell(\overline{T}_N, \overline{K})$ *is internally stable and* $\|\mathcal{F}_\ell(\overline{T}_N, \overline{K})\|_\infty < 1$ *if and only if* $L_D \neq \emptyset$, *where* L_D *is defined by (8.64) and the matrices* $A_N, B_1, B_2, C_1, C_2, D_{11}, D_{12}$ *and* D_{21} *in (8.65) and (8.66) are defined by (8.68) and (8.69), respectively.*

8.6. Control of uncertain systems guaranteeing gradient-like property   197

Furthermore, if $L_D \neq \emptyset$, then transfer function $\overline{K}$ can be obtained by the method given in Section 2.7 and be parameterized (see [Iwasaki and Skelton (1994)] in detail), and thus controller K can also be obtained by $K = (I - \overline{K}D)^{-1}\overline{K}$.

Moreover, if $P_{cl}(s) = (I+PK)^{-1}P$ is non-degenerate and $\det(P_{cl}(0)) \neq 0$, then closed-loop feedback system shown in Fig. 8.16 with $P_\Delta = P$ is gradient-like.

Example 8.3. Consider an autonomous pendulum-like system

$$\begin{cases} \dot{x} = -\dfrac{1}{\mu}x + (\beta - 1)\varphi(\sigma), \\ \dot{\sigma} = x - \beta\mu\varphi(\sigma), \end{cases} \quad (8.70)$$

where $\mu > 0$ and $\beta \in (0,1)$ are constants. The scalar nonlinearity $\varphi(\sigma)$ is T periodic and has finite number of isolated zeros on $[0, \Delta)$. The transfer function of the linear part of (8.70) from the input $-\varphi(\sigma)$ to the output $\dot{\sigma}$ is

$$P(s) = \mu\frac{1 + \beta\mu s}{1 + \mu s}. \quad (8.71)$$

This system describes an autonomous first-order phase-locked loop with proportional-integrating filter [Leonov et al. (1996); Shakhgil'dyan (1972)]. Let $\varphi(\sigma) = \sin(\sigma) - \gamma$ with $\gamma \in (0,1)$. It is shown that in [Leonov et al. (1996)] that the system is gradient-like if

$$\frac{4\sqrt{\beta}}{1+\beta} > \frac{\pi\gamma}{\gamma\arcsin\gamma + \sqrt{1-\gamma^2}} \quad (8.72)$$

Let $\beta = 0.2, \mu = 10, \gamma = 0.8$. The inequality (8.72) is not satisfied. Choosing an initial value $(x_0, \sigma_0) = (80, -80)$, the solution of the system (8.70) with above initial value is not convergent. The states $x(t)$ and $\sigma(t)$ are shown in Fig. 8.17 (a) and $\varphi(\sigma)$ is shown in Fig. 8.17 (b).

Now we impose a linear controller K such that the closed-loop feedback system shown in Fig. 8.16 is gradient-like. In terms of Corollary 8.3, choosing $\alpha = 1, \delta = 0.5, \kappa = -10$, LMIs in Corollary 8.3 are feasible and we obtain a linear controller:

$$K(s) = \frac{1.7500s + 0.3830}{s - 0.4840}.$$

It is easy to verify the conditions of Corollary 8.3 are satisfied and thus the closed-loop pendulum-like system:

$$\begin{cases} \dot{x}_c = A_c x_c + B_c\varphi(\sigma), \\ \dot{\sigma} = C_c x_c + D_c\varphi(\sigma) \end{cases} \quad (8.73)$$

is gradient-like, where

$$A_c = \begin{pmatrix} -1.5000 & -0.9920 & -1.4000 \\ 0.9920 & 0.4840 & 0.9920 \\ 0 & 0 & -0.1000 \end{pmatrix}, \quad B_c = \begin{pmatrix} 3.1305 \\ -2.2181 \\ -0.8944 \end{pmatrix},$$

$$C_c = (-2.2361 \quad -2.2181 \quad -2.2361), \quad D_c = 5.0000.$$

The states $x_c(t)$ and $\sigma(t)$ and $\varphi(\sigma)$ of the closed-loop pendulum-like system (8.73) with an initial value $x_c(0) = (80, -80, 80, -80)^T$ are shown in Fig. 8.18 (a) and Fig. 8.18 (b) respectively. The solution converges to the equilibrium point $(0, 0, 0, -456.4582)$ of the system (8.73).

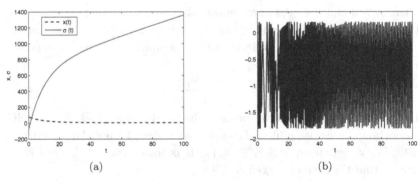

Fig. 8.17 The behavior of the pendulum-like system (8.70). (a) $x(t)$ and $\sigma(t)$ for the system (8.70). (b) $\varphi(\sigma(t))$ for the system (8.70).

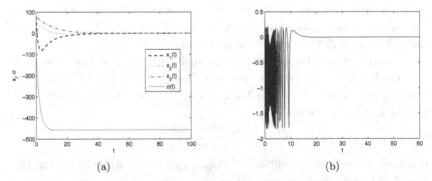

Fig. 8.18 The behavior of the closed-loop feedback system (8.73). (a) $x_c(t)$ and $\sigma(t)$ for the system (8.73). (b) $\varphi(\sigma(t))$ for the system (8.73).

8.7 Gradient-like property of systems with norm bounded uncertainties

This section discusses the gradient-like property of the feedback systems with norm bounded uncertainties. Here the feedback is a periodic nonlinear function which is local Lipschitz and possesses a finite number of zero on one period.

Consider an uncertain feedback system:

$$\begin{cases} \dot{x} = (A + \Delta A)x + (b + \Delta b)\varphi(\sigma), \\ \dot{\sigma} = cx + \rho\varphi(\sigma), \end{cases} \quad (8.74)$$

where $A \in \mathbb{R}^{n \times n}$ is a constant real matrix, $b \in \mathbb{R}^n$ and $c^T \in \mathbb{R}^n$ are real vectors and ρ is a real number. $x \in \mathbb{R}^n, \sigma \in \mathbb{R}$. Assume that parameter uncertainties ΔA and Δb can be expressed as the form of norm uncertainties

$$\Delta A = H_1 F E_1, \quad \Delta b = H_2 F E_2 \quad (8.75)$$

where $H_1 \in \mathbb{R}^{n \times m}, E_1 \in \mathbb{R}^{p \times n}, H_2 \in \mathbb{R}^{n \times m}, E_2 \in \mathbb{R}^{p \times 1}$ and $F \in \mathbb{R}^{m \times p}$ indicates uncertainty which satisfies $F^T F \leq I$. In this section we suppose that $A + \Delta A$ is invertible and $c^T(A + H_1 F E_1)^{-1}(b + H_2 F E_2) - \rho \neq 0$ which ensures that

$$\{(x_{eq}, \sigma_{eq}) : x_{eq} = 0, \varphi(\sigma_{eq}) = 0\}$$

is the equilibrium set of system (8.74). The nonlinear function $\varphi : \mathbb{R} \to \mathbb{R}$ is T-periodic, satisfies a local Lipschitz condition and possesses a finite number of zeros on $[0, \Delta)$.

When there are no uncertainties in (8.74), i.e., $\Delta A = 0, \Delta b = 0$, by Corollary 5.5, the conditions guaranteeing the gradient-like property for SISO system (8.74) are summarized as follows.

Lemma 8.13. *Suppose that A is a stable matrix. (A, b) is controllable and (A, c) is observable. Assume that there exist real numbers $\delta > 0, \varepsilon > 0$ and κ satisfying*

(i) there exists a matrix $P = P^T > 0$ such that

$$\begin{pmatrix} A^T P + PA + \varepsilon c^T c & Pb + (\dfrac{\kappa}{2} + \varepsilon\rho)c^T \\ b^T P + (\dfrac{\kappa}{2} + \varepsilon\rho)c & \delta + \varepsilon\rho^2 + \kappa\rho \end{pmatrix} < 0; \quad (8.76)$$

(ii) $2\sqrt{\varepsilon\delta} \int_0^\Delta |\varphi(\sigma)|d\sigma > |\kappa| |\int_0^\Delta \varphi(\sigma)d\sigma|.$

Then system (8.74) with $\Delta A = 0$ and $\Delta b = 0$ is gradient-like.

Using the bounded real lemma (Theorem 2.7) the H_∞ norm condition guaranteeing the property of gradient-like for system (8.74) with $\Delta A = 0$ and $\Delta b = 0$ is described as follows.

Lemma 8.14. *Suppose that A is a stable matrix . (A, b) is controllable and (A, c) is observable. If there exist real numbers $\varepsilon > 0$ and $|\kappa| > 2$ satisfying the condition (ii) of Lemma 8.13 and*

$$\|G_0\|_\infty = \|\varepsilon c(sI - A)^{-1}b + \mu\|_\infty < 1,$$

where $\mu = \dfrac{\kappa}{2} + \varepsilon\rho$ and δ in the condition (ii) of Lemma 8.13 is defined by $\delta = \dfrac{\kappa^2}{4\varepsilon} - \dfrac{1}{\varepsilon}$, then the system (8.74) with $\Delta A = 0$ and $\Delta b = 0$ is gradient-like.

Applying Lemma 8.13 to the uncertain system (8.74), we have

Theorem 8.14. *Suppose that there exist real numbers $\delta > 0, \varepsilon > 0, \nu > 0$, κ and matrix $P = P^T > 0$ such that*

(i)
$$\begin{pmatrix} A^T P + PA + \varepsilon c^T c + \lambda E_1^T E_1 & Pb + \mu c^T & PH_1 & PH_2 \\ b^T P + \mu c & h + \lambda E_2^T E_2 & 0 & 0 \\ H_1^T P & 0 & -\lambda I & 0 \\ H_2^T P & 0 & 0 & -\lambda I \end{pmatrix} < 0 \quad (8.77)$$

where

$$\mu = \frac{\kappa}{2} + \varepsilon\rho, \quad h = \delta + \varepsilon\rho^2 + \kappa\rho, \quad \lambda = \frac{1}{\nu}; \quad (8.78)$$

(ii) (A, b) is controllable and (A, c) is observable;
(iii)

$$\max\{m^{\frac{1}{2}}\|H_1\|_1\|E_1\|_1,\ m^{\frac{1}{2}}\|H_2\|_1\|E_2\|_1\} < \frac{1}{\|Q^{-1}(A,b)\|_1}$$

and

$$p\|H_1\|_\infty\|E_1\|_\infty < \frac{1}{\|H^{-1}(A,c)\|_1}$$

where $Q(A, b)$ and $H(A, c)$ are defined by (1.5) and (1.6) in Chapter 1, respectively;
(iv) the condition (ii) of Lemma 8.13 holds.

Then the uncertain pendulum-like system (8.74) is gradient-like.

8.7. Gradient-like property of systems with norm bounded uncertainties

Proof. Applying Lemma 8.13 to the uncertain feedback system (8.74), the corresponding inequality (8.76) is written as

$$\begin{pmatrix} P(A + H_1FE_1) + (A + H_1FE_1)^T P + \varepsilon c^T c & P(b + H_2FE_2) + \mu c^T \\ (b + H_2FE_2)^T P + \mu c & h \end{pmatrix} < 0 \quad (8.79)$$

where μ and h are defined by (8.78). Rewritting (8.79) as

$$\Omega + \begin{pmatrix} PH_1 & PH_2 \\ 0 & 0 \end{pmatrix} \begin{pmatrix} F & 0 \\ 0 & F \end{pmatrix} \begin{pmatrix} E_1 & 0 \\ 0 & E_2 \end{pmatrix}$$
$$+ \begin{pmatrix} E_1 & 0 \\ 0 & E_2 \end{pmatrix}^T \begin{pmatrix} F & 0 \\ 0 & F \end{pmatrix}^T \begin{pmatrix} PH_1 & PH_2 \\ 0 & 0 \end{pmatrix}^T < 0 \quad (8.80)$$

where

$$\Omega = \begin{pmatrix} PA + A^T P + \varepsilon c c^T & Pb + \mu c \\ b^T P + \mu c^T & h \end{pmatrix}.$$

From Lemma 8.3 and $F^T F \leq I$, (8.80) holds if and only if there exists a positive number ν such that

$$\begin{pmatrix} PA + A^T P + \varepsilon c^T c + \dfrac{1}{\nu} E_1^T E_1 + \nu P(H_1 H_1^T + H_2 H_2^T) P & Pb + \mu c^T \\ b^T P + \mu c & h + \dfrac{1}{\nu} E_2^T E_2 \end{pmatrix} < 0.$$

using the Schur complement it is clear that the above inequality holds if and only if (8.77) holds. From (8.79), one implies that $P(A + H_1 F E_1) + (A + H_1 F E_1)^T P < 0$ which means $A + \Delta A$ is quadratically stable for $\Delta A = H_1 F E_1$ where F satisfies $F^T F \leq I$.

As proof in Theorem 8.1, conditions (ii) and (iii) imply that $(A+\Delta A, b+\Delta b)$ is controllable and $(A + \Delta A, c)$ is observable. Combining with Lemma 8.13 completes the proof. $\square$

Remark 8.18. If there are no uncertainties in (8.74), that is, E_1, E_2, H_1, H_2 can be regarded as zero matrix, then the result of Theorem 8.14 is the same as the result of Lemma 8.13.

In the following, we use the results of Theorem 8.14 to analyze the gradient-like property of an autonomous phase-locked loop.

Example 8.4. Consider the system of autonomous phase-locked loop with a second order filter of type "2/2" [Shakhgil'dyan (1972); Leonov et al. (1996)]. This system can be written as

$$\begin{cases} \dfrac{dz}{dt} = \bar{A}z + \bar{b}\varphi(\sigma), \\ \dfrac{d\sigma}{dt} = cz + \rho\varphi(\sigma), \end{cases} \quad (8.81)$$

where

$$\bar{A} = \begin{pmatrix} 0 & -\dfrac{1}{\alpha_2} \\ 1 & -\dfrac{\alpha_1}{\alpha_2} \end{pmatrix}, \bar{b} = \begin{pmatrix} -\dfrac{1-\beta_2}{\alpha_2} \\ -\dfrac{\alpha_1(\beta_1-\beta_2)}{\alpha_2} \end{pmatrix}, c^T = \begin{pmatrix} 0 \\ 1 \end{pmatrix}, \rho = -\beta_2$$

and φ is T-periodic. $\alpha_1, \alpha_2, \beta_1, \beta_2$ are positive parameters and $\beta_1 < \beta_2 < 1$. The transfer function of the linear part of (8.81) from the input $-\varphi$ to the output $\dot{\sigma}$ is

$$P(s) = c(\bar{A} - sI)^{-1}\bar{b} - \rho = \frac{1 + \beta_1\alpha_1 s + \beta_2\alpha_2 s^2}{1 + \alpha_1 s + \alpha_2 s^2}.$$

It is easy to show that A is a stable matrix. $(\bar{A}, \bar{b})$ is controllable and $(\bar{A}, c)$ is observable. $P(0) = 1$. The equilibria set of

$$\{(z_{eq}, \sigma_{eq}) : z_{eq} = 0, \varphi(\sigma_{eq}) = 0\}.$$

Let $\varphi(\sigma) = \sin\sigma - \gamma, \gamma \in (0,1)$. In [Leonov et al. (1996)], it is shown that system (8.81) is gradient-like if the parameters are chosen as $\alpha_1 = 300, \beta_1 = 0.2, \alpha_2 = 5, \beta_2 = 0.5$ and $\gamma \le 0.54$.

In what follows we analyze the gradient-like property of (8.81) in terms of Theorem 8.14. If parameters $\beta_1, \alpha_2, \beta_2$ are chosen as the same in [Leonov et al. (1996)], that is, $\beta_1 = 0.2, \alpha_2 = 5, \beta_2 = 0.5$, one implies that system (8.81) is gradient-like for $288.96 \le \alpha_1 \le 310, \gamma \le 0.468$ by using Theorem 8.14.

In fact, system (8.81) can be rewritten as the form of (8.74) if α_1 is considered as an uncertain parameter, where

$$A = \begin{pmatrix} 0 & \alpha \\ 1 & \alpha_0\alpha \end{pmatrix}, \quad b = \begin{pmatrix} \alpha(1-\beta_2) \\ \alpha_0\alpha(\beta_1-\beta_2) \end{pmatrix}, \quad H_1 = H_2 = \begin{pmatrix} 0 \\ m\alpha \end{pmatrix}, \quad (8.82)$$

$$E_1 = (0 \quad 1), \quad E_2 = \beta_1 - \beta_2, \quad F = \frac{\alpha_1 - \alpha_0}{m} \quad (8.83)$$

$\alpha = -\dfrac{1}{\alpha_2}$, α_0 is a given positive number and $m > 0$ is an adjustable real number. Choosing parameters $\alpha_0 = 300, \beta_1 = 0.2, \alpha_2 = 5$ and $\beta_2 = 0.5$. Let $m = 0.08$. LMI (8.77) is feasible, $\delta = 19, \varepsilon = 130, \kappa = 150, \lambda = 0.3967$ and $P = \begin{pmatrix} 5.8644 & -0.0449 \\ -0.0449 & 4.2837 \end{pmatrix}$ is a set of solution of (8.77). The condition (iv) of Theorem 8.14 is equivalent to

$$\frac{0.5\pi\gamma}{\gamma\arcsin\gamma + \sqrt{1-\gamma^2}} < \frac{2\sqrt{\varepsilon\delta}}{\kappa} = 0.6627$$

8.7. Gradient-like property of systems with norm bounded uncertainties

which implies that $\gamma \leq 0.468$. Furthermore, it is easy to verify that other conditions of Theorem 8.14 are satisfied. Therefore by Theorem 8.14 the system (8.81) is gradient-like for $|\alpha_1 - \alpha_0| \leq m$, i.e., $299.92 \leq \alpha_1 \leq 300.08$. In order to enlarge the range of α_1 we propose the following algorithm in terms of Theorem 8.14.

Algorithm 8.1.

(i) For a given $\alpha_0 = \alpha_0^{(1)}, \alpha_2, \beta_1, \beta_2$, choosing a positive m as large as possible, solve LMI (8.77).

(ii) If LMI (8.77) is feasible, test the conditions (ii)-(iii) of Theorem 8.14 and determine the range of γ in terms of (iv) and $\delta, \varepsilon, \kappa$ obtained by solving LMI (8.77). If the conditions above are satisfied, go to step (iii). If not, choose a smaller $m > 0$ and return to step (i).

(iii) Enlarging the range of α_1 from the right hand of α_0. Let $\alpha_0 = \alpha_0^{(1)} + m$, test the conditions (ii)-(iii) of Theorem 8.14 and show that M is a negative definite matrix, where M is the matrix on the left of inequality (8.77) in which $\delta, \varepsilon, \kappa, \lambda, P$ is a set of solution of LMI (8.77) obtained by step (i) and $A, b, c, H_1, E_1, H_2, E_2$ are defined by (8.82) and (8.83).

(iv) If the conditions of step (iii) hold, let $\alpha_0 = \alpha_0^{(1)} + 2m$, return step (iii). Repeat this procedure until the conditions of step (iii) are satisfied for $\alpha_0 = \alpha_0^{(1)} + jm$ but are not satisfied for $\alpha_0 = \alpha_0^{(1)} + (j+1)m$, where j is a positive integer. Therefore the range of α_1 can be enlarged to $\alpha_1 \leq \alpha_0^{(1)} + jm$ from the right hand.

(v) Enlarging the range of α_1 from the left hand of α_0. Let $\alpha_0 = \alpha_0^{(1)} - m$ and $\alpha_0 = \alpha_0^{(1)} - km$ in step (iii) and step (iv) respectively, where k is a positive integer. By the same methods of step (iii) and (iv), the range of α_1 can be enlarged to $\alpha_0^{(1)} - km \leq \alpha_1$ from the left hand.

Finally, the range of α_1 can be enlarged to $\alpha_0^{(1)} - km \leq \alpha_1 \leq \alpha_0^{(1)} + jm$.

In this example, we note that $\bar{A}$ is a stable matrix and the system (8.81) is non-degenerate, i.e., $(\bar{A}, b)$ is controllable and $(\bar{A}, c)$ is observable for $\alpha_1 > 0, \alpha_2 > 0, \beta_1 > 0, \beta_2 > 0$ and $\beta_1 < \beta_2 < 1$. Therefore, the conditions (ii) and (iii) of Theorem 8.14 need not be tested. In this case, the range of parameter α_1 can be enlarged to $288.96 \leq \alpha_1 \leq 310$.

Remark 8.19. Using the same way, we can also discuss the gradient-like property of this system when α_2 or β_1 is considered as a uncertain parameter.

The analysis property for the gradient-like behavior given in Theorem 8.14 can be used to discuss the corresponding control problems.

Consider a nonlinear control system given by

$$\begin{cases} \dot{x} = (A + \Delta A)x + (b + \Delta b)\varphi(\sigma) + b_1 u, \\ \dot{\sigma} = cx + \rho\varphi(\sigma), \end{cases} \quad (8.84)$$

where $A \in \mathbb{R}^{n \times n}$ is a constant real matrix, $b \in \mathbb{R}^n$ and $c^T \in \mathbb{R}^n$ are real vectors and ρ is a number. $x \in \mathbb{R}^n, \sigma \in \mathbb{R}$. The parameter uncertainties $\Delta A \in \mathbb{R}^{n \times n}$ and $\Delta b \in \mathbb{R}^{n \times 1}$ can be expressed as the form of (8.75). The nonlinear function $\varphi : \mathbb{R} \to \mathbb{R}$ is T-periodic, satisfies a local Lipschitz condition and possesses a finite number of zeros on $[0, \Delta)$.

In what follows, we design a state feedback $u = Kx$ such that the closed-loop feedback system:

$$\begin{cases} \dot{x} = (A + b_1 K + \Delta A)x + (b + \Delta b)\varphi(\sigma), \\ \dot{\sigma} = cx + \rho\varphi(\sigma) \end{cases} \quad (8.85)$$

is gradient-like. Applying Theorem 8.14 to the closed-loop system (8.85) and let $Q = P^{-1}$ and $KQ = Y$, combining with Schur complement it is not difficult to obtain the following results based on matrix inequality.

Theorem 8.15. *Suppose that there exist reals numbers $\delta > 0, \varepsilon > 0, \nu > 0$, κ and matrices $Q > 0, Y$ such that*

$$\begin{pmatrix} AQ + QA^T + b_1 Y + Y^T b_1^T + W & \mu Qc^T + b & 0 & Qc^T & QE_1^T \\ \mu cQ + b^T & h & E_2^T & 0 & 0 \\ 0 & E_2 & -\nu I & 0 & 0 \\ cQ & 0 & 0 & -\dfrac{1}{\varepsilon}I & 0 \\ E_1 Q & 0 & 0 & 0 & -\nu I \end{pmatrix} < 0$$

where $W = \nu H_1 H_1^T + \nu H_2 H_2^T$, $\mu = \dfrac{\kappa}{2} + \varepsilon\rho$, $h = \delta + \varepsilon\rho^2 + \kappa\rho$.

Then there exists a control matrix $K = YQ^{-1}$. Furthermore, if

(i) $(A + b_1 K, b)$ is controllable and $(A + b_1 K, c)$ is observable,
(ii)

$$\max\{m^{\frac{1}{2}}||H_1||_1||E_1||_1, m^{\frac{1}{2}}||H_2||_1||E_2||_1\} < \frac{1}{||Q^{-1}(A + b_1 K, b)||_1}$$

and

$$p||H_1||_\infty ||E_1||_\infty < \frac{1}{||H^{-1}(A + b_1 K, c)||_1},$$

where $Q(A + b_1 K, b)$ and $H(A + b_1 k, c)$ are defined by (1.5) and (1.6) while A is substituted by $A + b_1 K$,

(iii) the condition (ii) of Lemma 8.13 holds, and
(iv) $c^T(A + b_1k + H_1FE_1)^{-1}(b + H_2FE_2) - \rho \neq 0$,

then the closed-loop nonlinear system (8.85) is gradient-like.

8.8 Notes and references

The nonlinear feedback systems studied in this chapter include two classes of systems. One class contains the systems with finite equilibria while the nonlinear function has a bounded derivative such as Chua's circuit, see Section 8.3 and [Wang et al. (2006c)]. The other class contains the pendulum-like systems with periodic nonlinearities and multiple equilibria such as phase-locked loops and synchronous machines in engineering. Dichotomy, Lagrange stability and the gradient-like property are important global properties of solutions for pendulum-like systems. The results on analysis and control for these global properties apply to both certain systems and uncertain systems. A series of frequency-domain inequalities conditions guaranteeing various global properties were given in [Leonov et al. (1992a,b, 1996); Rasvan (1998)] and in Chapter 5 of this book. These results are the basis for this chapter. The problems of controller design for Lagrange stability and dichotomy for the pendulum-like systems without uncertainties were considered in [Wang et al. (2004); Wang et al. (2006a)], see also Chapter 6. The robust analysis and robust control on the global properties of solutions for nonlinear feedback systems with uncertainties were discussed in [Wang et al. (2006b, 2007); Yang and Huang (2003); Yang et al. (2004); Ao et al. (2008)]. Lagrange stabilizability for uncertainty-free systems and the systems with norm bounded uncertainty in the linear part within the framework of the H_∞ control were presented in [Yang and Huang (2003)]. The results of robust dichotomy for Lur'e system with structured uncertainties were given in [Ao et al. (2008)]. For results different from this chapter, and based on LMI for the gradient-like property of nonlinear feedback systems with norm bounded uncertainty can be found in [Yang et al. (2004)]. Parts of the results in this chapter are adapted from [Wang et al. (2006b, 2007)] in which the robust analysis and robust control on dichotomy, Lagrange stability and the gradient-like property for nonlinear feedback systems with addictive, multiplicatice and H_∞ uncertainties were discussed.

Chapter 9

Control of Periodic Oscillations in Nonlinear Systems

This chapter is devoted to presenting robust analysis and synthesis results on guaranteeing the nonexistence of periodic solutions in nonlinear Lur'e systems. Particularly, a class of periodic oscillations, which are qualitatively different from auto-oscillation are introduced and the control design problems for the absence of such kind of oscillations in a certain frequency range in pendulum-like systems are discussed. In synchronization systems, cycle slipping is a frequently observed nonlinear phenomenon. It is also one of the most significant attributes in the transient mode of nonlinear phase controlled systems. A time domain method is proposed for the estimation of numbers of cycles slipped before the convergence of a solution, which is of great importance to the transient performance and synchronization of the system.

9.1 Periodic solutions in systems with cylindrical phase space

One of the important properties that nonlinear systems differentiate from linear systems is that nonlinear systems may have isolated periodic solutions. Periodic oscillation is a more complicated dynamic process than equilibrium position, which has wide applications in modern mechanisms, aerospace and communication engineering. A main issue of the nonlinear oscillation theory is the existence and stability of periodic solutions. Generally speaking, there are three kinds of periodic oscillations in nonlinear systems, auto-oscillations (i.e. limit cycles, also called cycles of the first kind), cycles of the second kind and forced oscillations. There are numerous work on the analysis of auto-oscillations and forced oscillations, and most researches are concerned with low dimensional systems. As an important dynamic property of nonlinear systems, cycles of the second kind has received less attention. This nonlinear effect can be observed in dynamical systems with cylindrical phase surface. Note that a broad class of differential equations describing different pendulums, systems of synchronization, Josephson junctions can be written in the form

$$\dot{x} = f(x) \tag{9.1}$$

with the vector field $f : \mathbb{R}^n \to \mathbb{R}^n$ having the property $f(x) = f(x+d)$ for every $d \in \Gamma$ and $x \in \mathbb{R}^n$, where Γ is a subgroup of $\mathbb{R}^n$ defined by $\Gamma := \{\sum_{i=1}^m k_i d_i | k_i \in \mathbb{Z}\}, m \le n$ and $\{d_i\}_{i=1}^m$ are linearly independent vectors in $\mathbb{R}^n$. Such systems have a cylindrical phase space and are also called cylinder system in the oscillating theory of applied sciences. The simplest example of systems with cylindrical phase space is the equation of mathematical pendulum in the form of a mass point suspended on an inextensible weightless thread to a fixed point, the motion of which can be represented by the equation

$$\ddot{\theta} + a\dot{\theta} + b\sin\theta = L.$$

The equation can be written into the form

$$\dot{\theta} = \eta, \dot{\eta} = -a\eta - b\sin\theta + L. \tag{9.2}$$

It is noted that if $(\theta(t), \eta(t))$ is a solution of system (9.2), then for any integer k the function $(\theta(t) + 2k\pi, \eta(t))$ is also a solution of the system. The natural requirement on the phase space of a mathematical model of a real system is that to every physical state of the system there should correspond one and only one point of the space. Since in a plane (θ, η) there are infinitely many points corresponding to the same physical state of a pendulum, the plane (θ, η) cannot be used for such a phase space of system (9.2). For ensuring the requirement of uniqueness, a cylindrical phase space $\{(\theta \bmod 2\pi, \eta)\}$ is introduced. The trajectories of system (9.2) can be represented on the phase cylinder $\mathbb{C} \times \mathbb{R}^1$ by piling up a phase portrait of trajectories in the covering space $\mathbb{R}^2 = \{(\theta, \eta)\}$ on it. It is important to note that in cylindrical phase space there are two kinds of closed paths. Closed path of the first kind encircles an equilibrium point and closes on itself without making a full net rotation around the cylinder, see Figure 9.1 (a). Closed path of the second kind does not encircle any equilibrium point and makes one or more full net rotations about the cylinder before closing on itself, see Figure 9.1 (b). A cycle of the first kind in cylindrical phase space is topologically equivalent to that in covering phase space, see Figure 9.2 (a), but this is not the case for cycles of the second kind. A cycle of the second kind remains closed in cylindrical phase space but lose the property of closure under transition to covering phase space, see Figure 9.2 (b). In the engineering practice, those two types of cycles correspond to different physical motions. A cycle of the first kind corresponds to the pendulum motion of periodic undamped oscillations around an equilibrium when $a = 0$, and a cycle of the second kind corresponds to circular motions around the point of suspension with periodically repeated instantaneous angular velocity when $a > 0$. Another example of this class of systems is phase-locked loops (PLLs). A cycle of the first kind corresponds to wobbly synchronization of a PLL while a cycle of the second kind corresponds to asynchronized operation of a PLL, which is an undesirable regime of the acting of the system.

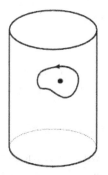 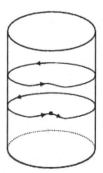

(a) Cycle of the first kind.   (b) Cyclee of the second kind.

Fig. 9.1 Two kinds of cycles on the phase cylinder.

A class of pendulum-like systems can be described by the following Lur'e-Postnikov form

$$\dot{x} = Ax + B\varphi(\sigma)$$
$$\dot{\sigma} = C^T x + D\varphi(\sigma)$$
(9.3)

where $A \in \mathbb{R}^{n \times n}, B, C \in \mathbb{R}^{n \times m}, D \in \mathbb{R}^{m \times m}, x = (x_1, \cdots, x_n)^T, \sigma = (\sigma_1, \cdots, \sigma_m)^T$ and $\varphi(\sigma) = (\varphi_1(\sigma_1), \cdots, \varphi_m(\sigma_m))^T$. Note that for system (9.3), it is possible to introduce a cylindrical phase space $\{(\sigma_1 \bmod \Delta, \cdots, \sigma_m \bmod \Delta, x_1, \cdots, x_n)\}$ where there can exist closed trajectories that lose the property of closure under transition to the covering space $\mathbb{R}^{n+m}$. In the following we formulate definitions of these two kinds of cycles for system (9.3).

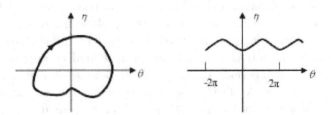

(a) Cycle of the first kind.   (b) Cycle of the second kind.

Fig. 9.2   Two kinds of cycles on the phase (θ, η).

Definition 9.1. The solution $(x(t), \sigma(t))$ of (9.3) is called the cycle of the first kind if there exist a number $T \neq 0$ such that

$$x(t+T) = x(t), \quad \sigma_k(t+T) = \sigma_k(t).$$

Definition 9.2. The solution $(x(t), \sigma(t))$ of (9.3) is called the cycle of the second kind if there exist a number $T \neq 0$ and integers $j_k, k = 1, 2, \cdots, m$ such that

$$x(t+T) = x(t), \quad \sigma_k(t+T) = \sigma_k(t) + j_k \Delta.$$

Previous work for oscillations in systems with cylindrical phase space concentrates on two or three dimensional systems. However, a rough idealization and the complication of practical systems require the investigation of systems of higher dimension. In the classical nonlinear theory of oscillation, the main methods for oscillation analysis in high dimensional systems are harmonic balance method and various asymptotic methods, including averaging method and method of slowly varying energy, etc. Although these methods are very effective in practice, generally they are not rigorous in mathematics. A rigorous mathematical method solving this problem was stated in [Leonov (1974, 1976, 1985); Leonov et al. (1992a,b)], where the Lyapunov direct method was used together with the construction of various comparison systems. In this way frequency domain criteria for the existence and the nonexistence of various types of periodic solutions were established and the results for two dimensional systems were extended to multidimensional systems.

From the control theory point of view, it is difficult to analyze the periodic oscillations based on frequency domain conditions when systems are subject to parameter uncertainties or external perturbations, as well as design controllers guaranteeing the existence or the nonexistence of various types of cycles. In the following sections, a time-domain framework is proposed and the robust analysis and synthesis results on the nonexistence of oscillations for nonlinear Lur'e systems are obtained based on the linear matrix inequality method.

9.2 Nonexistence of periodic solutions in Lur'e systems

Periodic solutions or limit cycles exemplify the complex dynamic behavior of certain nonlinear systems. In most practical dynamical systems, such property should be suppressed because it may lead to instability. If there exist periodic oscillations in communication systems or electronic devices, it will shorten the equipment life and result in mistriggering in digital circuits [Lindsey (1972); Fridman et al. (1997)]. Therefore, the research on the nonexistence of periodic solutions in nonlinear systems is certainly significant.

9.2.1 LMI-based conditions for nonexistence of periodic solutions

Consider the following nonlinear system

$$\dot{x} = Ax + B\varphi(\sigma)$$
$$\sigma = C^T x \qquad (9.4)$$

where $A \in \mathbb{R}^{n \times n}$ has no eigenvalues on the imaginary axis, $B, C \in \mathbb{R}^n$, and $\varphi : \mathbb{R} \to \mathbb{R}$ is a continuous function satisfying a Lipschitz condition with

$$\mu_1 \sigma^2 \leq \varphi(\sigma)\sigma \leq \mu_2 \sigma^2. \qquad (9.5)$$

The transfer function of the linear part of (9.4) from the input φ to the output $-\sigma$ is given by

$$G(s) = C^T(A - sI)^{-1}B.$$

Definition 9.3. The solution $x(t, x_0)$ of the system (9.4) is called a periodic solution if the following relations are satisfied for some time $\tau > 0$

$$x(\tau, x_0) - x_0 = 0, \quad \sigma(\tau, x_0) - \sigma_0 = 0,$$

where $\sigma_0 = C^T x_0$.

The following lemma from [Leonov et al. (1992b)] gives the frequency domain conditions of the nonexistence of periodic solutions with frequency $\omega > \omega_0$.

Lemma 9.1. *Suppose that there exists a number ν, such that the following requirement is satisfied for $\forall \omega > \omega_0$:*

$$\mathbf{Re}\{[1+\mu_1 G(j\omega)]^*[1+\mu_2 G(j\omega)]\} + \nu\,\mathbf{Re}\{j\omega K(j\omega)\} > 0. \qquad (9.6)$$

Then, system (9.4) has no periodic solutions with frequency $\omega > \omega_0$.

The following theorem gives the LMI-based conditions for the nonexistence of periodic solutions in a certain frequency range in system (9.4).

Theorem 9.1. *System (9.4) has no periodic solutions with frequency $\omega > \omega_0$ if one of the following requirements is satisfied.*

(i) There exist matrices $P = P^T, Q > 0$ and number ν such that

$$\begin{pmatrix} PA + A^T P + A^T QA - \omega_0^2 Q - \mu_1\mu_2 CC^T & A^T QB + \Gamma \\ B^T QA + \Gamma^T & B^T QB + r \end{pmatrix} < 0, \qquad (9.7)$$

where

$$\Gamma = PB + \frac{1}{2}(\mu_1+\mu_2)C + \frac{1}{2}\nu A^T C, \quad r = \nu\,\mathbf{Re}\{C^T B\} - 1. \qquad (9.8)$$

(ii) There exist matrices $P = P^T, Q > 0, G, F$, and number ν such that

$$\begin{pmatrix} -(G+G^T) & GA+P-F^T & GB & 0 \\ A^T G^T + P - F & \Lambda_{11} & \Lambda_{12} & A^T Q \\ B^T G^T & \Lambda_{12}^T & r & B^T Q \\ 0 & QA & QB & -Q \end{pmatrix} < 0, \qquad (9.9)$$

where

$$\Lambda_{11} = FA + A^T F^T - \mu_1\mu_2 CC^T - \omega_0^2 Q,$$

$$\Lambda_{12} = FB + \frac{1}{2}(\mu_1+\mu_2)C + \frac{1}{2}\nu A^T C.$$

Proof. Through simple computation, the frequency inequality (9.6) can be written as follows

$$\begin{pmatrix} (j\omega I - A)^{-1}B \\ I \end{pmatrix}^* \Theta \begin{pmatrix} (j\omega I - A)^{-1}B \\ I \end{pmatrix} < 0 \qquad (9.10)$$

where

$$\Theta = \begin{pmatrix} -\mu_1\mu_2 CC^T & \frac{1}{2}(\mu_1+\mu_2)C + \frac{1}{2}\nu A^T C \\ \frac{1}{2}(\mu_1+\mu_2)C^T + \frac{1}{2}\nu C^T A & \nu\,\mathbf{Re}\{C^T B\} - 1 \end{pmatrix}.$$

9.2. Nonexistence of periodic solutions in Lur'e systems

In virtue of the generalized KYP Lemma in Chapter 1, it can be proved that (9.7) is equivalent to (9.6). Using Schur complement, (9.9) is equivalent to

$$\begin{pmatrix} -(G+G^T) & GA+P-F^T & GB \\ A^TG^T+P-F & \Lambda_{11}+A^TQA & \Lambda_{12}+A^TQB \\ B^TG^T & \Lambda_{12}^T+B^TQA & \nu\,\mathbf{Re}\{C^TB\}-1+B^TQB \end{pmatrix} < 0.$$
(9.11)

By the projection lemma, it can be easily verified that (9.11) is feasible for G if and only if (9.7) is feasible for P. □

9.2.2 Robustness analysis

Assume that the system matrices A and B of (9.4) belong to the polytopic uncertainty domain Ω defined by

$$\Omega = \{[A,B] : [A,B] = \sum_{i=1}^{n} \alpha_i [A_i, B_i], \sum_{i=1}^{n} \alpha_i = 1, \alpha_i \geq 0\}.$$
(9.12)

Then we state the following theorem which guarantees the nonexistence of periodic solutions for system (9.4) with uncertainties (9.12).

Theorem 9.2. *Suppose that there exist* $P_i = P_i^T (i = 1, 2, ..., n), Q > 0$, G, F, *and number* ν *such that*

$$\begin{pmatrix} -(G+G^T) & GA_i+P_i-F^T & GB_i & 0 \\ A_i^TG^T+P_i-F & \Lambda_{11}^i & \Lambda_{12}^i & A_i^TQ \\ B_i^TG^T & (\Lambda_{12}^i)^T & \nu\,\mathbf{Re}\{C^TB_i\}-1 & B_i^TQ \\ 0 & QA_i & QB_i & -Q \end{pmatrix} < 0, \quad (9.13)$$

where

$$\Lambda_{11}^i = FA_i + A_i^T F^T - \mu_1\mu_2 CC^T - \omega_0^2 Q,$$

$$\Lambda_{12}^i = FB_i + \frac{1}{2}(\mu_1+\mu_2)C + \frac{1}{2}\nu A_i^T C.$$

Then system (9.4) with uncertainties (9.12) has no periodic solutions with frequency $\omega > \omega_0$.

Remark 9.1. It is shown that the inequality condition (9.13) involves only the vertices of the polytope domain. The LMI characterization provides a kind of decoupling between the Lyapunov matrix and the system matrices which makes it less conservative than single Lyapunov function method.

It is noted that when $\omega_0 = 0$, system (9.37) has no periodic solutions in the whole frequency ranges. The following corollary is a special case of Theorem 9.1 with $\nu = 0$, which enables one using the result of Theorem 9.1 to deal with uncertainties in all system matrices and design feedback controllers to ensure the nonexistence of periodic solutions for the system.

Corollary 9.1. *Suppose there exist matrices* $P = P^T, G, F$ *such that*

$$\begin{pmatrix} -G - G^T & GA + P - F^T & GB \\ A^T G^T + P - F & FA + A^T F^T - \mu_1 \mu_2 CC^T & \Lambda_{12} \\ B^T G^T & \Lambda_{12}^T & -1 \end{pmatrix} < 0, \quad (9.14)$$

where Λ_{12} *is as in Theorem 9.1. Then system (9.4) has no periodic solutions.*

9.2.3 Robust synthesis

It is usually difficult to design a static output feedback controller using the convex optimization method due to the variables coupling. Here we consider the design of a dynamic output feedback controller guaranteeing the nonexistence of periodic solutions for the nonlinear control system

$$\begin{aligned} \dot{x} &= Ax + B\varphi(u) \\ w &= C^T x \end{aligned} \quad (9.15)$$

where x is the system state, u is the control input and w is the system output. Assuming that the system matrices lie in the polytope

$$\Omega = \{[A, B, C] : [A, B, C] = \sum_{i=1}^{n} \alpha_i [A^i, B^i, C^i], \sum_{i=1}^{n} \alpha_i = 1, \alpha_i \geq 0\}, \quad (9.16)$$

and the controller G_k to be designed is

$$\begin{aligned} \dot{x}_k &= A_k x_k + B_k w \\ u &= C_k x_k + D_k w \end{aligned} \quad (9.17)$$

then the closed-loop nonlinear feedback system can be described by

$$\begin{aligned} \dot{x}_{cl} &= A_{cl} x_{cl} + B_{cl} \varphi(u) \\ u &= C_{cl}^T x_{cl}, \end{aligned} \quad (9.18)$$

with

$$A_{cl} = \begin{pmatrix} A & 0 \\ B_k C^T & A_k \end{pmatrix}, B_{cl} = \begin{pmatrix} B \\ 0 \end{pmatrix}, C_{cl} = \begin{pmatrix} CD_k^T \\ C_k^T \end{pmatrix}.$$

9.2. Nonexistence of periodic solutions in Lur'e systems   215

We obtain the following theorem which establishes the nonexistence conditions of periodic solutions for the closed-loop system (9.18).

Theorem 9.3. Given the system (9.15) with uncertainties (9.16) and $\mu_1\mu_2 < 0$, there exists a controller $G_k(s)$ such that the closed-loop system (9.18) has no periodic solutions if there are matrices $P_{11}^i, P_{12}^i, P_{22}^i, i = 1, 2, ..., n, G_{11}, G_{21}, G_2, F_{11}, F_{21}, T_A, T_B, T_C, T_D$, and numbers ρ_1, ρ_2 such that the following inequalities are satisfied for $i = 1, 2, \cdots, n$

$$\begin{pmatrix} -(G_{11} + G_{11}^T) & -(G_2 + G_{21}^T) & \Pi_{11}^i & \Pi_{12}^i & G_{11}B^i & 0 \\ * & -(G_2 + G_2^T) & \Pi_{21}^i & \Pi_{22}^i & G_{22}B^i & 0 \\ * & * & \Pi_{31}^i & \Pi_{32}^i & \Pi_{33}^i & C^i T_D^T \\ * & * & * & \rho_2(T_A + T_A^T) & \Pi_{43}^i & T_C^T \\ * & * & * & * & -1 & 0 \\ * & * & * & * & * & \mu_1^{-1}\mu_2^{-1} \end{pmatrix} < 0,$$
(9.19)

where

$$\Pi_{11}^i = G_{11}A^i + T_B(C^i)^T + P_{11}^i - F_{11}^T,$$

$$\Pi_{12}^i = T_A + P_{12}^i - F_{21}^T,$$

$$\Pi_{21}^i = G_{21}A^i + T_B(C^i)^T + P_{12}^i - \rho_1 G_2^T,$$

$$\Pi_{22}^i = T_A + P_{22}^i - \rho_2 G_2^T,$$

$$\Pi_{31}^i = F_{11}A^i + (A^i)^T F_{11}^T + \rho_1(T_B(C^i)^T + C^i T_B^T),$$

$$\Pi_{32}^i = \rho_1 T_A + (A^i)^T F_{21}^T + \rho_2 C^i T_B^T,$$

$$\Pi_{33}^i = F_{11}B^i + \frac{1}{2}(\mu_1 + \mu_2)C^i T_D^T,$$

$$\Pi_{43}^i = F_{21}B^i + \frac{1}{2}(\mu_1 + \mu_2)T_C^T.$$

Then, from T_A, T_B, T_C, T_D, we have the controller parameters:

$$A_k = G_2^{-1}T_A, B_k = G_2^{-1}T_B, C_k = T_C, D_k = T_D.$$

Proof. Exploiting Corollary 9.1, the closed-loop system (9.18) has no periodic solutions if there exist matrices $P = P^T, G, F$, such that the following inequality is satisfied

$$\begin{pmatrix} -(G + G^T) & GA_{cl} + P - F^T & GB_{cl} \\ A_{cl}^T G^T + P - F & FA_{cl} + A_{cl}^T F^T - \mu_1\mu_2 C_{cl}^T C_{cl} & FB_{cl} + \frac{1}{2}(\mu_1 + \mu_2)C_{cl} \\ B_{cl}^T G^T & B_{cl}^T F^T + \frac{1}{2}(\mu_1 + \mu_2)C_{cl}^T & -1 \end{pmatrix} < 0.$$
(9.20)

From the above inequality, $-(G + G^T) < 0$ implies that G is nonsingular. We need partitions of G, F and P in the form

$$G = \begin{pmatrix} G_{11} & G_{12} \\ G_{21} & G_{22} \end{pmatrix}, F = \begin{pmatrix} F_{11} & F_{12} \\ F_{21} & F_{22} \end{pmatrix}, P = \begin{pmatrix} P_{11} & P_{12} \\ P_{12}^T & P_{22} \end{pmatrix}.$$

Without loss of generality, we can assume that

$$G = \begin{pmatrix} G_{11} & G_2 \\ G_{21} & G_2 \end{pmatrix}, F = \begin{pmatrix} F_{11} & \rho_1 G_2 \\ F_{21} & \rho_2 G_2 \end{pmatrix}$$

where ρ_1, ρ_2 are numbers to be searched. Performing the linearization transformations

$$T_A = G_2 A_k, T_B = G_2 B_k, T_C = C_k, T_D = D_k,$$

then the inequality (9.19) is testified by putting the above matrices G, F, P and A_{cl}, B_{cl}, C_{cl} into (9.20). □

Example 9.1. A dynamic output feedback controller is designed to guarantee the absence of periodic oscillations for Chua's circuit shown in Figure 9.3. The differential equation of the Chua's system can be written into

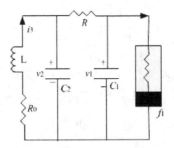

Fig. 9.3 Chua's Circuit.

the following form

$$\frac{dx_1}{d\tau} = \alpha(x_2 - x_1 - \phi(x_1))$$

$$\frac{dx_2}{d\tau} = x_1 - x_2 + x_3 \qquad (9.21)$$

$$\frac{dx_3}{d\tau} = -\beta x_2 - \gamma x_3$$

with

$$\phi(x_1) = m_2 x_1 + \frac{1}{2}(m_1 - m_2)[|x_1 + 1| - |x_1 - 1|].$$

9.2. Nonexistence of periodic solutions in Lur'e systems

Given $\alpha, \beta, \gamma, m_1, m_2$ as

$$\alpha = -1.4151, \beta = -0.0256, \gamma = 0.0297,$$

$$m_1 = 0.2290817, m_2 = -2.4467822,$$

and let

$$\mu_1 = \min\{m_1, m_2\}, \mu_2 = \max\{m_1, m_2\},$$

it is easy to test that the LMI condition in Theorem 9.1 for (9.21) is broken. Such Chua's circuit has periodic solutions which are shown in Figure 9.4 with initial value $x(0) = (1, -1, 0.8)^T$.

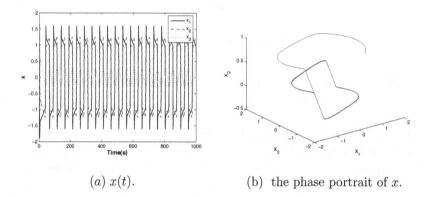

(a) $x(t)$. (b) the phase portrait of x.

Fig. 9.4 Solutions of Chua's circuit (9.21).

Consider the system matrices

$$A = \begin{pmatrix} -(\alpha+a) & \alpha+a & 0 \\ 1 & -1 & 1 \\ 0 & -(\beta+b) & -(\gamma+c) \end{pmatrix}, B = \begin{pmatrix} -(\alpha+a) \\ 0 \\ 0 \end{pmatrix}, C = \begin{pmatrix} 1 \\ 0 \\ 0 \end{pmatrix}$$

where a, b, c are uncertain parameters. The uncertainty polytope includes 8 vertices. In virtue of Theorem 9.3, we obtain the output feedback controller:

$$G_k(s) = \left(\begin{array}{c|c} A_k & B_k \\ \hline C_k & D_k \end{array}\right) = \left(\begin{array}{ccc|c} -3.8171 & -0.8830 & 0.0201 & -4.4570 \\ -0.4637 & -0.8981 & 0.3978 & -1.3089 \\ -0.2826 & 0.1987 & -0.6582 & -0.1412 \\ \hline 0.2008 & -0.3479 & -0.7587 & -0.4045 \end{array}\right)$$

for $|a| \leq 0.428, |b| \leq 0.0199, |c| \leq 0.0197$. Simulation results for one of the vertex systems are shown in Figure 9.5 with initial value $x(0) = (1, -1, 0.8)^T$.

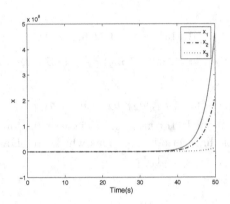

Fig. 9.5 Solutions of Chua's circuit with controller.

9.3 Nonexistence of cycles of the second kind in interconnected systems

A significant part of theories for large-scale systems is concerned with the problem of control design in a decentralized way. Decentralized control schemes present a practical and efficient means for designing control algorithms that utilize just the state of each subsystem without information from other subsystems. On the other hand, due to the characteristics of decentralized control method, interconnections among subsystems are not utilized sufficiently, and in some cases, interconnections are even viewed as disadvantages. Some of our recent work studied the effects of interconnections among independent subsystems and the results show that unstable subsystems can possibly generate a stable interconnected system through some effective interconnections [Duan et al. (2003, 2004c); Yang et al. (2006)].

Consider the following nonlinear feedback systems

$$\dot{x} = Ax + B\varphi(y)$$
$$\dot{\sigma} = C^T x + D\varphi(y)$$
(9.22)

where $A \in \mathbb{R}^{n \times n}, B, C \in \mathbb{R}^{n \times m}, D \in \mathbb{R}^{m \times m}, x = (x_1, \cdots, x_n)^T, \sigma = (\sigma_1, \cdots, \sigma_m)^T$ and $\varphi(y) = (\varphi_1(y_1), \cdots, \varphi_m(y_m))^T$ with $y_1 = a_{11}\sigma_1 + \cdots + a_{1m}\sigma_m, \cdots, y_m = a_{m1}\sigma_1 + \cdots + a_{mm}\sigma_m$. The matrix

$$\Xi \triangleq (a_{ij}), \; i,j = 1,2,\cdots,m \qquad (9.23)$$

defines a particular class of nonlinearly interconnected systems satisfying $y = \Xi\sigma$. The dimension of the nonlinear interconnection matrix Ξ is equal

9.3. Nonexistence of cycles of the second kind in interconnected systems

to the number of subsystems. The transfer function of the linear part of (9.22) from the input φ to the output $-\dot{\sigma}$ is given by

$$G(s) = C^T(A - sI)^{-1}B - D. \qquad (9.24)$$

The following assumptions are made on system (9.22).

Assumption 9.1. (A, B) is controllable, A has no eigenvalues on the imaginary axis, Ξ and $G(0)$ are nonsingular.

Assumption 9.2. $\varphi_i : \mathbb{R} \to \mathbb{R}$ is Δ periodic and satisfies

$$\varphi_i^2(\sigma) + [\varphi_i'(\sigma)]^2 \neq 0, \quad \forall \sigma \in \mathbb{R}.$$

Assumption 9.3. The derivative of $\varphi(\sigma)$ is Lipschitz continuous and there exist numbers $\mu_{1j}, \mu_{2j}, j = 1, 2, \cdots, m$ such that

$$\mu_{1j} \leq \frac{d\varphi_j(\sigma)}{d\sigma} \leq \mu_{2j}, \quad \forall \sigma \in \mathbb{R}. \qquad (9.25)$$

System (9.22) can be viewed as an interconnected system composed of m single input single output isolated subsystems through some linear and nonlinear interconnections. If we introduce the vector $d_j = (0, \cdots, 0, \Delta, 0, \cdots, 0)^T$, where Δ is the $(n+j)$-th component of d_j, then the vector field of system (9.22) is periodic with respect to $\Gamma = \{\sum_{j=1}^m k_j d_j, k_j \in \mathbb{Z}\}$. From the periodicity of φ_i we know $\mu_{1j}\mu_{2j} < 0$ and any equilibrium of system (9.22) satisfies

$$Ax_{eq} = -B\varphi(y_{eq}), \quad C^T x_{eq} = -D\varphi(y_{eq}), \quad y_{eq} = \Xi\sigma_{eq}$$

Consequently we have

$$(D - C^T A^{-1} B)\varphi(y_{eq}) = 0.$$

By the assumption above, we can get that $\varphi(y_{eq}) = 0$, i.e. $\varphi_j(y_{jeq}) = 0, j = 1, 2, \cdots, m$. It follows that $x_{eq} = 0$ and $\sigma_{eq} = \Xi^{-1}y_{eq}$. Since φ_j is Δ-periodic, system (9.22) has infinitely many isolated equilibria. The equilibria set of (9.22) is

$$\Lambda = \{(x_{eq}, \sigma_{eq}) \mid x_{eq} = 0, \sigma_{eq} = \Xi^{-1}y_{eq}, \varphi(y_{eq}) = 0\}.$$

In the following, we introduce a number of definitions and notations which will be used in the next subsection to derive our main results. Define

$$\nu_j = \int_0^\Delta \varphi_j(\sigma)d\sigma \Big/ \int_0^\Delta |\varphi_j(\sigma)|d\sigma, \quad j = 1, 2, \cdots, m. \qquad (9.26)$$

Denotes $\nu = \text{diag}(\nu_1, \nu_2, \cdots, \nu_m)$ and

$$\mu_1 = \text{diag}(\mu_{11}, \cdots, \mu_{1m}), \quad \mu_2 = \text{diag}(\mu_{21}, \cdots, \mu_{2m})$$

with the numbers μ_{1j} and μ_{2j} from (9.25). Then we introduce the function

$$F(y) = \varphi(y) - \nu|\varphi(y)| \qquad (9.27)$$

with $|\varphi(y)| = (|\varphi_1(y_1)|, \cdots, |\varphi_m(y_m)|)^T$.

9.3.1 Nonexistence of cycles of the second kind in interconnected systems

First, we consider the problem of the absence of cycles of the second kind in system (9.22).

Theorem 9.4. *Suppose there exist number $\omega_0 \geq 0$ and diagonal matrices $\epsilon > 0, \delta > 0, \tau \geq 0, \kappa$ such that the following requirements are satisfied*

(i) $4\epsilon\delta > (\nu\kappa)^2$;
(ii) $G^T(0)\Xi^T(\epsilon + \tau)\Xi G(0) - \mathrm{Re}\,\kappa\Xi G(0) + \delta \leq 0$;
(iii) $\mathrm{Re}\{\kappa\Xi G(j\omega) - [j\mu_1^{-1}\omega + \Xi G(j\omega)]^*\tau[\Xi G(j\omega) + j\mu_2^{-1}\omega]\} - G^*(j\omega)\Xi^T\epsilon\Xi G(j\omega) - \delta > 0, \forall \omega > \omega_0$.

Then system (9.22) has no cycle of the second kind with the frequency $\omega > \omega_0$. If $\omega_0 = 0$, the system has no cycles of the second kind.

Proof. Suppose that system (9.22) has a cycle of the second kind $(x(t), \sigma(t))$ and the period of $x(t)$ is $T = 2\pi/\omega$, where $\omega > \omega_0$. From the first equation of (9.22) and the T-periodicity of $x(t)$ it follows that $\varphi(y(t))$ is also a T-periodic function of t. Then $\varphi(y(t))$ can be expanded into a Fourier series that converges to the function for all $t \in (-\infty, +\infty)$

$$\varphi(y(t)) = \sum_{k=-\infty}^{k=+\infty} C_k e^{jk\omega t}. \qquad (9.28)$$

Therefore

$$\dot\varphi(y(t)) = \sum_{k=-\infty}^{k=+\infty} jk\omega C_k e^{jk\omega t}. \qquad (9.29)$$

From (9.24) we have

$$\dot\sigma(t) = -\sum_{k=-\infty}^{k=+\infty} G(jk\omega) C_k e^{jk\omega t}. \qquad (9.30)$$

Consider the functions
$J_1(t) = \dot\sigma^*(t)\Xi^T\epsilon\Xi\dot\sigma(t) + [\varphi(y(t)) - F(y(t))]^*\kappa\Xi\dot\sigma(t) + \varphi^*(y(t))\delta\varphi(y(t))$
$J_2(t) = -[\Xi\dot\sigma(t) - \mu_1^{-1}\dot\varphi(y(t))]^*\tau[\mu_2^{-1}\dot\varphi(y(t)) - \Xi\dot\sigma(t)]$

Let

$$J(t) = J_1(t) + J_2(t).$$

From (9.27) we have

$$J_1(t) = \dot\sigma^*(t)\Xi^T\epsilon\Xi\dot\sigma(t) + |\varphi^*(y(t))|\nu\kappa\Xi\dot\sigma(t) + \varphi^*(y(t))\delta\varphi(y(t)).$$

9.3. Nonexistence of cycles of the second kind in interconnected systems

So $J_1(t)$ is non-negative in virtue of hypothesis (i) of the theorem. In addition,

$$J_2(t) = \sum_{j=1}^{m} \tau_j \mu_{1j}^{-1} \mu_{2j}^{-1} [\dot{y}_j(t)]^2 [\varphi_j'(y_j(t)) - \mu_{1j}][\varphi_j'(y_j(t)) - \mu_{2j}].$$

Since $\varphi_j'(y) \in [\mu_{1j}, \mu_{2j}]$ and $\mu_{1j}\mu_{2j} < 0$ we conclude that $J_2(t) \geq 0$. As a result $J(t) \geq 0$ for $t \in \mathbb{R}$ and consequently

$$\int_0^T J(t)dt \geq 0, \qquad (9.31)$$

where T is the period of the function $x(t)$.

Let us calculate $\int_0^T J(t)dt$ and replace $\varphi(y(t)), \dot{\varphi}(y(t))$ and $\dot{\sigma}(t)$ by their expressions from (9.28),(9.29) and (9.30).

$$\int_0^T F_j(y_j(t))\dot{y}_j(t)dt = \int_{y_j(0)}^{y_j(T)} F_j(y_j)dy_j = 0, \quad j = 1, 2, \cdots, m.$$

Hence

$$\int_0^T F^*(y(t))\kappa\Xi\dot{\sigma}(t)dt = 0.$$

Denote $C_k = (c_{k1}, \cdots, c_{km})$, where $c_{ki}, i = 1, 2, \cdots, m$ is the column of C_k. Using the fact

$$c_{(-k)l} = c_{kl}^T \quad (k > 0, l = 1, 2, \cdots, m),$$

and the known relations

$$\int_0^T e^{j(k_1-k_2)\omega t}dt = \begin{cases} T & \text{for } k_1 = k_2 \\ 0 & \text{for } k_1 \neq k_2 \end{cases},$$

then we get

$$\int_0^T J(t)dt = \int_0^T J_1(t)dt + \int_0^T J_2(t)dt$$

$$= TC_0^T[G^T(0)\Xi^T(\epsilon + \tau)\Xi G(0) - \mathbf{Re}\{\kappa\Xi G(0)\} + \delta]C_0$$

$$+ 2T\sum_{k=1}^{\infty} C_k^T\{G^*(j\omega)\Xi^T\epsilon\Xi G(j\omega) - \mathbf{Re}\{\kappa\Xi G(j\omega) - (\Xi G(jk\omega)$$

$$+ \mu_1^{-1}jk\omega)^*\tau(\Xi G(jk\omega) + \mu_2^{-1}jk\omega)\} + \delta\}C_k.$$

So if we assume that system (9.22) has a cycle of the second kind with the frequency $\omega > \omega_0$ then according to the hypotheses (ii) and (iii) of the theorem

$$\int_0^T J(t) < 0.$$

This contradicts inequality (9.31). Consequently, there exists no cycle of the second kind which has the frequency greater than ω_0. Thus the theorem is proved. □

Remark 9.2. A special form of Ξ is $\Xi = I$ (the identity matrix), which indicates that the nonlinear function $\varphi(y)$ has decoupled components and every σ_i is only input to φ_i. Such decoupled nonlinearities are usually adopted in nonlinear Lur'e systems in order to derive the multivariable extensions of Popov criterion [Haddad and Bernstein (1993)]. In this case, (9.22) can be viewed as an interconnected system composed of single input single output subsystems through linear interconnections. Since the interconnection matrix Ξ characterizes the nonlinear interconnection structure of the interconnected system, when the interconnection structure of the system is allowed to be chosen arbitrarily, we can design the interconnection matrix Ξ to ensure the absence of cycles of the second kind in the entire system.

Remark 9.3. Another special form of Ξ is $m \times m$ permutation matrix got by exchanging the columns of the identity matrix I_m. Let T denote the permutation matrix and
$$(i_1, \cdots, i_m)^T \triangleq T(1, \cdots, m)^T,$$
then
$$\varphi(y) = (\varphi_1(\sigma_{i_1}), \cdots, \varphi_m(\sigma_{i_m}))^T.$$
In this case, the interconnection matrix $\Xi = T$ indicates a class of input and output intercross in the interconnected system.

Based on the KYP lemma, the conditions in Theorem 9.4 can be transformed into the following LMI-based criterion.

Theorem 9.5. *Suppose there exist number* $\omega_0 \geq 0, P = P^T$ *and diagonal matrices* $\epsilon > 0, \delta > 0, \tau \geq 0, \kappa$ *such that the following linear matrix inequalities holds*

$$\begin{pmatrix} 2\epsilon & \nu\kappa \\ \kappa\nu & 2\delta \end{pmatrix} > 0, \tag{9.32a}$$

$$\begin{pmatrix} \Pi_1(\epsilon, \tau) + A^T P + PA & \Pi_2(\kappa, \epsilon, \tau) + PB \\ \Pi_2^T(\kappa, \epsilon, \tau) + B^T P & \Pi_3(\delta, \kappa, \epsilon, \tau) + \Pi_4(\tau, \omega_0) \end{pmatrix} < 0, \tag{9.32b}$$

$$G^T(0)\Xi^T(\epsilon + \tau)\Xi G(0) - \mathbf{Re}\{\kappa \Xi G(0)\} + \delta \leq 0, \tag{9.32c}$$

where
$$\Pi_1(\epsilon, \tau) = C\Xi^T(\epsilon + \tau)\Xi C^T,$$
$$\Pi_2(\kappa, \epsilon, \tau) = \frac{1}{2}[C\Xi^T\kappa + A^T C\Xi^T \tau(\mu_1^{-1} - \mu_2^{-1})] + C\Xi^T(\epsilon + \tau)\Xi D,$$
$$\Pi_3(\delta, \kappa, \epsilon, \tau) = \delta + 2\mathbf{Re}\{(\mu_1^{-1} - \mu_2^{-1})\tau \Xi C^T B + \kappa \Xi D\} + 2D^T \Xi^T(\epsilon + \tau)\Xi D,$$
$$\Pi_4(\tau, \omega_0) = \mu_1^{-1}\tau\mu_2^{-1}\omega_0^2.$$

9.3. Nonexistence of cycles of the second kind in interconnected systems

Then system (9.22) has no cycle of the second kind with the frequency $\omega > \omega_0$. If $\omega_0 = 0$, the system has no cycles of the second kind.

Proof. The inequality in condition (iii) of Theorem 9.4 can be written into the form of

$$G^*(j\omega)\Xi^T(\epsilon + \tau)\Xi G(j\omega) + \delta + \mu_1^{-1}\tau\mu_2^{-1}\omega^2 -$$
$$\mathbf{Re}\{[j\omega G(j\omega)]^*\tau\mu_2^{-1} + \mu_1^{-1}\tau[j\omega G(j\omega)] + \kappa G(j\omega)\} < 0, \ \forall \omega > \omega_0.$$

In virtue of the fact $\mu_1^{-1}\tau\mu_2^{-1} \leq 0$, the above inequality holds if

$$G^*(j\omega)\Xi^*(\epsilon + \tau)\Xi G(j\omega) + \delta + \mu_1^{-1}\tau\mu_2^{-1}\omega_0^2 -$$
$$\mathbf{Re}\{[j\omega G(j\omega)]^*\tau\mu_2^{-1} + \mu_1^{-1}\tau[j\omega G(j\omega)] + \kappa G(j\omega)\} < 0, \ \forall \omega \in \mathbb{R} \cup \{\infty\}.$$
(9.33)

Since
$$sG(s) = C^T A(A - sI)^{-1}B - C^T B - sD,$$
then
$$\mathbf{Re}\{j\omega G(j\omega)\} = \mathbf{Re}\{C^T A(A - j\omega I)^{-1}B - C^T B\}.$$

Let
$$M = \begin{pmatrix} \Pi_1(\epsilon, \tau) & \Pi_2(\kappa, \epsilon, \tau) \\ \Pi_2^T(\kappa, \epsilon, \tau) & \Pi_3(\delta, \kappa, \epsilon, \tau) + \Pi_4(\tau, \omega_0) \end{pmatrix}.$$

According to the KYP Lemma, it can be proved that (9.33) holds if and only if

$$M + \begin{pmatrix} A^T P + PA & PB \\ B^T P & 0 \end{pmatrix} < 0.$$
□

From the above results, we can also derive the following nonexistence conditions based on the determination of the largest bound of the frequency of the cycles.

Corollary 9.2. *System (9.22) has no cycle of the second kind with the frequency $\omega > \sqrt{\lambda}$, where λ is the global minimum of the following generalized eigenvalue minimization problem with respect to the matrix variables $P = P^T, \kappa, \epsilon > 0, \delta > 0, \tau \geq 0$:*

$$\min \lambda$$

$$s.t. \begin{pmatrix} 2\epsilon & \nu\kappa \\ \kappa\nu & 2\delta \end{pmatrix} > 0,$$

$$\begin{pmatrix} \Pi_1(\epsilon, \tau) + A^T P + PA & \Pi_2(\kappa, \epsilon, \tau) + PB \\ \Pi_2^T(\kappa, \epsilon, \tau) + B^T P & \Pi_3(\delta, \kappa, \epsilon, \tau) + \lambda\mu_1^{-1}\tau\mu_2^{-1} \end{pmatrix} < 0,$$
(9.34)

$$G^T(0)\Xi^T(\epsilon + \tau)\Xi G(0) - \mathbf{Re}\{\kappa\Xi G(0)\} + \delta \leq 0,$$

where $\Pi_i(\cdot)$ are given in Theorem 9.5.

Corollary 9.3. *Suppose there exist number $\omega_0 \geq 0$ and diagonal matrices $\epsilon > 0, \delta > 0, \kappa$ satisfying the following conditions:*

(i) $4\epsilon\delta > (\nu\kappa)^2$;
(ii) $G^T(0)\Xi^T\epsilon\Xi G(0) - \mathbf{Re}\{\kappa\Xi G(0)\} + \delta \leq 0$;
(iii) $\mathbf{Re}\{\kappa\Xi G(j\omega)\} - G^*(j\omega)\Xi^T\epsilon\Xi G(j\omega) - \delta > 0, \forall \omega > \omega_0$.

Then system (9.22) has no cycle of the second kind with the frequency $\omega > \omega_0$. If $\omega_0 = 0$, the system has no cycles of the second kind.

Remark 9.4. Corollary 9.3 is a special case of Theorem 9.4 with $\tau = 0$. It is shown that by utilizing the slope restrictions on $\varphi(y)$, one can get a less conservative condition to ensure nonexistence of cycles of the second kind. But the conditions in Corollary 9.3 has a simpler form, so is easier to use.

9.3.2 Nonlinear interconnection design

In this subsection, we propose an LMI-based approach on designing the interconnection matrix Ξ to ensure the absence of cycles of the second kind in system (9.22). To simplify the design problem, first we give the following result which is deduced from Corollary 9.3.

Corollary 9.4. *If there exist a matrix Ξ and diagonal matrices $\epsilon > 0, \delta > 0$ such that*

(i) $4\epsilon\delta > \nu^2$,
(ii) $\mathbf{Re}\{\Xi G(j\omega)\} + G^*(j\omega)\Xi^T\epsilon\Xi G(j\omega) + \delta < 0, \forall \omega \in \mathbb{R}$,

then for this Ξ, the corresponding system (9.22) has no cycle of the second kind.

The following theorem shows that the existence of the interconnection matrix Ξ can be characterized in terms of the solutions of a set of LMIs.

Theorem 9.6. *If there exist $P = P^T$ and diagonal matrices $\tau > 0, \delta > 0$ such that*

$$4\delta > \tau\nu^2, \tag{9.35a}$$

$$\begin{pmatrix} PA + A^TP & PB \\ B^TP & \delta - \tau/4 \end{pmatrix} < 0, \tag{9.35b}$$

$$\begin{pmatrix} C \\ D^T \end{pmatrix}^\perp \begin{pmatrix} PA + A^TP & PB \\ B^TP & \delta \end{pmatrix} \begin{pmatrix} C \\ D^T \end{pmatrix}^{\perp T} < 0, \tag{9.35c}$$

9.3. Nonexistence of cycles of the second kind in interconnected systems

then there exists an interconnection matrix Ξ as defined in (9.23) such that the corresponding system (9.22) has no cycles of the second kind.

Proof. Based on the KYP Lemma and the Schur complement, the frequency-domain inequality in condition (ii) of Corollary 9.4 can be reduced to the existence of a matrix $P = P^T$ such that the following linear matrix inequality holds

$$\begin{pmatrix} PA + A^T P & PB + \frac{1}{2}C\Xi^T & C\Xi^T \\ B^T P + \frac{1}{2}\Xi C^T & \delta + \mathbf{Re}\{\Xi D\} & D^T \Xi^T \\ \Xi C^T & \Xi D & -\tau \end{pmatrix} < 0,$$

where $\tau = \epsilon^{-1}$. Since the above inequality can be written as

$$Q + \mathbf{He}(U\Xi V) < 0, \qquad (9.36)$$

where

$$Q = \begin{pmatrix} PA + A^T P & PB & 0 \\ B^T P & \delta & 0 \\ 0 & 0 & -\tau \end{pmatrix}, \quad U = \begin{pmatrix} 0 \\ I/2 \\ I \end{pmatrix}, \quad V = \begin{pmatrix} C^T & D & 0 \end{pmatrix}.$$

Using the projection lemma on inequality (9.36) we can obtain the following equivalent conditions where the parameter Ξ is eliminated,

$$U^{\perp} Q U^{\perp T} < 0, \quad V^{T\perp} Q V^{T\perp T} < 0.$$

Then using the standard LMI method given in [Iwasaki and Skelton (1994)] we get (9.35). □

Remark 9.5. When the linear matrix inequalities (9.35) have feasible solutions, the matrix Ξ can be parameterized by taking

$$U^+ = \begin{pmatrix} 0 & 0 & I \end{pmatrix}, \quad U^{\perp} = \begin{pmatrix} I & 0 & 0 \\ 0 & I & -I/2 \end{pmatrix}$$

as follows by the method in Chapter 1 (see also [Helmersson (1995)])

$$\Xi = G_1 + G_2 L G_3, \quad \|L\|_2 < 1,$$

where

$$G_1 = (-X_1 X_2^{-1} X_3^T)(X_3 X_2^{-1} X_3^T)^{-1},$$
$$G_2 = (X_1 X_2^{-1} X_1^T + \tau + G_1 G_3^{-2} G_1^T)^{1/2},$$
$$G_3 = (-X_3 X_2^{-1} X_3^T)^{-1/2},$$
$$X_1 = \begin{pmatrix} 0 & \tau/2 \end{pmatrix},$$
$$X_2 = U^{\perp} Q U^{\perp T},$$
$$X_3 = \begin{pmatrix} C^T & D \end{pmatrix}.$$

Corollary 9.4 gives a sufficient condition for the absence of cycles of the second kind in the system with the nonlinear interconnections specified by Ξ. Observe that the second condition of the corollary also ensures the strictly positive realness of the system with $\Xi G(j\infty) + G^T(-j\infty)\Xi^T < 0$ provided that A is Hurwitz. We state the results in the following corollary.

Corollary 9.5. *If A is Hurwitz stable, then the following statements are equivalent:*

(i) There exist a matrix Ξ and diagonal matrices $\epsilon > 0, \delta > 0$ such that

$$\mathbf{Re}\{\Xi G(j\omega)\} + G^*(j\omega)\Xi^T \epsilon \Xi G(j\omega) + \delta < 0, \forall \omega \in \mathbb{R};$$

(ii) There exists Ξ such that $-\Xi G(s)$ is SPR and $\Xi G(j\infty)+G^T(-j\infty)\Xi^T < 0$;

(iii) There exist Ξ and $P = P^T > 0$ such that

$$\begin{pmatrix} PA + A^T P & C\Xi^T - PB \\ \Xi C^T - B^T P & -\Xi D - D^T \Xi^T \end{pmatrix} < 0;$$

(iv) There exists $P = P^T > 0$ such that

$$PA + A^T P < 0 \quad and \quad \begin{pmatrix} C \\ D^T \end{pmatrix}^\perp \begin{pmatrix} PA + A^T P & PB \\ B^T P & 0 \end{pmatrix} \begin{pmatrix} C \\ D^T \end{pmatrix}^{\perp T} < 0.$$

Example 9.2. Let us consider system (9.22) with

$$A = \begin{pmatrix} -0.5 & 0 & 0.8 \\ 0 & 0 & 1 \\ 0 & -2.5 & -1 \end{pmatrix}, B = \begin{pmatrix} 1 & 0 \\ 0 & 0 \\ 0 & -1 \end{pmatrix},$$

$$C^T = \begin{pmatrix} 0.3 & 0 & 0 \\ 0 & 0.2 & 0 \end{pmatrix}, D = \begin{pmatrix} -0.1 & 3 \\ 5 & -0.1 \end{pmatrix},$$

and the nonlinear feedback $\varphi_1(y_1) = \sin(y_1) - 0.3$, $\varphi_2(y_2) = \sin(y_2) - 0.2$. Clearly, A has no eigenvalues on imaginary axis and (A, B) is controllable. Solving the linear matrix inequalities in Theorem 9.6 we get a set of feasible solutions as follows

$$P = \begin{pmatrix} 1.6822 & -1.0997 & -0.4274 \\ -1.0997 & 1.2247 & 0.4507 \\ -0.4274 & 0.4507 & 0.3630 \end{pmatrix},$$

$$\tau = \begin{pmatrix} 12.4981 & 0 \\ 0 & 11.5044 \end{pmatrix}, \delta = \begin{pmatrix} 0.3189 & 0 \\ 0 & 0.1408 \end{pmatrix}.$$

9.3. Nonexistence of cycles of the second kind in interconnected systems

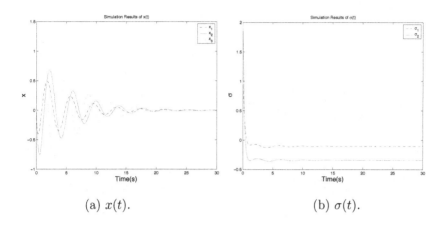

(a) $x(t)$. (b) $\sigma(t)$.

Fig. 9.6 Gradient-like behavior with $\Xi = (-0.3177 \ -0.7872; -1.4716 \ -0.1305)$.

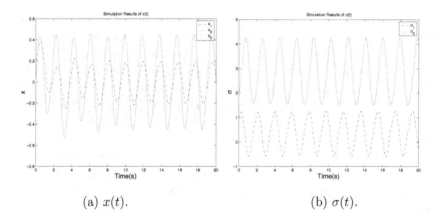

(a) $x(t)$. (b) $\sigma(t)$.

Fig. 9.7 Periodic oscillations with $\Xi = I_2$.

From the above solutions we get

$$\Xi = \begin{pmatrix} -0.3177 & -0.7872 \\ -1.4716 & -0.1305 \end{pmatrix}.$$

Thus the system has no cycles of the second kind. Furthermore, since A is Hurwitz stable, the conditions in Theorem 9.4 actually guarantee the gradient-like behavior of the system and every trajectory tends to an equilibrium as $t \to +\infty$. For the purpose of verification, a simulation result with initial value $\mathbf{x}_0 = (-0.2, 0, 0.1)^T, \sigma_0 = (1, 2)^T$ is given in Figure 9.6.

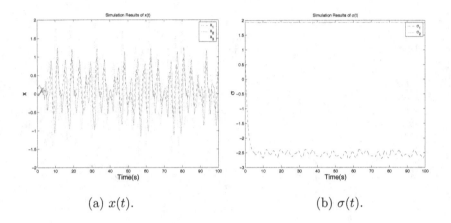

(a) $x(t)$. (b) $\sigma(t)$.

Fig. 9.8 Chaotic oscillations with $\Xi = (-0.3\ 50; 50\ -0.3)$.

This illustrative result coincides with Theorem 9.4 and Theorem 9.5 and confirms the nonexistence of cycles of the second kind for the system.

In order to demonstrate the effect of nonlinear interconnections on system behavior, the same experiment was repeated with varied connection structure by taking different Ξ. Simulation results for $\Xi = I_2$ and $\Xi = (\ -0.3\ 50; 50\ -0.3)$ are respectively shown in Figure 9.7 and Figure 9.8. When $\Xi = I_2$, i. e., there is no nonlinear interaction in the system, the system has periodic motions. When $\Xi = (-0.3\ 50; 50\ -0.3)$, chaotic oscillations appear in the variables x. From these simulation results, it can be clearly seen that with respect to different interconnection matrix the system has a rich dynamic behavior from gradient-like behavior to chaotic oscillations.

9.4 Cycle slipping in phase synchronization systems

Phase synchronization systems include a broad class of carrier tracking loops, suppressed carrier tracking loops, phase coherent demodulators of the phase-locked loop or automatic phase control type, and various closed loop symbol synchronization systems etc., which can be described by the model of pendulum-like systems. A typical behavior of phase synchronization systems is cycle slipping, which indicates that the absolute value of phase error increases by $k\Delta$, where k is an integer and Δ is the period of the nonlinearity, when t tends to infinity. In physical terms the generator to be synchronized drops or adds one or several cycles of oscillation relative to the input phase. Analogous phenomenon occurs in alternating current machinery when it skips a pole or in a pendulum when it rotates through 2π radians which can have disastrous effects. Outside the area of electrical engineering, it is found that *extrasystole* or *premature beat*, the most common disorder of the heart rhythm, is an analogous phenomenon of cycle slipping in phase synchronization systems. A series of premature beats occurs in succession at a rapid rate, which is analogous to the loss of phase synchronization in a PSS where several cycles are slipped in rapid succession before phase-lock is achieved [Lindsey (1972)]. Since cycle slipping may lead to frequency and phase diffusion in the generator to be synchronized, therefore analysis of cycle slipping and estimation of the number of cycles slipped is of interest.

Consider the following feedback system

$$\dot{z} = Az + b\varphi(\sigma)$$
$$\dot{\sigma} = cz + \rho\varphi(\sigma)$$
(9.37)

where $A \in \mathbb{R}^{n \times n}, b, c^T \in \mathbb{R}^n, \rho \in \mathbb{R}$. The Δ- periodic function $\varphi : \mathbb{R} \to \mathbb{R}$ is assumed to be differentiable and has exactly two zeros on $[0, \Delta)$. We suppose also that $[\varphi'(\sigma)]^2 + \varphi^2(\sigma) \neq 0$ on $\mathbb{R}$. The transfer function of the linear part of the system from the input φ to the output $-\dot{\sigma}$ is characterized by

$$G(s) = c(A - sI)^{-1}b - \rho$$

and $G(s)$ is nondegenerate.

Definition 9.4. We say that (z, σ) is a solution of (9.37) with k slipped cycles if there exists a $T > 0$ such that

$$|\sigma(T) - \sigma(0)| = k\Delta \quad \text{and} \quad |\sigma(t) - \sigma(0)| < (k+1)\Delta$$

for all $t \geq 0$.

In reference [Leonov et al. (1992a)], frequency domain conditions for estimation of the number of cycles slipped in system (9.37) was established.

Lemma 9.2. *Suppose that the matrix A is Hurwitz stable and for a symmetric $n \times n$ matrix $H = H^T$ and certain numbers $\epsilon > 0, \delta > 0$ and a positive integer k the following relations are true:*

(i) $\mathbf{Re}\{G(j\omega)\} - \epsilon|G(j\omega)|^2 \geq \delta, \ \forall \omega \in \mathbb{R};$

(ii) $4\epsilon\delta \geq [\nu_j(k, z(0)^T H z(0))]^2$ for $j = 1, 2$, where $\nu_j : \mathbb{N} \times \mathbb{R} \twoheadrightarrow \mathbb{R}$ are defined by

$$\nu_j(m, \xi) = \left[\int_0^\Delta \varphi(\sigma)d\sigma + (-1)^j m^{-1}\xi\right] \bigg/ \int_0^\Delta |\varphi(\sigma)|d\sigma$$

and H satisfies $2\operatorname{Re}\{z^T H(Az + b\xi)\} - F(z, \xi) \leq 0$, where

$$F(z, \xi) = -[\operatorname{\mathbf{Re}}\{\xi^T(cz + \rho\xi)\} + \epsilon|cz + \rho\xi|^2 + \delta|\xi|^2].$$

Then for any solution (z, σ) of system (9.37) starting from $(z(0), \sigma(0))$ with $\sigma(0) \in [0, 2\pi)$ it follows that

$$|\sigma(t) - \sigma(0)| < k\Delta$$

for all $t \geq 0$.

Although system (9.37) has a form of Lur'e type, it allows to have multiple isolated equilibria due to the periodic nonlinear function, which is different from the canonical Lur'e system with single equilibrium in absolute stability theory. Thus the above result can be used to analyze nonlinear systems with multiple equilibria described by the differential equation (9.37), such as phase-locked loops [Best (1984)], synchronization motors [Yakubovich et al. (2004)] and some mechanical systems [Leonov et al. (1992b)].

Theorem 9.7. *Suppose that the matrix A is Hurwitz stable and there exist a matrix $H = H^T \geq 0$, numbers $\epsilon > 0, \delta > 0$ and a positive integer k such that*

$$\begin{pmatrix} \epsilon c^T c + A^T H + HA & (\epsilon\rho + \frac{1}{2})c^T + Hb \\ (\epsilon\rho + \frac{1}{2})c + b^T H & \delta + \epsilon\rho^2 + \rho \end{pmatrix} \leq 0, \qquad (9.38a)$$

$$\begin{pmatrix} 2\epsilon & \nu_j(k, z(0)^T H z(0)) \\ \nu_j(k, z(0)^T H z(0)) & 2\delta \end{pmatrix} \geq 0, \qquad (9.38b)$$

where $\nu_j(\cdot, \cdot), j = 1, 2$ are as defined in Lemma 9.2, then for any solution (z, σ) of system (9.37) starting from $(z(0), \sigma(0))$ with $\sigma(0) \in [0, \Delta)$ it follows that

$$|\sigma(t) - \sigma(0)| < k\Delta$$

for all $t \geq 0$.

Proof. Similar to the proof of Theorem 9.5, the result can be easily obtained if we let

$$M = \begin{pmatrix} \epsilon c^T c & (\epsilon\rho + \frac{1}{2})c^T \\ (\epsilon\rho + \frac{1}{2})c & \delta + \epsilon\rho^2 + \rho \end{pmatrix}.$$

□

The symmetric matrix H in (9.38) of the theorem can be also determined according to the algorithm as follows

Step 1 Define the functions
$$d(s) = \det(sI - A),$$
$$Q(s) = d(s)(sI - A)^{-1}b,$$
$$F(w) = -w^T Gw, \ w \in \mathbb{R}^{n+1},$$
and polynomial Ξ by
$$\Xi(s) =: \begin{pmatrix} Q(-s^*) \\ d(-s^*) \end{pmatrix}^* G \begin{pmatrix} Q(s) \\ d(s) \end{pmatrix}$$
with
$$G = \begin{pmatrix} -\epsilon c^T c & -(\epsilon\rho + \frac{1}{2})c \\ -(\epsilon\rho + \frac{1}{2})c^T & -\rho - \epsilon\rho^2 - \delta \end{pmatrix}.$$
Assume that $F(0, \xi) := -\Gamma\xi^2, \xi \in \mathbb{R}$.

Step 2 Find a polynomial Ψ of degree n which satisfies the factorization equation
$$\Xi(s) = \Psi(-s)\Psi(s).$$

Step 3 Find an n-vector h and a number τ such that
$$\Psi(s) = h^T Q(s) + \tau d(s)$$
which is equivalent to the linear problem
$$\tau = \sqrt{\Gamma}, \quad h^T q_i = \Psi_i - \tau d_i, \text{ for } i = 0, 1, \cdots, n-1,$$
where Ψ_i, d_i, q_i are defined by the representations
$$\Psi(s) = \sqrt{\Gamma} s^n + \Psi_{n-1} s^{n-1} + \cdots + \Psi_0,$$
$$d(s) = s^n + d_{n-1} s^{n-1} + \cdots + d_0,$$
$$Q(s) = q_{n-1} s^{n-1} + \cdots + q_0.$$

Step 4 Find a matrix $H = H^T$ satisfying $HA + A^T H = D$, where D is defined by $z^T Dz := -F(z, 0) - |h^T z|$. H exists and is unique.

Define
$$\mu := \int_0^\Delta \varphi(\sigma) d\sigma \Big/ \int_0^\Delta |\varphi(\sigma)| d\sigma$$
and we state the following result according to the sign of μ.

Corollary 9.6. *Suppose that the matrix A is Hurwitz stable and there exist a matrix $H = H^T \geq 0$, numbers $\epsilon > 0, \delta > 0$ and a positive integer k such that (9.38a) and*

$$\begin{pmatrix} 2\epsilon & \nu_1(k,z(0)^T H z(0)) \\ \nu_1(k,z(0)^T H z(0)) & 2\delta \end{pmatrix} \geq 0, \text{ if } \mu < 0,$$

or

$$\begin{pmatrix} 2\epsilon & \nu_2(k,z(0)^T H z(0)) \\ \nu_2(k,z(0)^T H z(0)) & 2\delta \end{pmatrix} \geq 0, \text{ if } \mu \geq 0$$

hold, then for any solution (z,σ) of system (9.37) starting from $(z(0),\sigma(0))$ with $\sigma(0) \in [0,\Delta)$ it follows that

$$|\sigma(t) - \sigma(0)| < k\Delta$$

for all $t \geq 0$.

Remark 9.6. For the minimum $k(k \in \mathbb{Z}, k > 1)$ such that the conditions in Corollary 9.6 are satisfied, there must exists $T > 0$ such that

$$|\sigma(T) - \sigma(0)| = (k-1)\Delta \text{ and } |\sigma(t) - \sigma(0)| < k\Delta,$$

i.e. the solution (z,σ) starting from $(z(0),\sigma(0))$ with $\sigma(0) \in [0,\Delta)$ is a solution of (9.37) with $k-1$ slipped cycles. If the conditions of the corollary hold with $k = 1$, then there is no cycle slipping for the solution starting from $(z(0),\sigma(0))$. Based on this point, it is possible to design feedback controller guaranteeing the nonexistence of cycle slippings in the system.

Without loss of generality, we suppose that $\mu < 0$ and derive the following conditions estimating from above the numbers of cycles that are slipped by a solution of system (9.37) starting from $(z(0),\sigma(0))$. The result shows that the estimation problem can be recast into a generalized eigenvalue problem and solved by LMI techniques.

Corollary 9.7. *The solution (z,σ) starting from $(z(0),\sigma(0))$ with $\sigma(0) \in [0,\Delta)$ is a solution of (9.37) with $\text{int}(-1/\lambda)$ slipped cycles, where $\text{int}(\cdot)$ is the integer part of a real number and λ is the global minimum of the following generalized eigenvalue minimization problem with respect to $H = H^T > 0, \epsilon > 0, \delta > 0$*

$$\min \lambda$$

$$\text{s.t. } (9.38a), \begin{pmatrix} 2\epsilon & \widetilde{\nu}(\lambda, z(0)^T H z(0)) \\ \widetilde{\nu}(\lambda, z(0)^T H z(0)) & 2\delta \end{pmatrix} > 0,$$

where

$$\widetilde{\nu}(\lambda,\xi) = \left[\int_0^\Delta \varphi(\sigma)d\sigma + \lambda\xi\right] \bigg/ \int_0^\Delta |\varphi(\sigma)|d\sigma$$

9.4. Cycle slipping in phase synchronization systems

Example 9.3. The block diagram shown in Figure 9.9 is a simple communication system with a phase-locked loop. The harmonic oscillations of the signal generator $s(t)$ and the voltage control oscillator (VCO) output $r(t)$ act on the multiplier (phase detector) to produce a voltage $\epsilon(t)$. This voltage $\epsilon(t)$ depends on the difference between the phase and frequency of the signals $r(t)$ and $s(t)$, and on the additive noise $n(t)$. The resulting signal $\epsilon(t)$, after being filtered by the loop filter, changes the output frequency and phase of the VCO, causing them to coincide (in the absence of noise) with the frequency and phase of the signal generator.

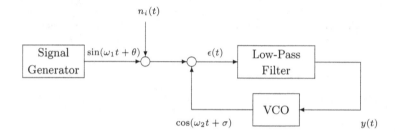

Fig. 9.9   A simple communication system with a phase-locked loop.

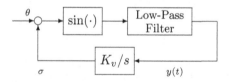

Fig. 9.10   Model of Phase-locked loop.

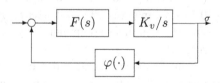

Fig. 9.11   Rearrangement into a linear component with a feedback nonlinear component.

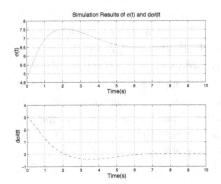

Fig. 9.12 Simulation results of $\sigma(t)$ and $\dot{\sigma}(t)$ with no cycle slipped.

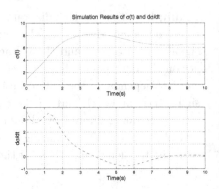

Fig. 9.13 Simulation results of $\sigma(t)$ and $\dot{\sigma}(t)$ with 1 cycle slipped.

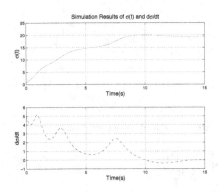

Fig. 9.14 Simulation results of $\sigma(t)$ and $\dot{\sigma}(t)$ with 3 cycles slipped.

Fig. 9.15 Simulation results of $\sigma(t)$ and $\dot{\sigma}(t)$ with 5 cycles slipped.

In case of the absence of noise, with some assumptions the PLL model in Figure 9.9 can be simplified to the one in Figure 9.10 and the voltage controlled oscillator (VCO) can be considered as an integrator. The model shown in Figure 9.10 is rearranged so that all linear components appear in the forward path and the nonlinear component appears in the feedback path as shown in Figure 9.11.

Let us consider an autonomous second order phase-locked loop with proportionally integrating filter, which is of greatest interest in practice and described by transfer function from $\varphi(\sigma)$ to $-\dot{\sigma}$

$$G(s) = T\frac{1+mTs}{1+Ts}, \qquad (9.39)$$

9.4. Cycle slipping in phase synchronization systems   235

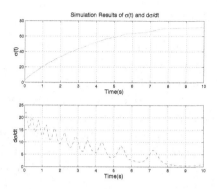

Fig. 9.16   Simulation results of $\sigma(t)$ and $\dot{\sigma}(t)$ with 10 cycles slipped.

	T	m	γ	ε	δ	H	$(z(0), \sigma(0))$
0	1	0.1	0.3	0.9091	0.0909	0.4545	$(3, 3\pi/2)$
1	2.5	0.3	0.3	0.3077	0.5769	0.3846	$(4, \pi/4)$
3	3	0.3	0.5	0.2564	0.6923	0.3846	$(5, \pi/6)$
5	8	0.1	0.1	0.1136	0.7273	0.4545	$(6, \pi/8)$
10	3.5	0.8	0.6	0.1587	1.5556	0.2778	$(18, \pi)$

where $T > 0, m \in (0,1)$ are constants and the nonlinearity $\varphi(\sigma) = \sin \sigma - \gamma, \gamma \in (0,1)$. Note that system (9.39) may be written as

$$\dot{z} = -T^{-1}z - (1-m)\varphi(\sigma),$$
$$\dot{\sigma} = z - mT\varphi(\sigma).$$
(9.40)

It is obvious that

$$A = -T^{-1}, \ b = -(1-m), \ c = 1, \ \rho = -mT.$$

Choosing different parameters T, m, γ and initial values $z(0), \sigma(0)$, then solving the linear matrix inequalities and the generalized eigenvalue problem in Corollary 9.6 we get the feasible solutions and the numbers of cycles slipped by the solution of (9.37) with the given initial conditions. Simulation results of the solution $\sigma(t)$ and its derivative $\dot{\sigma}(t)$ are shown in Figure 9.12 to Figure 9.16 corresponding to different parameter values given in Table 9.1 with the number of cycles slipped in the first column, which demonstrate the validity of the proposed method.

9.5 Notes and references

The mathematical concepts of auto-oscillation and a self-oscillating system originate from [Andronov (1955)], where the property of nonlinear dynamic systems of generating undamped oscillations was firstly connected with the concept of a Poincare cycle. From then on, various methods were developed for investigation of periodic solutions in nonlinear systems. Although there are many qualitative analysis results for low dimensional systems, analysis for high dimensional systems often relies on numerical computation and only approximate analytic results can be obtained (Lindsey, 1972; Liu and Chen, 2001; Mitropolsky and Nguyen, 1997; Nayfeh and Mook, 1995; Strzhinski, 1977). Pioneering work for oscillations in two dimensional systems with cylindrical phase space can be found in (Tricomi, 1933; Hayes, 1953; Serfert, 1952), and the work of [Leonov et al. (1996, 1992b,a)] extend the results for low dimensional systems to high dimensional systems in an analytic way. Previous results devoted to theoretical as well as experimental studies of cycle slipping can be seen in [Ascheid and Meyr (1982); Sannemann and Rowbotham (1964); Tausworthe (1967)]. Recent contributions to this problem are given in [Leonov et al. (1992a)], where frequency domain methods are used and estimation of the number of cycles slipped is characterized by frequency domain inequalities. More recently, robust analysis and synthesis results on oscillations of nonlinear Lur'e systems within the time domain framework are established by the authors and their co-authors. [Lu et al. (2007)] considers the nonexistence of periodic solutions in a certain frequency range in nonlinear Lur'e systems with polytopic uncertainties; [Lu et al. (2008a)] is concerned with the existence of cycles of the second in a class of pendulum-like systems which describe some models of synchronous machines; [Yang et al. (2005)] establishes frequency-domain conditions as well as linear matrix inequalities conditions for the nonexistence of cycles of the second kind in multi-input multi-output pendulum-like systems; [Yang et al. (2006)] proposes a novel methodology for scheming interconnection structure to guarantee the absence of cycles of the second kind in a class of interconnected nonlinear systems; [Yang and Huang (2007)] presents an LMI-based approach for the estimation of the number of cycles slipped in a class of phase synchronization systems; [Lu et al. (2008b)] addresses the design of a dynamic output feedback controller guaranteeing the nonexistence of cycle slipping in such systems. The main content of this chapter are based on the results of the above mentioned papers.

Chapter 10

Interconnected Systems

One of the greatest challenges in system theory is to deal with the area increasing size and complexity of system models for application such as electric power systems, the internet, and biological systems, etc. Since the 1960s, the large-scale systems theory has been developed to deal with such a challenge, and many interesting results have been established on some basic issues such as stability analysis, decomposition and aggregation, decentralized control, etc. Large-scale system methods have been successfully applied to study economics, multispecies communities, spacecraft control systems and power systems. Generally in the study of large-scale systems, in order to get acceptable stability results, the whole system must be connectively stable, i.e., the whole system is still stable when some connections among subsystems are broken. From the viewpoint of stability, this translates to reliability and robustness of stability with respect to structural perturbations, but the connections among subsystems do not play any useful roles. In this chapter, we mainly discuss interconnected systems composed of two subsystems and focus on the effects of interconnections. Section 10.1 introduces linearly interconnected systems and discusses the effects of unstable subsystem and applicability of small gain theorem. Section 10.2 discusses interconnected Lur'e systems. Section 10.3 discusses Lagrange stability of a generalized smooth Chua circuit from an interconnected system point of view. Section 10.4 introduces a class of input and output coupled pendulum-like systems. Section 10.5 concludes the chapter.

10.1 Linearly interconnected systems

Consider the following interconnected system composed of two subsystems,
$$\begin{cases} \dot{x}_1 = A_1 x_1 + A_{12} x_2 + B_1 u_1 \\ \dot{x}_2 = A_2 x_2 + A_{21} x_1 + B_2 u_2 \end{cases} \tag{10.1}$$
$u_1 = K_1 x_1, u_2 = K_2 x_2, A_{12}, A_{21}$ are matrices with compatible dimensions. We say system (10.1) can be decentrally stabilizable, i.e., there exist K_1 and K_2 such that the state matrix of the closed-loop system

$$A_{cl} = \begin{pmatrix} A_1 + B_1 K_1 & A_{12} \\ A_{21} & A_2 + B_2 K_2 \end{pmatrix}$$

is stable. Obviously, the decentralized control problem is completely different from the traditional centralized control problem such as designing a controller K such that $A + BK$ is stable. In decentralized control framework, the controller is structurally constrained. This constraint generally makes the controller design problem hard. In this kind of problems, an important concept is the notion of decentrally fixed mode, i.e., system mode or eigenvalue which cannot be changed by decentralized controllers, which is a generalization of an uncontrollable mode or unobservable mode in the centralized control problems, refer to [Siljak (1978, 1991)] and references therein for more details.

10.1.1 The effect of the unstable subsystem

First, we consider a simple example. In system (10.1) if
$$A_1 = \begin{pmatrix} 0 & 1 \\ 0 & 0 \end{pmatrix}, A_{12} = \begin{pmatrix} 0 & \alpha \\ 0 & 0 \end{pmatrix}, A_{21} = \begin{pmatrix} 0 & \beta \\ 0 & 0 \end{pmatrix}, A_2 = \begin{pmatrix} 0 & 1 \\ 0 & 0 \end{pmatrix},$$

$$B_1 = B_2 = \begin{pmatrix} 0 \\ 1 \end{pmatrix}, K_1 = -(k_1 \quad k_2), K_2 = -(k_3 \quad k_4),$$

then
$$A_{cl} = \begin{pmatrix} 0 & 1 & 0 & \alpha \\ -k_1 & -k_2 & 0 & 0 \\ 0 & \beta & 0 & 1 \\ 0 & 0 & -k_3 & -k_4 \end{pmatrix}.$$

Obviously, at this time, $det(sI - A_{cl}) = (s^2 + k_2 s + k_1)(s^2 + k_4 s + k_3) - \alpha \beta k_1 k_3$. For this simple case, one can get the following results easily:

(i) when $\alpha\beta = 1$, 0 is a fixed mode;
(ii) when $\alpha\beta > 1$, for any $k_i > 0, i = 1,2,3,4$, i.e., $A_1 + B_1K_1$ and $A_2 + B_2K_2$ are stable, A_{cl} cannot be stable (the constant term of its characteristic polynomial is less than 0 at this time);
(iii) when $\alpha\beta < 1$, it is possible that the interconnected system and two subsystems can be stabilized simultaneously.

An example as above with $\alpha = 2$ and $\beta = 1$ was also introduced by [Siljak (1991)] to show the complexity in interconnected systems. But with the changes of α and β we can see more in the above example. And according to the above analysis, we can give a special structural property of interconnections under which subsystems and the overall system cannot be simultaneously stabilized.

Theorem 10.1. *If the interconnected system in (10.1) satisfies that:*

(i) there exists A'_{12} such that $A_{12} = A'_{12}A_2$, and $A'_{12}B_2 = 0$, and
(ii) there exists A'_{21} such that $A_{21} = A'_{21}A_1$, and $A'_{21}B_1 = 0$,

then when $det(I - A'_{21}A'_{12}) < 0$, there are no K_1 and K_2 such that $A_1 + B_1K_1$, $A_2 + B_2K_2$ and A_{cl} are Hurwitz stable simultaneously.

Proof. Computing the determinant of A_{cl}, one can get

$$det(A_{cl}) = det(A_1 + B_1K_1)det(A_2 + B_2K_2 - A'_{21}A_1(A_1 + B_1K_1)^{-1}A'_{12}A_2).$$

Noticing conditions (i) and (ii),

$det(A_{cl}) = det(A_1 + B_1K_1)det(A_2 + B_2K_2 - A'_{21}(A_1 + B_1K_1)(A_1 + B_1K_1)^{-1}A'_{12}(A_2 + B_2K_2))$,

that is,

$$det(A_{cl}) = det(A_1 + B_1K_1)det(A_2 + B_2K_2)det(I - A'_{21}A'_{12}).$$

When $A_1 + B_1K_1, A_2 + B_2K_2$ are stable and $det(I - A'_{21}A'_{12}) < 0$, one gets that the constant term of the characteristic polynomial of A_{cl} is less than zero. Therefore, A_{cl} must be unstable. □

Remark 10.1. Obviously, under the conditions in Theorem 10.1, when $det(I - A'_{21}A'_{12}) = 0$, 0 is a fixed mode. When $det(I - A'_{21}A'_{12}) > 0$, it is possible that there exist K_1 and K_2 such that $A_1 + B_1K_1$, $A_2 + B_2K_2$ and A_{cl} are stable simultaneously. And if A_{cl} is stable, one of $A_1 + B_1K_1$ and $A_2 + B_2K_2$ must be unstable when $det(I - A'_{21}A'_{12}) < 0$. This also shows the complexity of interconnected large-scale systems. For the study of the effects of interconnections in large scale systems, it is important to design decentralized controllers when some subsystems must be unstable. At this time, the interconnections play real roles for stability of large-scale systems.

Remark 10.2. Because of the existence of unstable subsystem under stability of interconnected systems, the traditional structural disturbance [Siljak (1991)], i.e., communication failure between subsystems, is not allowed for the interconnected structure discussed in Theorem 10.1.

Corollary 10.1. *For any interconnected matrix* $A = \begin{pmatrix} A_1 & A_{12} \\ A_{21} & A_2 \end{pmatrix}$, *if* A_{12} *and* A_{21} *can be written as* $A_{12} = A'_{12}A_2$, $A_{21} = A'_{21}A_1$, *and* $det(I - A'_{21}A'_{12}) < 0$, *then* A_1, A_2 *and* A *cannot be stable simultaneously.*

When A_1 and A_2 are nonsingular, we can directly take $A'_{12} = A_{12}A_2^{-1}$ and $A'_{21} = A_{21}A_1^{-1}$.

Obviously, the above results can be generalized to cases of multiple subsystems. For example, for an interconnected system composed of three subsystems, its closed-loop system matrix is given by

$$A_{cl} = \begin{pmatrix} A_1 + B_1K_1 & A_{12} & A_{13} \\ A_{21} & A_2 + B_2K_2 & A_{23} \\ A_{31} & A_{32} & A_3 + B_3K_3 \end{pmatrix}.$$

Let $\overline{A}_1 = \begin{pmatrix} A_1 & A_{12} \\ A_{21} & A_2 \end{pmatrix}$, $\overline{B}_1 = diag(B_1, B_2)$, $\overline{A}_{13} = \begin{pmatrix} A_{13} \\ A_{23} \end{pmatrix}$, $\overline{A}_{31} = (A_{31} \quad A_{32})$. If the following conditions are satisfied:

(i) there exists A'_{13} such that $\overline{A}_{13} = A'_{13}A_3$, and $A'_{13}B_3 = 0$, and
(ii) there exists A'_{31} such that $\overline{A}_{31} = A'_{31}\overline{A}_1$, and $A'_{31}\overline{B}_1 = 0$,

then there are not K_1, K_2 and K_3 such that $\overline{A}_1 + \overline{B}_1 diag(K_1, K_2)$, $A_3 + B_3K_3$ and A_{cl} are stable simultaneously when $det(I - A'_{31}A'_{13}) < 0$.

Similarly to the case of decentralized state feedbacks, we can also study the problem of decentralized dynamic output feedback stabilization. Suppose the outputs of two subsystems in (10.1) are $y_1 = C_1x_1$ and $y_2 = C_2x_2$ respectively. The decentralized dynamic controllers are given as

$$\begin{cases} x_{k1} = A_{k1}x_{k1} + B_{k1}y_1, \\ u_1 = C_{k1}x_{k1} + D_{k1}y_1, \end{cases} \quad \begin{cases} x_{k2} = A_{k2}x_{k2} + B_{k2}y_2, \\ u_2 = C_{k2}x_{k2} + D_{k2}y_2. \end{cases} \quad (10.2)$$

Then the state matrix of the closed-loop system is

$$\tilde{A}_{cl} = \begin{pmatrix} \tilde{A}_1 + \tilde{B}_1\tilde{K}_1\tilde{C}_1 & \tilde{A}_{12} \\ \tilde{A}_{21} & \tilde{A}_2 + \tilde{B}_2\tilde{K}_2\tilde{C}_2 \end{pmatrix},$$

10.1. Linearly interconnected systems

where

$$\tilde{A}_1 = \begin{pmatrix} A_1 & 0 \\ 0 & 0 \end{pmatrix}, \tilde{B}_1 = \begin{pmatrix} B_1 & 0 \\ 0 & I \end{pmatrix}, \tilde{C}_1 = \begin{pmatrix} C_1 & 0 \\ 0 & I \end{pmatrix}, \tilde{A}_{12} = \begin{pmatrix} A_{12} & 0 \\ 0 & 0 \end{pmatrix},$$

$$\tilde{A}_2 = \begin{pmatrix} A_2 & 0 \\ 0 & 0 \end{pmatrix}, \tilde{B}_2 = \begin{pmatrix} B_2 & 0 \\ 0 & I \end{pmatrix}, \tilde{C}_2 = \begin{pmatrix} C_2 & 0 \\ 0 & I \end{pmatrix}, \tilde{A}_{21} = \begin{pmatrix} A_{21} & 0 \\ 0 & 0 \end{pmatrix},$$

$$\tilde{K}_1 = \begin{pmatrix} D_{k1} & C_{k1} \\ B_{k1} & A_{k1} \end{pmatrix}, \tilde{K}_2 = \begin{pmatrix} D_{k2} & C_{k2} \\ B_{k2} & A_{k2} \end{pmatrix}.$$

At this time, the problem of decentralized dynamic output feedback stabilization can be viewed as a special problem of decentralized static output feedback stabilization. Obviously, if A_{12} and A_{21} satisfy the conditions of Theorem 10.1, then there are not decentralized controllers (10.2), i.e. $\tilde{K}_1$ and $\tilde{K}_2$, such that $\tilde{A}_1 + \tilde{B}_1 \tilde{F}_1 \tilde{C}_1$, $\tilde{A}_2 + \tilde{B}_2 \tilde{F}_2 \tilde{C}_2$ and $\tilde{A}_{cl}$ are stable simultaneously.

The above results show that in some cases the stability of interconnected systems is closely dependent on the interconnections. In order to study the actions of the interconnections between subsystems further, we study a special kind of decentralized control problem which can be viewed as harmonic stability problem of subsystems.

10.1.2 Interconnected feedbacks

Consider the following interconnected system

$$\begin{cases} \dot{x}_1 = A_1 x_1 + b_{12} u_{12}, \\ \dot{x}_2 = A_2 x_2 + b_{21} u_{21}, \end{cases} \quad (10.3)$$

where $u_{12} = k_{12} x_2$, $u_{21} = k_{21} x_1$, b_{12} and b_{21} are given real column vectors. k_{12} and k_{21} are real row vectors to be designed. In system (10.3), there is information interchange between two subsystems. It means that two systems are cooperating. For this special decentralized control problem, one can get some simple result for its stabilizability with the following lemma.

Lemma 10.1. *Given a real monic polynomial $f(\lambda)$ with degree n, $f(\lambda)$ has no real root if and only if $f(x) > 0$, $\forall x \in \mathbf{R}$.*

One can prove this lemma easily by writing $f(\lambda)$ as

$$f(\lambda) = (\lambda - \lambda_1)(\lambda - \lambda_2) \cdots (\lambda - \lambda_n)$$

where $\lambda_1, \lambda_2, \cdots, \lambda_n$ are roots of $f(\lambda)$.

Theorem 10.2. *If (A_1, b_{12}) and (A_2, b_{21}) are controllable, and $a = trace(A_1) + trace(A_2) < 0$, where $trace(\cdot)$ denotes the trace of the corresponding matrix, then there are k_{12} and k_{21} such that $A_{cl} = \begin{pmatrix} A_1 & b_{12}k_{12} \\ b_{21}k_{21} & A_2 \end{pmatrix}$ is stable.*

Proof. Suppose (A_1, b_{12}) and (A_2, b_{21}) are standard controllable canonical form. Let the orders of A_1 and A_2 be n and m, respectively. Set $H_1(s) = k_{21}(sI - A_1)^{-1}b_{12}$, $H_2(s) = k_{12}(sI - A_2)^{-1}b_{21}$, $k_{12} = (\beta_0, \beta_1, \cdots, \beta_{m-1})$, $k_{21} = (\alpha_0, \alpha_1, \cdots, \alpha_{n-1})$, $d_1(s) = det(sI - A_1)$, $d_2(s) = det(sI - A_2)$, $k_1(s) = \alpha_0 + \alpha_1 s + \cdots + \alpha_{n-1}s^{n-1}$ and $k_2(s) = \beta_0 + \beta_1 s + \cdots + \beta_{m-1}s^{m-1}$, then A_{cl} is stable if and only if the feedback system shown in Fig. 10.1 is stable, i.e., the polynomial $d_{cl}(s) = d_1(s)d_2(s) - k_1(s)k_2(s)$ is stable. Obviously, $d_{cl}(s)$ is a monic polynomial and the coefficient of s^{n+m-1} in $d_{cl}(s)$ is $-a = -(trace(A_1) + trace(A_2)) > 0$. Let $d_1(s)d_2(s) = s^{n+m} - as^{n+m-1} + d_0(s)$, then the stability of $d_{cl}(s)$ is completely determined by $d(s) = d_0(s) - k_1(s)k_2(s)$. When at least one of n and m is odd, one can choose $d(s)$ arbitrarily such that $d_{cl}(s)$ is stable, and decompose $d(s) - d_0(s)$ into the product of real polynomials $k_1(s)$ and $k_2(s)$. This means that we find real vectors k_{12} and k_{21} such that A_{cl} is stable. When both n and m are even, the degrees of $k_1(s)$ and $k_2(s)$ are odd, but the degree of $k_1(s)k_2(s)$ is even. At this time, one chooses $d(s)$ suitably such that d_{cl} is stable and $d(s) - d_0(s)$ has a real root in order to decompose $d(s) - d_0(s)$ into the product of two real polynomials. Actually this can be done as follows. First choose a real number x such that $d_{cl}(x) < 0$ by adjusting the roots of $d_{cl}(s)$ in the left half plane with the constraint that the sum of all its roots is a, and $d_{cl}(x) - x^{n+m} + ax^{n+m-1} - d_0(x) < 0$ can be guaranteed by enlarging the imaginary parts of the roots of $d_{cl}(s)$. Then one knows that $d_{cl}(s) - s^{n+m} + as^{n+m-1} - d_0(s)$ has a real root by Lemma 10.1. So one can decompose $d_{cl}(s) - s^{n+m} + as^{n+m-1} - d_0(s)$ into the product of two real polynomials $k_1(s)$ and $k_2(s)$. This completes the proof. □

Remark 10.3. From the proof of the theorem, we know that the eigenvalues of A_{cl} can be assigned arbitrarily with the only constraint $a = \lambda_1 + \cdots + \lambda_{n+m}$ when one of n and m is odd. When n and m are even simultaneously, the eigenvalues of A_{cl} can also be assigned properly with the constraints $a = \lambda_1 + \cdots + \lambda_{n+m}$ and k_{12}, k_{21} being real vectors.

10.1. Linearly interconnected systems

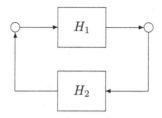

Fig. 10.1 Feedback system.

Remark 10.4. System (10.3) can be viewed as cooperative behavior between two subsystems. Two subsystems can be unstable themselves, but they can realize a stable system overall through intercrossed feedback. Each subsystem does not use its own information, but it uses the other subsystem's information. That is, they help each other to achieve stability. Of course, there may be self-feedback in any one subsystem. We can imagine that in such cooperation behavior, the subsystems need not to be controllable or stabilizable themselves, see the following system.

If there is self-feedback in subsystem, system (10.3) can be stated as follows,

$$\begin{cases} \dot{x}_1 = A_1 x_1 + b_1 u_1 + b_{12} u_{12}, \\ \dot{x}_2 = A_2 x_2 + b_2 u_2 + b_{21} u_{21}, \end{cases} \quad (10.4)$$

where $u_{12}, u_{21}, b_{12}, b_{21}$ are given as in system (10.3), b_1 and b_2 are real vectors with compatible dimensions, $u_1 = k_1 x_1$ and $u_2 = k_2 x_2$. By using Theorem 10.2, one can get the following result easily.

Theorem 10.3. If $(A_1, [b_1 \ b_{12}])$ and $(A_2, [b_2 \ b_{21}])$ are controllable, and b_1 and b_2 are not zero vectors simultaneously, then there are real vectors k_1, k_{12}, k_2 and k_{21} such that system (10.4) is asymptotically stable, i.e.,

$$A_{cl} = \begin{pmatrix} A_1 + b_1 k_1 & b_{12} k_{12} \\ b_{21} k_{21} & A_2 + b_2 k_2 \end{pmatrix} \text{ is stable.}$$

Remark 10.5. One can see clearly in Theorem 10.3, (A_1, b_1) and (A_2, b_2) need not to be controllable or stabilizable. This theorem shows that two subsystems with effective control can cooperate easily for sake of stability. The actions of interconnections are shown here to some degree. Obviously, the framework in Theorems 10.1, 10.2 and 10.3 can be generalized to multi-subsystem cases.

10.1.3 Decentralized controller design

Although we have discussed some special decentralized control problems in the above sections, it is still hard to design decentralized controllers. Obviously, the traditional diagonal Lyapunov function method does not work for the case that the subsystem is not stable. In the following, we introduce the parameter-dependent Lyapunov function method. By using this method we present an LMI-based design method for the problems discussed above. First we introduce the following lemma to begin this section.

Lemma 10.2. *Given a real matrix $A \in \mathbb{R}^{n \times n}$, A is stable if and only if there exist a matrix $P = P^T > 0$ and any matrix V such that*

$$\begin{pmatrix} -V - V^T & V^T A^T + P & V^T \\ AV + P & -P & 0 \\ V & 0 & -P \end{pmatrix} < 0. \qquad (10.5)$$

We can turn (10.5) into Lyapunov inequality $A^T P + PA < 0$ easily by using the well known projection lemma in Chapter 1. By introducing a new variable V, the products of PA and $A^T P$ are relaxed to new products AV and $V^T A^T$, where V needs not to be symmetric and positive definite. In this way Lyapunov matrix P can be parameter-dependent for the study of robust stability and robust performances [de Oliveira et al. (1999, 2002b)]. The case of diagonal blocked matrix V for decentralized control of discrete-time systems was considered in [de Oliveira et al. (2002b)]. Here we discuss the upper trigonal constraint of V for system (10.1) as follows, the lower trigonal constraint can be considered similarly. Sometimes, the upper trigonal constraint is less conservative than the diagonal constraint. Corresponding to system (10.1), we suppose

$$V = \begin{pmatrix} V_1 & \lambda V_1 \\ 0 & V_2 \end{pmatrix}, \quad X = \begin{pmatrix} X_1 & \lambda X_1 \\ 0 & X_2 \end{pmatrix}, \qquad (10.6)$$

where the dimensions of V_1 and V_2 are identical with the ones of A_1 and A_2, the dimensions of X_1 and X_2 are identical with the ones of K_1 and K_2, respectively, and λ is a parameter to be searched. The parameter λ can provide one more degree of freedom to reduce the conservativeness of the above method, see [Duan et al. (2006b)] for the similar method. For simplicity, we assumed that V and X acquire the upper trigonal structure as in (10.6). In fact, it is only fit for the case of $order(A_1) = order(A_2)$, where $order(\cdot)$ is the order of a square matrix. If $order(A_1) \neq order(A_2)$,

10.1. Linearly interconnected systems

for example, $order(A_1) < order(A_2)$, we can add zero blocks in V and X as follows to meet such cases,

$$V = \begin{pmatrix} V_1 & V_{12} \\ 0 & V_2 \end{pmatrix}, \quad X = \begin{pmatrix} X_1 & X_{12} \\ 0 & X_2 \end{pmatrix},$$

where $V_{12} = (\lambda V_1 \quad 0), X_{12} = (\lambda X_1 \quad 0)$. When $order(A_1) > order(A_2)$, we assume V and X acquire the following lower trigonal structure

$$V = \begin{pmatrix} V_1 & 0 \\ V_{21} & V_2 \end{pmatrix}, \quad X = \begin{pmatrix} X_1 & 0 \\ X_{21} & X_2 \end{pmatrix},$$

where $V_{21} = (\lambda V_2 \quad 0), X_{21} = (\lambda X_2 \quad 0)$, then we can get the similar result as in the following theorem.

Let

$$A = \begin{pmatrix} A_1 & A_{12} \\ A_{21} & A_2 \end{pmatrix}, \quad B = \begin{pmatrix} B_1 & 0 \\ 0 & B_2 \end{pmatrix}, \quad K = \begin{pmatrix} K_1 & 0 \\ 0 & K_2 \end{pmatrix} \quad (10.7)$$

in system (10.1). By Lemma 10.2, we can get the following result for stability of system (10.1).

Theorem 10.4. *If there are $P = P^T$ and V, X with form (10.6) such that*

$$\begin{pmatrix} -V - V^T & V^T A^T + X^T B^T + P & V^T \\ AV + BX + P & -P & 0 \\ V & 0 & -P \end{pmatrix} < 0,$$

then there exist diagonal blocked matrix K as in (10.7) such that $A_{cl} = A + BK$ is stable, and the decentralized controller can be obtained as $K_1 = X_1 V_1^{-1}$ and $K_2 = X_2 V_2^{-1}$.

Remark 10.6. From Theorem 10.4, we can see that P is not blocked, and V_1, V_2 are generally not symmetric, of course not positive definite. Intuitively, we can imagine that at this time, $A_1 + B_1 K_1$ or $A_2 + B_2 K_2$ can be unstable under stability of A_{cl}. We can also see this from the forthcoming examples. Similar results for systems (10.3) and (10.4) can be established.

Example 10.1. In the simple example studied in section 10.1.1, let $\alpha = \beta = 2$. By Theorem 10.4 ($\lambda = 1$), one can get a controller $K = \begin{pmatrix} -2.62 & -5.69 & 0 & 0 \\ 0 & 0 & 1.13 & -11.48 \end{pmatrix}$ such that A_{cl} is stable, but $A_2 + B_2 K_2$ is not stable at this time.

Example 10.2. Consider system (10.3) defined by matrices

$$A_1 = \begin{pmatrix} 0 & 1 & 0 & 0 \\ 0 & 0 & 1 & 0 \\ 0 & 0 & 0 & 1 \\ -3 & -3 & -8 & -5 \end{pmatrix}, \quad A_2 = \begin{pmatrix} 0 & 1 & 0 & 0 \\ 0 & 0 & 1 & 0 \\ 0 & 0 & 0 & 1 \\ -3 & -2 & -5 & 0 \end{pmatrix},$$

$$b_{12} = \begin{pmatrix} 0 \\ 0 \\ 0 \\ 1 \end{pmatrix}, \quad b_{21} = \begin{pmatrix} 0 \\ 0 \\ 0 \\ 1 \end{pmatrix}.$$

Obviously, the conditions in Theorem 10.2 are satisfied. A_2 is not stable here. Using the similar method as in Theorem 10.4 ($\lambda = 1$), one can get a decentralized controller $k_{21} = (\,-1.7976\ -3.0015\ -5.4645\ -4.6861\,)$ and $k_{12} = (\,0.1507\ 1.0410\ 0.8293\ 1.3124\,)$ such that A_{cl} is stable in (10.3).

From the above examples we can see that subsystems need not to be stable, even in some special cases some subsystems must be unstable. This shows the special effects of interconnections.

10.1.4 *The effect of small gain theorem*

The small gain theorem plays an important role in robust control theory [Zhou et al. (1996)], also see Chapter 2. By this lemma, we know that the system shown in Fig. 10.1 is internally stable, that is

$$\begin{bmatrix} (I - H_2 H_1)^{-1} & H_2(I - H_1 H_2)^{-1} \\ H_1(I - H_2 H_1)^{-1} & (I - H_1 H_2)^{-1} \end{bmatrix} \in RH_\infty,$$

if $H_1, H_2 \in RH_\infty$ and $\|H_1\|_\infty \|H_2\|_\infty < 1$.

In what follows, we discuss its applications to interconnected systems when subsystems are stable. Consider the following system composed of two subsystems,

$$\begin{cases} \dot{x}_1 = A_1 x_1 + A_{12} x_2 + B_{11} u_1, \\ \dot{x}_2 = A_2 x_2 + A_{21} x_1 + B_{22} u_2, \end{cases} \quad (10.8)$$

where $u_1 = K_1 x_1$ and $u_2 = K_2 x_2$. The objective is to design K_1 and K_2 such that the closed-loop system matrix A_{cl} is stable, where

$$A_{cl} = A + \begin{pmatrix} B_{11} K_1 & 0 \\ 0 & B_{22} K_2 \end{pmatrix}, \quad A = \begin{pmatrix} A_1 & A_{12} \\ A_{21} & A_2 \end{pmatrix}.$$

10.1. Linearly interconnected systems

Decompose A_{12} and A_{21} as $A_{12} = B_1 C_2$ and $A_{21} = B_2 C_1$. Let $H_1(s) = C_1(sI - A_1)^{-1} B_1$ and $H_2(s) = C_2(sI - A_2)^{-1} B_2$. Then the system matrix A of (10.8) can be viewed as the state matrix of the closed-loop feedback system as shown in Fig. 10.1. In this way, system (10.8) can be viewed as an interconnected system composed of two subsystems (A_1, B_1, C_1) and (A_2, B_2, C_2) under local feedback.

Remark 10.7. Since A_{12} represents the action from the second subsystem to the first subsystem, we get the decomposition $A_{12} = B_1 C_2$, where B_1 can be viewed as the matrix of the first subsystem accepting the interconnected control, $C_2 x_2$ can be viewed as the output of the second subsystem. The decomposition $A_{21} = B_2 C_1$ can be viewed analogously. The main idea here is to decompose the off-diagonal components of matrix A such that the whole system appear to be an interconnection of two subsystems through input/output connection. Then small gain theorem can be used to dicuss its stability problem.

Using small gain theorem, one can get the following simple result,

Proposition 10.1. *If there exist K_1 and K_2 such that*

$$\|C_1(sI - A_1 - B_{11} K_1)^{-1} B_1\|_\infty \|C_2(sI - A_2 - B_{22} K_2)^{-1} B_2\|_\infty < 1, \quad (10.9)$$

then (10.8) can be stabilized by local feedback.

In Proposition 10.1, the decompositions $A_{12} = B_1 C_2$ and $A_{21} = B_2 C_1$ are generally not unique. Hence, we will look for the suitable decompositions such that (10.9) holds by LMI method.

Based on bounded real lemma in H_∞ control theory (Chapter 2), one can get the following result

Theorem 10.5. *Given a matrix* $A = \begin{pmatrix} A_1 & A_{12} \\ A_{21} & A_2 \end{pmatrix}$, *and for any fixed full rank decompositions* $A_{12} = B_1^0 C_2^0$ *and* $A_{21} = B_2^0 C_1^0$, *there exist decompositions* $A_{12} = B_1 C_2$ *and* $A_{21} = B_2 C_1$ *such that*

$$\|C_1(sI - A_1)^{-1} B_1\|_\infty \|C_2(sI - A_2)^{-1} B_2\|_\infty < 1, \quad (10.10)$$

if and only if there exist positive definite matrices P_1, P_2, X and Y such that

$$\begin{pmatrix} P_1 A_1^T + A_1 P_1 + B_1^0 X B_1^{0T} & P_1 C_1^{0T} \\ C_1^0 P_1 & -Y \end{pmatrix} < 0,$$

$$\begin{pmatrix} P_2 A_2^T + A_2 P_2 + B_2^0 Y B_2^{0T} & P_2 C_2^{0T} \\ C_2^0 P_2 & -X \end{pmatrix} < 0. \quad (10.11)$$

Proof. If there exist positive definite matrices P_1, P_2, X, Y such that (10.11) holds, then let
$$B_1 = B_1^0 X^{\frac{1}{2}}, \ C_2 = X^{-\frac{1}{2}} C_2^0, \ B_2 = B_2^0 Y^{\frac{1}{2}}, \ C_1 = Y^{-\frac{1}{2}} C_1^0.$$
By (10.11), we have
$$\begin{pmatrix} P_1 A_1^T + A_1 P_1 + B_1 B_1^T & P_1 C_1^T \\ C_1 P_1 & -I \end{pmatrix} < 0$$
and
$$\begin{pmatrix} P_2 A_2^T + A_2 P_2 + B_2 B_2^T & P_2 C_2^T \\ C_2 P_2 & -I \end{pmatrix} < 0.$$
By using bounded real lemma and Schur complement, we know that the inequalities above hold, only if
$$\|C_1(sI - A_1)^{-1} B_1\|_\infty < 1 \text{ and } \|C_2(sI - A_2)^{-1} B_2\|_\infty < 1.$$
Then (10.10) holds.

Conversely, decompositions of A_{12} and A_{21}, except the trivial decompositions (not always full rank)
$$A_{12} = A_{12} \times I, \quad A_{21} = A_{21} \times I, \tag{10.12}$$
can be expressed as
$$A_{12} = (B_1^0 X_1) \times (X_1^+ C_2^0), \quad A_{21} = (B_2^0 Y_1) \times (Y_1^+ C_1^0), \tag{10.13}$$
where the superscript $+$ denotes pseudoinverse. Since $B_1^0 C_2^0$ and $B_2^0 C_1^0$ are full rank decompositions, X_1 and Y_1 are matrices with full row rank.

Moreover, the decompositions (10.12) can be replaced by
$$A_{12} = A_{12} \times (C_2^{0+} C_2^0), \quad A_{21} = A_{21} \times (C_1^{0+} C_1^0),$$
since it is easier for these decompositions to satisfy (10.10) than (10.12). Therefore, we only need to consider the decompositions in (10.13).

For any decompositions in (10.13), by bounded real lemma, we know that (10.10) holds, if and only if there exist positive definite matrices P_1, P_2 and a scalar γ such that
$$\begin{pmatrix} A_1 P_1 + P_1 A_1^T & B_1^0 X_1 & P_1 C_1^{0T} Y_1^{+T} \\ X_1^T B_1^{0T} & -\gamma I & 0 \\ Y_1^+ C_1^0 P_1 & 0 & -\gamma I \end{pmatrix} < 0$$
and
$$\begin{pmatrix} A_2 P_2 + P_2 A_2^T & B_2^0 Y_1 & P_2 C_2^{0T} X_1^{+T} \\ Y_1^T B_2^{0T} & -\frac{1}{\gamma} I & 0 \\ X_1^+ C_2^0 P_2 & 0 & -\frac{1}{\gamma} I \end{pmatrix} < 0.$$
Let $X = \frac{1}{\gamma} X_1 X_1^T, Y = \gamma Y_1 Y_1^T$, then $X^{-1} = \gamma X_1^{+T} X_1^+, Y^{-1} = \frac{1}{\gamma} Y_1^{+T} Y_1^+$. By Schur complement, we know that the above inequalities hold if and only if (10.11) holds. □

10.1. Linearly interconnected systems

Remark 10.8. From the above proof, we know that we only need to consider full rank decompositions among the different decompositions of A_{12}, A_{21}. When A_{12}, A_{21} are matrices with rank one, the decompositions of A_{12}, A_{21} are unique up to a constant.

Remark 10.9. It is clear, $\|C_1(sI_1 - A_1)^{-1}B_1\|_\infty \|C_2(sI_2 - A_2)^{-1}B_2\|_\infty$ can be minimized among different decompositions $A_{12} = B_1 C_2, A_{21} = B_2 C_1$ by the LMI method given above.

Clearly, (10.9) is only a sufficient condition for decentralized stabilizability of (10.8). But based on this condition, it is easy to establish an LMI algorithm for designing K_1 and K_2. In order to make (10.9) hold, we just need to minimize $\|C_i(sI_i - A_{ii} - B_{ii}K_i)^{-1}B_i\|_\infty$ by designing $K_i, i = 1, 2$. This problem can be solved by LMI method based on the result of Theorem 10.5.

Theorem 10.6. *There exist K_1 and K_2 and decompositions $A_{12} = B_1 C_2$ and $A_{21} = B_2 C_1$ such that (10.9) holds, if and only if there exist positive definite matrices P_1, P_2, X_1 and X_2, and any matrices Y_1 and Y_2 such that*

$$\begin{pmatrix} P_1 A_1^T + A_1 P_1 + B_1^0 X_1 B_1^{0T} + B_{11} Y_1 + Y_1^T B_{11}^T & P_1 C_1^{0T} \\ C_1^0 P_1 & -X_2 \end{pmatrix} < 0,$$
$$\begin{pmatrix} P_2 A_2^T + A_2 P_2 + B_2^0 X_2 B_2^{0T} + B_{22} Y_2 + Y_2^T B_{22}^T & P_2 C_2^{0T} \\ C_2^0 P_2 & -X_1 \end{pmatrix} < 0. \quad (10.14)$$

Then controllers are given by $K_1 = Y_1 P_1^{-1}$ and $K_2 = Y_2 P_2^{-1}$.

Proof. By Theorem 10.5, there exist K_1 and K_2, and decompositions $A_{12} = B_1 C_2$ and $A_{21} = B_2 C_1$ such that (10.9) holds, if and only if there exist positive definite matrices P_1, P_2, X_1 and X_2, and any matrices K_1 and K_2 such that

$$\begin{pmatrix} P_1(A_1 + B_{11}K_1)^T + (A_1 + B_{11}K_1)P_1 + B_1^0 X_1 B_1^{0T} & P_1 C_1^{0T} \\ C_1^0 P_1 & -X_2 \end{pmatrix} < 0,$$
$$\begin{pmatrix} P_2(A_2 + B_{22}K_2)^T + (A_2 + B_{22}K_2)P_2 + B_2^0 X_2 B_2^{0T} & P_2 C_2^{0T} \\ C_2^0 P_2 & -X_1 \end{pmatrix} < 0.$$

Let $Y_1 = K_1 P_1, Y_2 = K_2 P_2$, the result follows. □

In fact, by using the method in Chapter 1, we can get conditions for (10.14) which are independent of Y_i, i.e., independent of K_i, i=1,2. Using Schur complement, (10.14) is equivalent to

$$P_1 A_1^T + A_1 P_1 + B_1^0 X_1 B_1^{0T} + B_{11} Y_1 + Y_1^T B_{11}^T + P_1 C_1^{0T} X_2^{-1} C_1^0 P_1 < 0,$$
$$P_2 A_2^T + A_2 P_2 + B_2^0 X_2 B_2^{0T} + B_{22} Y_2 + Y_2^T B_{22}^T + P_2 C_2^{0T} X_1^{-1} C_2^0 P_2 < 0.$$

By the projection lemma in Chapter 1, there exist Y_1, Y_2 such that the above inequalities hold if and only if
$$B_{11}^{\perp}(P_1 A_1^T + A_1 P_1 + B_1^0 X_1 B_1^{0T} + P_1 C_1^{0T} X_2^{-1} C_1^0 P_1) B_{11}^{\perp T} < 0,$$
$$B_{22}^{\perp}(P_2 A_2^T + A_2 P_2 + B_2^0 X_2 B_2^{0T} + P_2 C_2^{0T} X_1^{-1} C_2^0 P_2) B_{22}^{\perp T} < 0.$$
By Lemma 1.3 in Chapter 1, we know that the inequalities above hold, if and only if there exist scalars $\mu_1, \mu_2 > 0$ such that
$$P_1 A_1^T + A_1 P_1 + B_1^0 X_1 B_1^{0T} + P_1 C_1^{0T} X_2^{-1} C_1^0 P_1 < \mu_1 B_{11} B_{11}^T,$$
$$P_2 A_2^T + A_2 P_2 + B_2^0 X_2 B_2^{0T} + P_2 C_2^{0T} X_1^{-1} C_2^0 P_2 < \mu_2 B_{22} B_{22}^T.$$

Remark 10.10. In fact, by Theorem 10.5, if A_1 and A_2 are stable, the negative definiteness of the left side of the inequalities above is equivalent to the condition that there exist decompositions $A_{12} = B_1 C_2$ and $A_{21} = B_2 C_1$ such that $\|C_1(sI - A_1)^{-1} B_1\|_\infty \|C_2(sI - A_2)^{-1} B_2\|_\infty < 1$. Further, by small gain theorem in Chapter 2, the smaller the $\|C_1(sI - A_1)^{-1} B_1\|_\infty \|C_2(sI - A_2)^{-1} B_2\|_\infty$ is, the better the robust stability of the feedback system in Fig. 10.1 is, i.e., the stronger the coupling of the interconnected system in Fig. 10.1 is. Therefore, the method presented above can be used to strengthen the connection of the interconnected system, as detailed in [Duan et al. (2004c)].

10.2 Interconnected Lur'e systems

The absolute stability of Lur'e systems has been studied in Chapter 4. In this section, consider the absolute stability of an interconnected Lur'e system. Given two Lur'e systems:
$$\begin{cases} \dot{x}_i = A_i x_i + B_i f_i(y), \\ y_i = C_i x_i, \quad i = 1, 2, \end{cases} \quad (10.15)$$
where $x_1 = (x_{11}, x_{12}, \cdots, x_{1n_1})^T$ and $x_2 = (x_{21}, x_{22}, \cdots, x_{2n_2})^T$ are the states of two Lur'e systems, $y_1 = (y_{11}, y_{12}, \cdots, y_{1m_1})^T$ and $y_2 = (y_{21}, y_{22}, \cdots, y_{2m_2})^T$ are the outputs, and $f_1(y_1) = (f_{11}(y_{11}), f_{12}(y_{12}), \cdots, f_{1m_1}(y_{1m_1}))^T$ and $f_2(y_2) = (f_{21}(y_{21}), f_{22}(y_{22}), \cdots, f_{2m_2}(y_{2m_2}))^T$ are nonlinear functions satisfying the following sector conditions:
$$0 \leq \frac{f_{1i}(y_{1i})}{y_{1i}} \leq \mu_{1i}, \quad 0 \leq \frac{f_{2k}(y_{2k})}{y_{2k}} \leq \mu_{2k}, \quad f_{1i}(0) = 0,$$
$$f_{2k}(0) = 0, i = 1, \cdots, m_1, k = 1, \cdots, m_2. \quad (10.16)$$
Consider the following system composing of two subsystems (10.15) through linear interconnections:
$$\begin{cases} \dot{x}_1 = A_1 x_1 + A_{12} x_2 + B_1 f_1(y_1), \\ \dot{x}_2 = A_2 x_2 + A_{21} x_1 + B_2 f_2(y_2), \end{cases} \quad (10.17)$$

10.2. Interconnected Lur'e systems

where A_{12} and A_{21} are given interconnection matrices with compatible dimensions. System (10.17) is said to be absolutely stable if it is globally stable for all f_1 and f_2 satisfying condition (10.16).

For simplicity, the sector conditions are taken as in (10.16) without loss of generality. For a general sector constraint, one can transform it into the form (10.16) by a loop transformation, see Chapter 4 for details.

Let $\mu_1 = diag(\mu_{11}, \cdots, \mu_{1m_1})$ and $\mu_2 = diag(\mu_{21}, \cdots, \mu_{2m_2})$. Similarly to Theorem 10.1, one has the following result for local instability.

Theorem 10.7. *Assume that there exist* $0 \leq \nu_{10} \leq \mu_1$, $0 \leq \nu_{20} \leq \mu_2$, *and* A'_{12}, A'_{21} *such that*

$$A_{12} = A'_{12}(A_2 + B_2\nu_{20}C_2), \quad A_{21} = A'_{21}(A_1 + B_1\nu_{10}C_1).$$

If $det(I - A'_{21}A'_{12}) < 0$, *then systems (10.15) and (10.17) cannot be absolutely stable simultaneously.*

Proof. Take $f_1(y_1) = \nu_1 y_1$ and $f_2(y_2) = \nu_2 y_2$, where ν_1 and ν_2 are diagonal matrices with $0 \leq \nu_1 \leq \mu_1$ and $0 \leq \nu_2 \leq \mu_2$. Denote $A_{f_1} = A_1 + B_1\nu_1 C_1$, $A_{f_2} = A_2 + B_2\nu_2 C_2$ and $A_f = \begin{pmatrix} A_1 + B_1\nu_1 C_1 & A_{12} \\ A_{21} & A_2 + B_2\nu_2 C_2 \end{pmatrix}$.

Obviously, the absolute stability of (10.15) and (10.17) implies the stability of A_{f_1}, A_{f_2} and A_f, respectively. Let $A_{f_{10}}, A_{f_{20}}$ and A_{f_0} be the resulting matrices obtained from A_{f_1}, A_{f_2} and A_f after substituting ν_1 and ν_2 by ν_{10} and ν_{20}, respectively. Computing the determinant of A_{f_0}, one can get $det(A_{f_0}) = det(A_1+B_1\nu_{10}C_1)det(A_2+B_2\nu_{20}C_2 - A_{21}(A_1+B_1\nu_{10}C_1)^{-1}A_{12})$. It then follows from the assumption of the theorem that $det(A_{f_0}) = det(A_{f_{10}})det(A_{f_{20}} - A'_{21}A_{f_{10}}A_{f_{10}}^{-1}A'_{12}A_{f_{20}}) = det(A_{f_{10}})det(A_{f_{20}})det(I - A'_{21}A'_{12})$. When $A_{f_{10}}$ and $A_{f_{20}}$ are Hurwitz stable and $det(I - A'_{21}A'_{12}) < 0$, the constant term of the characteristic polynomial of A_{f_0} is less than zero. This implies that A_{f_0} is unstable. Therefore, $A_{f_{10}}, A_{f_{20}}$ and A_{f_0} cannot be stable simultaneously. This completes the proof. □

Corollary 10.2. *If there exist* A'_{12} *and* A'_{21} *such that*

$$A_{12} = A'_{12}A_2, \quad A'_{12}B_2 = 0, \quad A_{21} = A'_{21}A_1, \quad A'_{21}B_1 = 0,$$

and $det(I - A'_{21}A'_{12}) < 0$, *then systems (10.15) and (10.17) cannot be absolutely stable simultaneously, and for any* $0 \leq \nu_1 \leq \mu_1$ *and* $0 \leq \nu_2 \leq \mu_2$, A_{f_1}, A_{f_2} *and* A_f *cannot be stable simultaneously.*

Obviously, the above result can be generalized to cases of multiple subsystems as linearly interconnected systems in Section 10.1.1. And complex networks with linear nodes or Lur'e nodes were studied in [Duan et al. (2008)]. In some special networks, some subsystem must be unstable to ensure the stability of the whole network. This shows the effects of the unstable subsystem in complex network systems. As a realizable circuit system, the coupled Chua circuit can be studied by the above method, see the following chapter for details.

10.3 Lagrange stability of a generalized smooth Chua circuit

The Chua circuit is an ideal paradigm for the study of chaos in nonlinear systems [Madan (1993)]. This section introduces a generalized smooth Chua circuit from a viewpoint of interconnected systems and studies its Lagrange stability. Consider the following dynamical system described by the differential equations

$$\begin{cases} \dot{x} = A_1 x + A_{12} y, \\ \dot{y} = A_{21} x + A_2 y + R f(y), \end{cases} \qquad (10.18)$$

where $x = (x_1, \cdots, x_n)^T \in \mathbf{R}^n$, $y = (y_1, \cdots, y_m)^T \in \mathbf{R}^m$, $f(y) = (y_1^{k_1}, \cdots, y_m^{k_m})^T$, A_1, A_{12}, A_{21}, A_2 and R are real matrices with compatible dimensions, k_i are odd numbers and $k_i \geq 3$, $1 \leq i \leq m$.

Obviously, system (10.18) can be viewed as a special interconnected system composed of a linear subsystem $\dot{x} = A_1 x$ and a nonlinear subsystem $\dot{y} = A_2 y + R f(y)$. And (10.18) can also be viewed as a Lur'e system since the nonlinear functions satisfy the general infinite sector constraints $y_i y_i^{k_i} \geq 0$, $i = 1, \cdots, m$.

Recall that system (10.18) is called to be Lagrange stable if every of its solutions is bounded. By the method of [Arcak et al. (2002)], we have the following result for Lagrange stability of (10.18).

Theorem 10.8. *If A_1 is Hurwitz stable, R is diagonal and $R < 0$, then system (10.18) is Lagrange stable.*

Proof. Take $V(x, y) = x^T(t) P_1 x(t) + y^T(t) P_2 y(t)$ as a Lyapunov function candidate, where P_1 and P_2 are symmetric matrices. The time derivative of $V(x, y)$ along any trajectory of system (10.18) is given by

$$\dot{V}(x, y) = 2x^T P_1 (A_1 x + A_{12} y) + 2y^T P_2 (A_{21} x + A_2 y + R f(y)). \qquad (10.19)$$

Obviously, for any scalar $\lambda > 0$,

$$2 x^T P_1 A_{12} y \leq \frac{1}{\lambda} x^T P_1 A_{12} A_{12}^T P_1 x + \lambda y^T y$$

10.3. Lagrange stability of a generalized smooth Chua circuit

and
$$2y^T P_2 A_{21} x \leq \frac{1}{\lambda} x^T A_{21}^T A_{21} x + \lambda y^T P_2 P_2 y.$$

Hence, we have
$$\dot{V}(x,y) \leq x^T (P_1 A_1 + A_1^T P_1 + \frac{1}{\lambda} P_1 A_{12} A_{12}^T P_1 + \frac{1}{\lambda} A_{21}^T A_{21}) x \\ + y^T (\lambda I + \lambda P_2 P_2 + P_2 A_2 + A_2^T P_2) y + 2y^T P_2 R f(y). \quad (10.20)$$

Since A_1 is Hurwitz stable, there exist $\lambda > 0$ and $P_1 > 0$ such that
$$P_1 A_1 + A_1^T P_1 + \frac{1}{\lambda} P_1 A_{12} A_{12}^T P_1 + \frac{1}{\lambda} A_{21}^T A_{21} < 0. \quad (10.21)$$

On the other hand, we can choose P_2 diagonal and positive definite such that $P_2 R < 0$ due to the assumption on R. Further, by the assumption on $f(y)$, there is a scalar M large enough such that
$$y^T (\lambda I + \lambda P_2 P_2 + P_2 A_2 + A_2^T P_2) y + 2y^T P_2 R f(y) < 0 \quad (10.22)$$

when there exists y_i with $|y_i| \geq M$, $1 \leq i \leq m$. By (10.21) and (10.22), we know
$$\dot{V}(x,y) < 0$$

when there exists y_i with $|y_i| \geq M$. It is also possible to estimate the attracting set by the above method. By (10.21), there also exist scalars $\epsilon > 0$ and $\lambda > 0$ such that
$$x^T (P_1 A_1 + A_1^T P_1 + \frac{1}{\lambda} P_1 A_{12} A_{12}^T P_1 + \frac{1}{\lambda} A_{21}^T A_{21}) x < -\epsilon x^T x. \quad (10.23)$$

Then we have
$$\dot{V}(x,y) < -\epsilon(x^T x + y^T y) + y^T (\lambda I + \epsilon I + \lambda P_2 P_2 + P_2 A_2 + A_2^T P_2) y \\ + 2y^T P_2 R f(y).$$
$$\quad (10.24)$$

Similarly to (10.22), there exists a scalar M^* large enough such that
$$y^T (\lambda I + \epsilon I + \lambda P_2 P_2 + P_2 A_2 + A_2^T P_2) y + 2y^T P_2 R f(y) < 0 \quad (10.25)$$

when there exists y_i with $|y_i| \geq M^*$, $1 \leq i \leq m$. Denoting
$$d := \max_{y_i \in [-M^*, M^*], 1 \leq i \leq m} (y^T (\lambda I + \epsilon I + \lambda P_2 P_2 + P_2 A_2 + A_2^T P_2) y + 2y^T P_2 R f(y)),$$
$$\quad (10.26)$$

where ϵ, λ and P_2 satisfy (10.23) and (10.25), then we get
$$\dot{V}(x,y) < -\epsilon(x^T x + y^T y) + d \quad (10.27)$$

which implies that $\dot{V} < 0$ outside the compact set
$$\Omega := \{(x,y) \mid \epsilon(x^T x + y^T y) \leq d\}. \quad (10.28)$$

Therefore, system (10.18) is Lagrange stable and every solution (x,y) converges to the smallest level set of V which includes Ω. □

Remark 10.11. Consider the following canonical smooth Chua equation [Huang et al. (1996)]
$$\dot{x}_1 = -\alpha c x_1 + \alpha x_2 - \alpha x_1^3,$$
$$\dot{x}_2 = x_1 - x_2 + x_3, \qquad (10.29)$$
$$\dot{x}_3 = -\beta x_2.$$

Obviously, (10.29) can be viewed as a special case of (10.18) with
$$A_1 = \begin{pmatrix} -1 & 1 \\ -\beta & 0 \end{pmatrix}, A_{12} = \begin{pmatrix} 1 \\ 0 \end{pmatrix}, A_{21} = (\alpha\ 0), x = \begin{pmatrix} x_2 \\ x_3 \end{pmatrix}, A_2 = -\alpha c$$
and $R = -\alpha, y = x_1$. By Theorem 10.8, when $\alpha > 0, \beta > 0$, (10.29) is Lagrange stable. According to [Huang et al. (1996)], there is a chaotic attractor in (10.29) with parameters $\alpha = 10, \beta = 16, c = -0.143$. Of course, every solution of (10.29) is bounded at this time. In Theorem 10.8, there are no any requirements on A_{12}, A_{21} and A_2. System (10.18) can be viewed as a general extension of the smooth Chua equation (10.29). According to (10.29), one can also see from the forthcoming examples that there are many chaotic solutions in (10.18) under Lagrange stability.

Remark 10.12. In fact, under the assumptions of the theorem, the Lagrange stability is mainly due to the stiffening property of $f(y)$ as studied in [Arcak et al. (2002)]. This special property determines the state y converges to an attracting set whenever $|y_i|$ is large enough. According to the proof of Theorem 10.8, we can also optimize the attracting set of system (10.18) when it is Lagrange stable by minimizing d in (10.26).

Besides the quadratic Lyapunov function method in Theorem 10.8, we can obtain the following less conservative theorem by taking a "quadratic plus integral" Lyapunov function.

Theorem 10.9. *Suppose A_1 is Hurwitz stable. If R is diagonal stable, i.e., there exists a diagonal and positive definite matrix Q such that $QR + R^T Q < 0$, then system (10.18) is Lagrange stable.*

Proof. Take $V(x,y) = x^T(t)P_1 x(t) + 2\sum_{i=1}^m q_i \int_0^{y_i} f_i(\tau)d\tau$ as a Lyapunov function candidate, where P_1 is a symmetric matrix. The time derivative of $V(x,y)$ along any trajectory of system (10.18) is given by
$$\dot{V}(x,y) = 2x^T P_1(A_1 x + A_{12} y) + 2f^T(y)Q(A_{21}x + A_2 y + Rf(y)), \quad (10.30)$$
where $Q =: diag(q_1, \cdots, q_m)$. Obviously, for any scalars $\lambda_1 > 0$ and $\lambda_2 > 0$, we have
$$2x^T P_1 A_{12} y \le \frac{1}{\lambda_1} x^T P_1 A_{12} A_{12}^T P_1 x + \lambda_1 y^T y$$

10.3. Lagrange stability of a generalized smooth Chua circuit

and

$$2f^T(y)QA_{21}x \leq \frac{1}{\lambda_2}x^T A_{21}^T A_{21}x + \lambda_2 f^T(y)QQf(y).$$

Hence, we have

$$\dot{V}(x,y) \leq x^T(P_1 A_1 + A_1^T P_1 + \frac{1}{\lambda_1} P_1 A_{12} A_{12}^T P_1 + \frac{1}{\lambda_2} A_{21}^T A_{21})x \\ + \lambda_1 y^T y + 2f^T(y)QA_2 y + f^T(y)(QR + R^T Q + \lambda_2 QQ)f(y). \quad (10.31)$$

By the assumption on R, there are a diagonal matrix $Q > 0$ and a scalar $\lambda_2 > 0$ such that

$$QR + R^T Q + \lambda_2 QQ < 0 \quad (10.32)$$

On the other hand, since A_1 is Hurwitz stable, there exist $\lambda_1 > 0$ and $P_1 > 0$ such that

$$P_1 A_1 + A_1^T P_1 + \frac{1}{\lambda_1} P_1 A_{12} A_{12}^T P_1 + \frac{1}{\lambda_2} A_{21}^T A_{21} < 0 \quad (10.33)$$

where λ_2 satisfies (10.32). According to (10.32) and the assumption on $f(y)$, there is a scalar M large enough such that

$$\lambda_1 y^T y + 2f^T(y)QA_2 y + f^T(y)(QR + R^T Q + \lambda_2 QQ)f(y) < 0 \quad (10.34)$$

when there exists y_i with $|y_i| \geq M$, $1 \leq i \leq m$. Combining (10.33), (10.34) with (10.31), we know

$$\dot{V}(x,y) < 0$$

when there exists y_i with $|y_i| \geq M$. Therefore, system (10.18) is Lagrange stable. □

Remark 10.13. One can also give an estimate of the attracting set by the method of Theorem 10.9. Although Theorem 10.8 can be viewed as a special case of Theorem 10.9, it is possible to give a less conservative estimate of the attracting set by the method of Theorem 10.8 when R is diagonal and negative definite.

Remark 10.14. We have discussed Lagrange stability of system (10.18) in the theorems above. In fact system (10.18) is a Lur'e system, we can also establish conditions of global stability for (10.18) by the canonical methods for absolute stability, see Chapter 4 for details.

The following Chua-like oscillator with 6 parameters was studied by [Tsuneda (2005)]

$$\dot{x}_1 = -k\alpha x_1 + k\alpha x_2 - k\alpha(ax_1^3 + bx_1), \\ \dot{x}_2 = kx_1 - kx_2 + kx_3, \\ \dot{x}_3 = -k\beta x_2 - k\gamma x_3. \quad (10.35)$$

Obviously, (10.35) can be viewed as a special case of (10.18) with

$$A_1 = \begin{pmatrix} -k & k \\ -k\beta & -k\gamma \end{pmatrix}, A_{12} = \begin{pmatrix} k \\ 0 \end{pmatrix}, A_{21} = \begin{pmatrix} k\alpha & 0 \end{pmatrix}, x = \begin{pmatrix} x_2 \\ x_3 \end{pmatrix}$$

and $A_2 = -k\alpha - k\alpha b$, $R = -k\alpha a$, $y = x_1$. 20 sets of parameter values for (10.35) were given in Table 1 of [Tsuneda (2005)]. Testing these parameter sets by Theorem 10.8, we know that (10.35) is Lagrange stable with parameter sets C-1, C-2, C-4, C-7, C-11, C-12, C-17, C-19 and C-20 (see Table 1 of [Tsuneda (2005)]). For example, for the parameter set C-4 (Table 1 of [Tsuneda (2005)]): $\alpha = 143.1037$, $\beta = 207.34198$, $\gamma = -3.8767721$, $k = -1$, $a = -0.0156316049$, $b = -0.897577333$, all solutions of (10.35) are bounded, see Fig. 10.2 for the boundedness of several solutions.

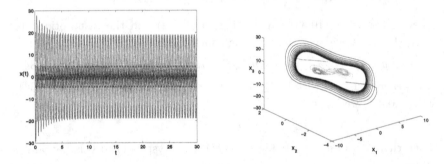

Fig. 10.2 The boundedness of several solutions of (10.35) for C-4 .

For other parameter sets C-3, C-5, C-6, C-8, C-9, C-10, C-13, C-14, C-15, C-16 and C-18 (see Table 1 of [Tsuneda (2005)]), although chaotic or periodic solutions were shown in [Tsuneda (2005)]), Theorem 10.8 fails for these parameter sets and computer simulation shows that system (10.35) is not Lagrange stable, i.e., there exists unbounded solutions. For example, for the parameter set C-9: $\alpha = -4.08685$, $\beta = -2$, $\gamma = 0$, $k = 1$, $a = 0.0345416029$, $b = -1.0936936418$, the solution of this system is bounded when the initial value is small, and the solution would be dramatically unbounded when the initial value becomes larger, see Fig. 10.3 for two solutions at $x(0) = (0.032, 0.013, 0.1)^T$ and $x(0) = (0.5, 0.13, 0.5)^T$.

A detailed analysis of parameter region of Lagrange stability for (10.35) was given in [Duan et al. (2007c)], which shows that Theorem 10.8 is very good for testing the Lagrange stability of (10.35). Lagrange stability of a class of nonlinearly coupled Chua circuits was also given in [Duan et al.

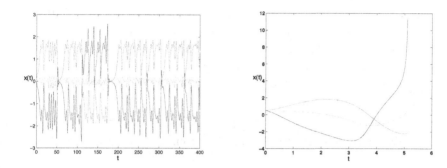

Fig. 10.3 Two solutions of (10.35) for C-9 with different initial values.

(2007c)]. Obviously, under the Lagrange stability, it allows the existence of multiple equilibria in system (10.18).

10.4 Input and output coupled nonlinear systems

Recall the pendulum-like system studied in Chapter 5,

$$\begin{cases} \frac{dx}{dt} = Ax + B\varphi(y), \\ \frac{dy}{dt} = Cx + D\varphi(y), \end{cases} \quad (10.36)$$

where $A \in \mathbb{R}^{n \times n}, B \in \mathbb{R}^{n \times m}, C \in \mathbb{R}^{m \times n}, D \in \mathbb{R}^{m \times m}, \varphi(x) = (\varphi_1(x_1), \cdots, \varphi_m(x_m))^T$. The transfer function from $\varphi(y)$ to $\dot{y}$ is

$$G(s) = C(sI - A)^{-1}B + D.$$

In system (10.36), φ_i only depends on x_i. In practical systems, the nonlinear functions φ_i may depend on $x_j, j \neq i$. As a special case, we consider the following system:

$$\begin{cases} \frac{dx}{dt} = Ax + B\varphi(z), \\ \frac{dy}{dt} = Cx + D\varphi(z), \end{cases} \quad (10.37)$$

where A, B, C and D are given as in (10.36), $\varphi(z) = (\varphi_1(z_1), \cdots, \varphi_m(z_m))^T$, $z_1 = \alpha_{11}y_1 + \cdots + \alpha_{1m}y_m, \cdots$, $z_m = \alpha_{m1}y_1 + \cdots + \alpha_{mm}y_m$. We denote the $m \times m$ matrix (α_{ij}) by M, so $z = My$. Then $\varphi(z)$ can also be denoted by $\varphi(My)$. Obviously, the transfer function from $\varphi(z)$ to $\dot{y}$ is still

$$G(s) = C(sI - A)^{-1}B + D.$$

In what follows, we consider the global convergence of system (10.37), where the interconnection matrix M will play an important role.

Suppose that the Assumptions 5.1, 5.2 and 5.3 in Chapter 5 hold. According to the equilibrium analysis of system (10.36) in Chapter 5, the number of equilibria of system (10.37) is infinite. If M is nonsingular, its equilibrium set can be written as

$$\Lambda = \{(x_{eq}, y_{eq}) \mid x_{eq} = 0, y_{eq} = M^{-1}z_{eq}, \varphi(z_{eq}) = 0\}.$$

Alternatively, system (10.37) can be written as

$$\begin{cases} \frac{dx}{dt} = Ax + B\varphi(z), \\ \frac{dz}{dt} = MCx + MD\varphi(z), \end{cases} \tag{10.38}$$

where A, B, C, D and M are given as in (10.37). Obviously, the transfer function from $\varphi(z)$ to $\dot{z}$ is $MG(s)$. About the global convergence of systems (10.37) and (10.38), we have the following statement.

Conclusion 10.1 *If M is nonsingular, then system (10.37) is gradient-like if and only if system (10.38) is gradient-like.*

Mathematically, system (10.37) is simple and its global convergence can be studied by following the argument for system (10.38). However, it is interesting from the viewpoint of input and output information coupling of systems. Combining Conclusion 10.1 and Theorems 5.8 and 5.9 in Chapter 5, one can get the following result [Duan et al. (2005a)].

Theorem 10.10. *Suppose A is Hurwitz stable. Under Assumptions 5.1, 5.2 and 5.3 in Chapter 5, if there exist real diagonal matrices P, Q and E with $\epsilon > 0, Q > 0$ such that*

$$\mathbf{Re}\{PMG(iw)\} + G^*(iw)M^T QMG(iw) + E < 0, \forall w \in \mathbb{R}$$

and the condition (ii) of Theorem 5.8 holds, then system (10.37) is gradient-like.

Given the coupling matrix M, we can analyze the gradient-like behavior of system (10.37) by testing the conditions of Theorem 10.10. In addition, we can also design such a coupling matrix such that (10.37) is gradient-like. For the design problem, first we rewrite Theorem 10.10 to reduce a matrix variable P [Duan et al. (2005a)].

Corollary 10.3. *Suppose A is Hurwitz stable. Under Assumptions 5.1, 5.2 and 5.3 in Chapter 5, if there exist a matrix M and diagonal matrices $E > 0$ and $Q > 0$ such that*

$$\mathbf{Re}\{MG(iw)\} + G^*(iw)M^T QMG(iw) + E < 0, \quad \forall w \in \mathbb{R} \cup \{\infty\}, \tag{10.39}$$

10.4. Input and output coupled nonlinear systems

and

$$4E > Q^{-1}V^2, \tag{10.40}$$

where V is defined as in Chapter 5, then for this M, the corresponding system (10.37) is gradient-like.

In order to design M by the canonical LMI method, we can convert (10.39) to an LMI by Corollary 1.4 and Schur complement.

Corollary 10.4. *The frequency-domain inequality (10.39) in Corollary 10.3 holds if and only if there is a symmetrical matrix W such that*

$$\begin{pmatrix} WA + A^T W & WB + \frac{1}{2}C^T M^T & C^T M^T \\ B^T P + \frac{1}{2}MC & E + \frac{1}{2}(MD + D^T M^T) & D^T M^T \\ MC & MD & -Q^{-1} \end{pmatrix} < 0.$$

Remark 10.15. Obviously, if the inequality in Corollary 10.4 holds, then M is nonsingular. Based on this corollary, the coupling matrix M can be designed easily by the popular LMI method (Chapter 1). By the LMI method given here, all the matrices M such that the conditions of Corollary 10.3 hold can be parameterized. When system (10.36) itself satisfies the conditions in Corollary 10.3 ($M = I$), one can also find some other M such that the corresponding systems as (10.37) are still globally convergent. Therefore, we can discuss the invariance of system properties with respect to the input and output coupling given here. It is also an interesting problem.

Example 10.3. We consider the following system

$$\begin{cases} \frac{dx}{dt} = Ax + B\varphi(y), \\ \frac{dy}{dt} = Cx + D\varphi(y), \end{cases} \tag{10.41}$$

where

$$A = \begin{pmatrix} -0.9 & 0 & 0.8 \\ 0 & 0 & 1.1 \\ 0 & -2.5 & -1 \end{pmatrix}, B = \begin{pmatrix} 1 & 0 \\ 0 & 0 \\ 0 & -1 \end{pmatrix}, C = \begin{pmatrix} 0.3 & 0 & 0 \\ 0 & 0.2 & 0 \end{pmatrix},$$

$$D = \begin{pmatrix} -0.1 & 3.2 \\ 6 & -0.1 \end{pmatrix}, y = \begin{pmatrix} y_1 \\ y_2 \end{pmatrix}, \varphi(y) = \begin{pmatrix} \varphi_1(y_1) \\ \varphi_2(y_2) \end{pmatrix},$$

$x = (x_1, x_2, x_3)^T$, $\varphi_1(t) = \sin(t) - 0.2$, $\varphi_2(t) = \sin(2t) - 0.1$.
It is easy to test that the conditions in Theorem 5.8 for system (10.41) are broken. Refer to Fig. 10.4 for the bounded oscillating solution of system (10.41) with initial value $x(0) = [0.2, -1, 0.1]^T, y(0) = [0, -1]^T$.

Solve the matrix inequality (10.40) and the inequality in Corollary 10.4 with system (10.41) by LMI method (Chapter 1), one can parameterize all solutions M. Take a simple one from this parameterized set, for example, take $M = \begin{pmatrix} 0.1 & 1 \\ 1.2 & 0.2 \end{pmatrix}$, system (10.41) becomes

$$\begin{cases} \frac{dx}{dt} = Ax + B\varphi(z), \\ \frac{dy}{dt} = Cx + D\varphi(z), \end{cases} \quad (10.42)$$

where A, B, C, D are as given in (10.41) and $z = My$. By testing the conditions in Theorem 10.10, we know that system (10.42) is globally convergent, refer to Fig. 10.5 for the global convergence of system (10.42), where the behavior of the solutions of (10.42) with three initial values $x(0) = [0.2, -1, 0.1]^T, y(0) = [0, -1]^T$; $x(0) = [0.8, -0.1, 0.2]^T, y(0) = [3, -1]^T$; $x(0) = [0.2, -0.1, 2]^T, y(0) = [-5, -1]^T$ are shown.

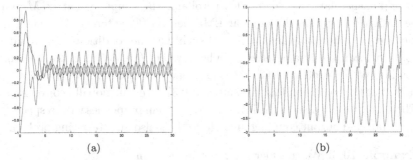

Fig. 10.4 The solution of system (10.41). (a) $x(t)$. (b) $y(t)$).

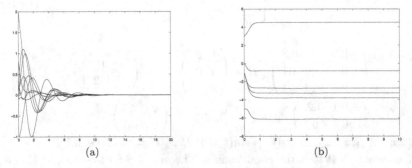

Fig. 10.5 The solutions of system (10.42). (a) $x(t)$. (b) $y(t)$).

10.4. Input and output coupled nonlinear systems

One can see from Fig. 10.5 that system (10.42) has more than one equilibrium and every solution $x(t) \to 0$, $y(t) \to y_{eq}$ which satisfies that $\varphi(z(t)) \to \varphi(z_{eq}) = 0$, $z_{eq} = Mx_{eq}$. From this example, we can see that the interconnection presented here can result in some large changes in nonlinear systems.

Similarly to gradient-like behavior, we can also study the property of dichotomy [Duan et al. (2005a)]. Using the result for dichotomy in Chapter 5, we can get

Theorem 10.11. *Suppose that A has no eigenvalues on the imaginary axis, (A, B) is controllable and (A, C) is observable. If there exist M and a scalar $\epsilon > 0$ such that*

$$\frac{1}{2}MG(iw) + \frac{1}{2}G^*(iw)M^T + \epsilon G^*(iw)M^T MG(iw) \leq 0, \ \forall w \in \mathbf{R} \cup \{\infty\}, \tag{10.43}$$

and with this M system (10.37) has isolated equilibria, then every bounded solution of (10.37) is convergent to a certain equilibrium.

Corollary 10.5. *If there exists a matrix M such that $MG(s)$ is SPR, then for this M, system (10.37) is dichotomous.*

Similarly to Corollary 10.4, we can also express the condition (10.43) in the form of LMI.

Corollary 10.6. *If (A, B) is controllable, (10.43) holds if and only if there exists a symmetric matrix W such that*

$$\begin{pmatrix} WA + A^T W & WB + \frac{1}{2}C^T M^T & C^T M^T \\ B^T W + \frac{1}{2}MC & \frac{1}{2}(MD + D^T M^T) & D^T M^T \\ MC & MD & -\epsilon^{-1}I \end{pmatrix} \leq 0. \tag{10.44}$$

Based on this corollary, the matrix M can be designed by the non-strict inequality method presented in Chapter 1.

The coupling matrix studied above is something like the input transformation in linear systems as discussed in Chapter 3. From the above corollaries, we can see the effects of nonlinear input and output coupling. Apart from the static coupling matrix discussed above, we can also design a dynamical coupling guaranteeing the property of dichotomy or gradient-like behavior for system (10.36). Take $\dot{y}$ as the input of the designed controller,

one can get the following controlled pendulum-like system:
$$\begin{cases} \frac{dx}{dt} = Ax + B\varphi(u), \\ \frac{dy}{dt} = Cx + D\varphi(u), \\ \frac{dx_k}{dt} = A_k x_k + B_k \dot{y}, \\ \frac{du}{dt} = C_k x_k + D_k \dot{y}, \end{cases} \tag{10.45}$$

Tidy up the system above as
$$\begin{cases} \frac{dx_{cl}}{dt} = A_{cl} x_{cl} + B_{cl} \varphi(u), \\ \frac{du}{dt} = C_{cl} x_{cl} + D_{cl} \varphi(u), \end{cases} \tag{10.46}$$

where
$$x_{cl} = \begin{pmatrix} y \\ x_k \end{pmatrix}, A_{cl} = \begin{pmatrix} A & 0 \\ B_k C & A_k \end{pmatrix}, B_{cl} = \begin{pmatrix} B \\ B_k D \end{pmatrix}, C_{cl} = (\, D_k C \;\; C_k \,)$$

and $D_{cl} = D_k D$. Apparently, the closed-loop transfer function from $\varphi(u)$ to $\dot{u}$ is $G_{cl}(s) = C_{cl}(sI - A_{cl})^{-1} B_{cl} + D_{cl} = K(s)G(s)$, where $K(s) = C_k(sI - A_k)^{-1} B_k + D_k$. The nonlinear feedback system (10.46) is shown in Fig. 10.6. Compared with the coupling matrix M discussed above, this controller can be viewed as a dynamic coupling. By the method in Chapter 3, we can design a multiplier $K(s)$ such that $K(s)G(s)$ is SPR, then the closed-loop system (10.46) must be dichotomous (Corollary 10.5).

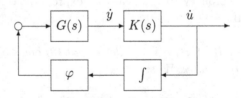

Fig. 10.6 Nonlinear feedback controlled system.

Example 10.4. Consider the following system,
$$\begin{cases} \frac{dx}{dt} = Ax + B\varphi(y), \\ \frac{dy}{dt} = Cx + D\varphi(y), \end{cases} \tag{10.47}$$

where $x = \begin{pmatrix} x_1 \\ x_2 \end{pmatrix}, y = \begin{pmatrix} y_1 \\ y_2 \end{pmatrix}, \varphi(y) = \begin{pmatrix} \varphi_1(y_1) \\ \varphi_2(y_2) \end{pmatrix}, A = \begin{pmatrix} -5 & 3 \\ -3 & 1 \end{pmatrix},$
$B = \begin{pmatrix} 5 & 0 \\ 0 & 1 \end{pmatrix}, C = \begin{pmatrix} 0.5 & 0 \\ 0 & 0.7 \end{pmatrix}, D = \begin{pmatrix} -0.4 & 3 \\ 3 & 0.3 \end{pmatrix}, \varphi_1(y_1) = \sin(y_1) - 0.3,$

10.4. Input and output coupled nonlinear systems

$\varphi_2(y_2) = \sin(2y_2) - 0.2$. Obviously (A, B) is controllable, (A, C) is observable and system (10.47) is a pendulum-like system. By computer simulation, we see that there is a bounded oscillating solution for system (10.47) with initial value $x(0) = [0.2, -3]^T, y(0) = [0.1, 20]^T$, see Fig. 10.7. So system (10.47) is not dichotomous or gradient-like.

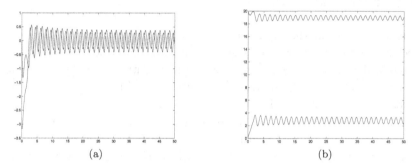

Fig. 10.7 The solution of system (10.47). (a) $x(t)$. (b) $y(t)$.

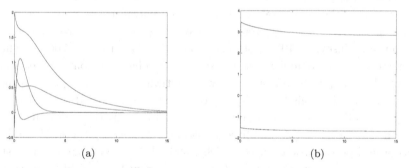

Fig. 10.8 The solution of the closed-loop system (10.46) with initial value $y(0) = [1, 2]^T, x_k(0) = [0.1, 0.5]^T, u(0) = [0.2, -0.1]^T$. (a) $x(t), x_k(t)$. (b) $u(t)$.

By the method given in Theorem 3.4 (Chapter 3), and noticing the order $G(s)$ and $K(s)$, one can get a controller $K(s)$ defined as in (10.45), where

$$A_k = \begin{pmatrix} -3.8770 & -0.5873 \\ 0.7540 & -1.0502 \end{pmatrix}, B_k = \begin{pmatrix} 1.4868 & -11.0411 \\ -1.0657 & 3.7262 \end{pmatrix},$$

$$C_k = \begin{pmatrix} 28.9955 & 71.8654 \\ 17.7084 & 31.5468 \end{pmatrix}, D_k = \begin{pmatrix} -6.5924 & 65.8415 \\ 65.8001 & 8.6869 \end{pmatrix}$$

such that $K(s)G(s)$ is SPR. Then the corresponding closed-loop system as given in (10.46) is dichotomous, see Fig. 10.8 (every bounded solution is convergent).

10.5 Notes and references

Large-scale system theory has been extensively studied in the last four decades, and many interesting results have been established, on such basic issues as decentrally fixed modes, decentralized controllers design, diagonal Lyapunov function method, M-matrix method, etc. ([Davison and Ozguner (1983); Gao and Huo (1994); Huang and Zhang (1998); Ikeda et al. (1983); Siljak (1978, 1991); Wang and Davison (1973); Wang (1978); Yang and Zhang (1995)]). Compared with centralized control methods, decentralized control has many advantages for its lower dimensionality, easier implementation, lower cost, etc., where interchange of state information among subsystems is generally not needed. Therefore, this control methodology has become quite popular in large-scale systems theory since the 1970s. However, due to the structural constraints in decentralized control, it is very difficult to develop a unified and effective design strategy [Davison (1974); Sezer and Huseyin (1978); Shi and Gao (1986); Zhai et al. (2001)]. This difficulty lies in the same problem just as the familiar static output-feedback control technique. As a result, many theoretical and practical problems remain unsolved in this field. Especially, a basic decentralized controller design method in large-scale systems is diagonal Laypunov function method. With this method, the stabilized closed-loop system is still stable under structural perturbation [Siljak (1978, 1991)]. From a viewpoint of robustness, this is a satisfying result. However, under such a control framework, the interconnected terms do not play any useful roles. In order to study the effects of interconnections in large-scale systems, an interconnected structure was given in [Duan et al. (2006a)] to show that under such a structure some subsystem must be unstable to stabilize the whole interconnected system, and a harmonic control method was also presented there. Correspondingly, discrete-time interconnected systems were studied in [Duan et al. (2007b)]. Small gain theorem was used to strengthen the stability degree of interconnected systems [Duan et al. (2004c)], and a class of special input-output coupled large-scale systems [Sezer and Huseyin (1978)] can also be studied [Duan et al. (2004c)] by this theorem. This interesting theorem was also used in [Siljak (1991)] for stability of large-scale systems.

10.5. Notes and references

The viewpoint of interconnected systems can be used to generalize smooth Chua circuit and study its Lagrange stability [Duan et al. (2007c)]. The similar method was first used in [Arcak et al. (2002)] for stiffening nonlinearities in Lur'e systems. The study of input and output coupled nonlinear systems can be found in [Duan et al. (2005a)].

In recent ten years, complex networks have got a great attention [Boccaletti et al. (2006); Duan et al. (2007d); Lü et al. (2004); Wang and Chen (2000)]. From a system theoretical viewpoint, complex dynamical networks can be viewed as large-scale systems. In fact, complex network systems are much similar to symmetric or similar large-scale systems [Kashlan and Geneidy (1996); Wang and Zhang (2000); Yang and Zhang (1995); Yan et al. (1998); Yang et al. (2001)]. Without requiring interchange of information among different nodes, the decentralized controller design methodology is obviously bringing an interesting research problem to the field of complex dynamical networks. Recently, stability analysis and decentralized control of complex dynamical networks were studied [Duan et al. (2008)], and a special structure was also found in networks under which sometimes some subsystem must be unstable to stabilize the whole network. Networks with Lur'e nodes or pendulum-like nodes were also studied in [Lu et al. (2008); Liu et al. (2007a)].

Chapter 11

Chua's Circuit

This chapter uses some of the theory established in the previous chapters and applies it to Chua's circuit, which is a simple circuit composed of a nonlinear diode and basic circuit elements such as capacitor, inductor and resistor. The only nonlinear element can be easily implemented by either a dual op amp package and six resistors or by an IC chip. This remarkable circuit, providing realization of a chaotic system, has been studied extensively by numerical, mathematical and experimental methods. It has become an ideal paradigm for the study of nonlinear dynamics. Section 11.1 introduces the canonical Chua's circuit and gives an analysis of dichotomy and absolute stability. Section 11.2 gives a chaos control method for Chua's circuit based on dichotomy. Section 11. 3 presents an application of Kalman conjecture to stabilization of Chua's circuit. Section 11.4 introduces an extended Chua circuit. Section 11.5 presents an analysis of coupled Chua circuits. Section 11.6 concludes the chapter.

11.1 Chua's circuit

Consider the canonical Chua's circuit in Fig. 11.1, it can be described by the following ordinary differential equation

$$\dot{v} = Av - bf_1(v_1), \qquad (11.1)$$

where

$$v = \begin{pmatrix} v_1 \\ v_2 \\ i_3 \end{pmatrix}, A = \begin{pmatrix} \frac{-1}{C_1 R_1} & \frac{1}{C_1 R_1} & 0 \\ \frac{1}{C_2 R_1} & \frac{-1}{C_2 R_1} & \frac{1}{C_2} \\ 0 & \frac{-1}{L_1} & \frac{-R_0}{L_1} \end{pmatrix}, b = \begin{pmatrix} \frac{1}{C_1} \\ 0 \\ 0 \end{pmatrix}$$

and

$$f_1(v_1) = G_2 v_1 + 0.5(G_1 - G_2)(|v_1 + 1| - |v_1 - 1|),$$

G_1 and G_2 are parameters to be chosen. It is well known that if we take $C_1 = 1, C_2 = 10, R_0 = 0, R_1 = 1, L_1 = 0.6725$ and $G_2 = -0.68, G_1 = -1.27$, then Chua's circuit (11.1) has a chaotic solution, see Fig. 11.2.

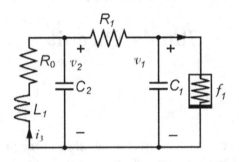

Fig. 11.1 Chua's circuit.

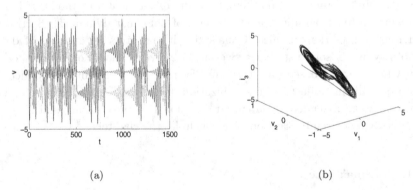

(a)   (b)

Fig. 11.2 The solution of Chua's circuit (11.1). (a) $v_1(t), v_2(t), i_3(t)$. (b) (v_1, v_2, i_3) space.

Rewrite Chua's circuit (11.1) as

$$\dot{v} = A_0 v - b_0 g_1(y_1), \quad y_1 = c_0 v, \tag{11.2}$$

where

$$A_0 = \begin{pmatrix} \frac{-1}{C_1 R_1} - \frac{G_2}{C_1} & \frac{1}{C_1 R_1} & 0 \\ \frac{1}{C_2 R_1} & \frac{-1}{C_2 R_1} & \frac{1}{C_2} \\ 0 & \frac{-1}{L_1} & \frac{-R_0}{L_1} \end{pmatrix}, \quad b_0 = \begin{pmatrix} \frac{G_1 - G_2}{C_1} \\ 0 \\ 0 \end{pmatrix}, \quad c_0 = (1, 0, 0),$$

11.1. Chua's circuit

$$g_1(v_1) = 0.5(|v_1 + 1| - |v_1 - 1|),$$

and v is given as in (11.1).

Obviously, system (11.2) can be viewed as a Lur'e system as in Chapter 4 with the sector constraint

$$0 \leq y_1 g_1(y_1) \leq y_1^2.$$

If A_0 is nonsingular, its equilibria can be analyzed as follows. Let $v_{eq} = (v_{1eq}, v_{2eq}, i_{3eq})^T$ be an equilibrium. Then, by equations (11.2), we have

$$v_{eq} = A_0^{-1} b_0 g_1(v_{1eq}), \quad v_{1eq} = c_0 v_{eq}.$$

So $v_{1eq} = c_0 A_0^{-1} b_0 g_1(v_{1eq})$. Therefore, it is clear that v_{1eq} is necessarily a point of intersection of the curve $y = g_1(v_{1eq})$ and the straight line $v_{1eq} = c_0 A_0^{-1} b_0 g_1$. For the function $g_1(v_1) = 0.5(|v_1 + 1| - |v_1 - 1|)$, obviously if $c_0 A_0^{-1} b_0 = 1$, then (11.2) has infinite equilibria with continuum.

If (11.2) has unique equilibrium, its absolute stability can be studied by the theory established in Chapter 4. For example, if we take $C_1 = 2, C_2 = 5, R_0 = 0, R_1 = 0.79, L_1 = 0.9$ and $G_2 = 6, G_1 = 10.7$, then by Theorem 4.2 in Chapter 4 system (11.2) is absolutely stable with the above sector condition.

On the other hand, we can also test the dichotomy property of system (11.2) to see when it has no bounded oscillating solutions. First, we know that the nonlinear function g_1 is piece-wise continuously differentiable and satisfies

$$0 \leq g_1'(y_1) \leq 1.$$

If we take $C_1 = 0.5, C_2 = 5, R_0 = -0.1, R_1 = 2, L_1 = 1$ and $G_2 = -0.68, G_1 = -1.27$, then by Theorem 4.6 in Chapter 4 system (11.2) is dichotomous. In fact, at this time, we do not need the bounded derivative condition for testing its dichotomy property.

Further, obviously g_1 is a bounded nonlinear function. Hence if A_0 is stable, then (11.2) is Lagrange stable. So this simple circuit can illustrate most of global properties of nonlinear systems discussed in the previous chapters.

In addition, for convenient numerical and theoretical analysis of Chua's circuit (11.1), by rescaling the parameters and variables and replacing the piece-wise linear nonlinearity by the cubic nonlinearity (11.1) can be normalized as the following mathematical equations

$$\begin{aligned} \dot{x}_1 &= -\alpha c x_1 + \alpha x_2 - \alpha x_1^3, \\ \dot{x}_2 &= x_1 - x_2 + x_3, \\ \dot{x}_3 &= -\beta x_2, \end{aligned} \quad (11.3)$$

which is well known as smooth Chua's circuit. Obviously, the cubic function $f(x_1) = x_1^3$ can be viewed as a nonlinearity with infinite sector constraint

$$0 \leq x_1 x_1^3 \leq +\infty.$$

So the absolute stability and dichotomy of (11.3) can be studied by the methods in Chapter 4 and the Lagrange stability can be studied by the method in Chapter 10.

11.2 Dichotomy: application to chaos control for Chua's circuit system

The property of dichotomy can ensure that all the bounded solutions of systems are convergent which indicates chaos oscillations cannot happen in the systems. This section studies the problem of designing a feedback control such that Chua's circuit system is dichotomous by injecting an external signal in the circuit, and therefore strange attractors in the system disappear. For this purpose we first discuss a simple problem of controller design for a class of special nonlinear systems.

Consider a nonlinear control system:

$$\begin{cases} \dot{x} = Ax + b\varphi(\sigma) + b_1 u, \\ \dot{\sigma} = cx + \rho\varphi(\sigma), \end{cases} \quad (11.4)$$

where $A \in \mathbb{R}^{n \times n}$ is a constant real matrix, $b \in \mathbb{R}^n, b_1 \in \mathbb{R}^n$ and $c^T \in \mathbb{R}^n$ are real vectors and ρ is a real number. $x \in \mathbb{R}^n, \sigma \in \mathbb{R}$ and u is control variable. Nonlinear function $\varphi : \mathbb{R} \to \mathbb{R}$ is piece-wise continuously differentiable on $\mathbb{R}$ and

$$-\infty < \mu_1 \leq \frac{d\varphi}{d\sigma} \leq \mu_2 < +\infty \quad (11.5)$$

for all $\sigma \in \mathbb{R}$. Consider the problem of designing a state feedback $u = Kx$ such that the closed-loop nonlinear system:

$$\begin{cases} \dot{x} = (A + b_1 K)x + b\varphi(\sigma), \\ \dot{\sigma} = cx + \rho\varphi(\sigma) \end{cases} \quad (11.6)$$

is dichotomous. Applying Lemma 8.2 to the closed-loop system (11.6) and supposing $Q = P^{-1}$ and $KQ = Y$, it is easy to obtain the following result.

Theorem 11.1. *Suppose that there exist real numbers $\eta, \varepsilon > 0, \tau \geq 0$ and*

11.2. Dichotomy: application to chaos control for Chua's circuit system   271

matrices $Q = Q^T$ and Y such that

$$\begin{bmatrix} AQ + QA^T + b_1Y + Y^Tb_1 & QJ - \frac{1}{2}r_2Y^Tb_1^Tc^T + b & Qc^T \\ J^TQ - \frac{1}{2}r_2cb_1Y + b^T & \rho^2r_1 - r_2cb + \eta\rho & 0 \\ cQ & 0 & -\frac{1}{r_1} \end{bmatrix} \leq 0 \quad (11.7)$$

where $r_1 = \varepsilon - \tau\mu_1\mu_2, r_2 = \tau(\mu_1 + \mu_2)$ and $J = \frac{\eta}{2}c^T + \rho r_1 c^T - \frac{1}{2}r_2A^Tc^T$.
Then there exists a control matrix $K = YQ^{-1}$. Furthermore, if

(i) $(A + b_1K, b)$ is controllable and $(A + b_1K, c)$ is observable;
(ii) $\varphi(\sigma)$ has a finite number of isolated zeros, or the matrix $Q > 0$,

then the closed-loop nonlinear system (11.6) is dichotomous.

Consider the canonical Chua's circuit system:

$$\begin{cases} \dot{v}_1 = \frac{1}{C_1}\left[\frac{v_2 - v_1}{R} - g(v_1)\right], \\ \dot{v}_2 = \frac{1}{C_2}\left[\frac{v_1 - v_2}{R} + i_3\right], \\ \dot{i}_3 = \frac{1}{L}[-v_2 - R_0i_3], \end{cases} \quad (11.8)$$

where v_1 and v_2 are the voltages across the capacitors C_1 and C_2 respectively, i_3 is the current through the inductor L. R, R_0 are resistors, and $g(v_1)$ is the current through the nonlinear resistor which is defined as

$$g(v_1) = G_bv_1 + \frac{1}{2}(G_a - G_b)(|v_1 + B_p| - |v_1 - B_p|) \quad (11.9)$$

In order to control chaos oscillations we inject an external current signal i_u in the RCL-node as shown in Fig. 11.3. The state equations describing the circuit are as follows

$$\begin{cases} \frac{dv_1}{dt} = \frac{1}{C_1}\left[\frac{v_2 - v_1}{R} - g(v_1)\right], \\ \frac{dv_2}{dt} = \frac{1}{C_2}\left[\frac{v_1 - v_2}{R} + i_3 + i_u\right], \\ \frac{di_3}{dt} = \frac{1}{L}[-v_2 - R_0i_3]. \end{cases} \quad (11.10)$$

Let

$$x = \frac{v_1}{B_p}, \quad y = \frac{v_2}{B_p}, \quad z = \frac{i_3R}{B_p}, \quad \tau = \pm\frac{t}{C_2R}, \quad u = \pm\frac{R}{B_p}i_u,$$

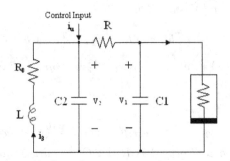

Fig. 11.3 Chua's circuit injected control signal i_u.

$$m_0 = G_a R, \quad m_1 = G_b R, \quad \alpha = \frac{C_2}{C_1}, \quad \beta = \frac{C_2 R^2}{L}, \quad \gamma = \frac{C_2 R_0}{L}.$$

Then the system (11.10) can be transformed into the following form

$$\begin{cases} \dfrac{dx}{d\tau} = k\alpha(y - x - f(x)), \\ \dfrac{dy}{d\tau} = k(x - y + z) + u, \\ \dfrac{dz}{d\tau} = k(-\beta y - \gamma z), \end{cases} \quad (11.11)$$

where u is injected control, $k = \pm 1$. The nonlinear function $f(x)$ is defined as

$$f(x) = m_1 x + \frac{1}{2}(m_0 - m_1)[|x+1| - |x-1|]. \quad (11.12)$$

We write the system (11.11) as the form of (11.4), i.e.,

$$\begin{cases} \dot{\bar{x}} = A\bar{x} + bf(\sigma) + b_1 u, \\ \dot{\sigma} = c\bar{x} + \rho f(\sigma), \end{cases} \quad (11.13)$$

where

$$A = \begin{pmatrix} -k\alpha & k\alpha & 0 \\ k & -k & k \\ 0 & -k\beta & -k\gamma \end{pmatrix}, \quad b = \begin{pmatrix} -k\alpha \\ 0 \\ 0 \end{pmatrix}, \quad \bar{x} = \begin{pmatrix} x \\ y \\ z \end{pmatrix}, \quad b_1 = \begin{pmatrix} 0 \\ 1 \\ 0 \end{pmatrix},$$

$$c = \begin{pmatrix} -k\alpha & k\alpha & 0 \end{pmatrix}, \quad \rho = -k\alpha, \quad \sigma = x.$$

It is obvious that $\mu_1 = \min\{m_0, m_1\} \le f'(x) \le \mu_2 = \max\{m_0, m_1\}$.

Chaotic oscillations of Chua's system (11.11) without control, i.e., $u = 0$, are possible when special parameter values of the system are chosen.

11.2. Dichotomy: application to chaos control for Chua's circuit system

Table 11.1 Parameters of the system and initial values for the gallery of strange attractors.

Figure	Fig.11.4 (a)	Fig. 11.5 (a)	Fig.11.6 (a)
α	-1.083	-1.461	-1.301
β	0.0000969	-0.09422	-0.0136
γ	0.007327	-0.3230	-0.02969
m_0	-0.09411	1.219	0.1690
m_1	0.0001899	-0.5141	-0.4767
k	-1	-1	1
initial value	$[0.1, 0.1, 0.1]^T$	$[-2.5, 0.2, 0.2]^T$	$[-2.5, 0.2, 0.2]^T$

Figure	Fig.11.7 (a)	Fig. 11.8 (a)	Fig.11.9 (a)
α	-1.609	-17.38	9.0
β	0.0704	1.413	14.28
γ	0.2973	1.045	0.0
m_0	-0.1392	-1.525	-1.142
m_1	-0.2175	-0.4576	-0.7142
k	1	-1	1
initial value	$[-0.4, -0.2, 0.4]^T$	$[0.1, 0.6, -0.4]^T$	$[0.1, 0.1, -0.4]^T$

Strange attractors are shown in Fig. 11.4 (a) — Fig. 11.9 (a) corresponding to different parameter values given in Table 11.1.

Applying Theorem 11.1 to Chua's system (11.13). We can design a control matrix K such that the system (11.13) with $u = K\bar{x}$ is dichotomous. Table 11.2 shows the obtained control matrices corresponding to the parameters given in Table 11.1. The states of system (11.13) injected control signal $u = K\bar{x}$ are shown in Fig. 11.4 (b) — Fig. 11.9 (b). It is obvious that the chaotic oscillations of the system disappear.

Remark 11.1. The result of Theorem 11.1 for controller design guaranteeing dichotomy of Chua's circuit (11.13) with $b_1 = (0, 0, 1)^T$ is also true.

In what follows we consider a modified Chua circuit described by four coupled first-order differential equations [Thamilmaran and Lakshmanan (2004); Liu et al. (2007a)] :

$$\begin{cases} C_1 \dfrac{dv_1}{dt} = i_1 - g(v_1), \\ C_2 \dfrac{dv_2}{dt} = -G_1 v_2 - i_1 - i_2, \\ L_1 \dfrac{di_1}{dt} = v_2 - v_1 - Ri_1, \\ L_2 \dfrac{di_2}{dt} = v_2, \end{cases} \quad (11.14)$$

Table 11.2 Parameter values and controllers obtained by Theorem 11.1.

Figure	Fig. 11.4 (b)	Fig. 11.5 (b)	Fig. 11.6 (b)
$u=0$	Fig. 11.4 (a)	Fig. 11.5 (a)	Fig. 11.6 (a)
η	50	10	-10
r_1	0.05	1	0.5
r_2	-0.6	0.6	-0.6
controller	$K_2 = \begin{pmatrix} 83.4289 \\ -83.4289 \\ 1.0000 \end{pmatrix}^T$	$K_3 = \begin{pmatrix} 45.0412 \\ -45.0412 \\ 1.0000 \end{pmatrix}^T$	$K_4 = \begin{pmatrix} -18.6638 \\ 18.6638 \\ -1.0000 \end{pmatrix}^T$

Figure	Fig. 11.7 (b)	Fig. 11.8 (b)	Fig. 11.9 (b)
$u=0$	Fig. 11.7 (a)	Fig. 11.8 (a)	Fig. 11.9 (a)
η	-8	200	100
r_1	0.1	0.1	0.1
r_2	-0.06	-6	-0.6
controller	$K_5 = \begin{pmatrix} 387.1546 \\ -387.1546 \\ -1.0000 \end{pmatrix}^T$	$K_6 = \begin{pmatrix} -16.7292 \\ -57.0452 \\ -53.5607 \end{pmatrix}^T$	$K_7 = \begin{pmatrix} 159.5567 \\ -159.5567 \\ -1.0000 \end{pmatrix}^T$

where v_1 and v_2 are the voltages across the capacitors C_1 and C_2 and i_1 and i_2 denotes the currents through the inductances L_1 and L_2 respectively. R is resistor, and $g(v_1)$ is the current through the nonlinear resistor which is defined by (11.9)

In order to control chaos and hyperchaos we inject an external current signal i_u in the circuit as shown in Fig. 11.10. The state equations describing the circuit are as follows

$$\begin{cases} C_1 \dfrac{dv_1}{dt} = i_1 - g(v_1), \\ C_2 \dfrac{dv_2}{dt} = -G_1 v_2 - i_1 - i_2 + i_u, \\ L_1 \dfrac{di_1}{dt} = v_2 - v_1 - R i_1, \\ L_2 \dfrac{di_2}{dt} = v_2. \end{cases} \quad (11.15)$$

Let

$$x_1 = \frac{v_1}{B_p}, \quad x_2 = \frac{v_2}{B_p}, \quad x_3 = \frac{i_1}{GB_p}, \quad x_4 = \frac{i_2}{GB_p}, \quad t = \tau \frac{C_2}{G},$$

$$\alpha_1 = \frac{C_2}{C_1}, \quad \alpha_2 = \frac{G_1}{G}, \quad \beta_1 = \frac{C_2}{L_1 G^2}, \quad \beta_2 = \frac{C_2}{L_2 G^2},$$

$$a = \frac{G_a}{G}, \quad b = \frac{G_b}{G}, \quad G = \frac{1}{R}, \quad u = \frac{1}{GB_p} i_u.$$

11.2. Dichotomy: application to chaos control for Chua's circuit system 275

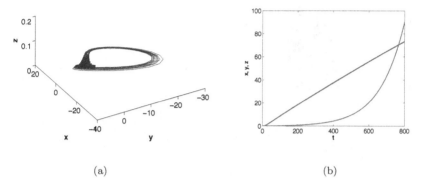

(a)   (b)

Fig. 11.4 (a) Chaotic attractor of Chua's equation (11.11) with $u = 0$. (b) The states of (11.11) injected control $u = K_2 \bar{x}$.

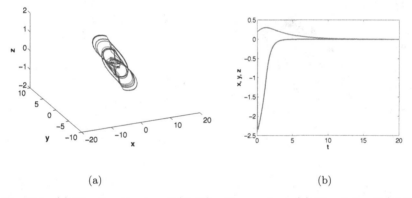

(a)   (b)

Fig. 11.5 (a) Chaotic attractor of (11.11) with $u = 0$. (b) The states of (11.11) injected $u = K_3 \bar{x}$.

Then the system (11.15) can be transformed into the following form

$$\begin{cases} \dfrac{dx_1}{d\tau} = \alpha_1(x_3 - f(x_1)), \\ \dfrac{dx_2}{d\tau} = -\alpha_2 x_2 - x_3 - x_4 + u, \\ \dfrac{dx_3}{d\tau} = \beta_1(x_2 - x_1 - x_3), \\ \dfrac{dx_4}{d\tau} = \beta_2 x_2, \end{cases} \quad (11.16)$$

where u is an injected control variable. The nonlinear function $f(x_1)$ is

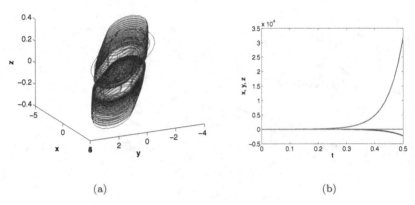

Fig. 11.6 (a) Chaotic attractor of (11.11) with $u = 0$. (b) The states of (11.11) injected $u = K_4\bar{x}$.

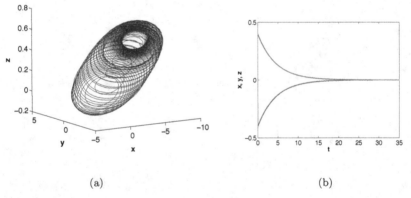

Fig. 11.7 (a) Chaotic attractor of (11.11) with $u = 0$. (b) The states of (11.11) injected $u = K_5\bar{x}$.

defined as
$$f(x_1) = bx_1 + \frac{1}{2}(a-b)[|x_1+1| - |x_1-1|]. \tag{11.17}$$
Writing the system (11.16) as the form of (11.4), i.e.,
$$\begin{cases} \dot{\bar{x}} = A\bar{x} + bf(\sigma) + b_1 u, \\ \dot{\sigma} = c\bar{x} + \rho f(\sigma), \end{cases} \tag{11.18}$$
where
$$A = \begin{pmatrix} 0 & 0 & \alpha_1 & 0 \\ 0 & -\alpha_2 & -1 & -1 \\ -\beta_1 & \beta_1 & -\beta_1 & 0 \\ 0 & \beta_2 & 0 & 0 \end{pmatrix}, \quad b = \begin{pmatrix} -\alpha_1 \\ 0 \\ 0 \\ 0 \end{pmatrix}, \quad x = \begin{pmatrix} x_1 \\ x_2 \\ x_3 \\ x_4 \end{pmatrix}, \quad b_1 = \begin{pmatrix} 0 \\ 1 \\ 0 \\ 0 \end{pmatrix},$$

11.2. Dichotomy: application to chaos control for Chua's circuit system

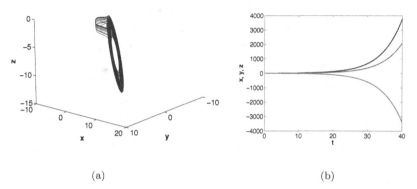

Fig. 11.8 (a) Chaotic attractor of (11.11) with $u = 0$. (b) The states of (11.11) injected $u = K_6 \bar{x}$.

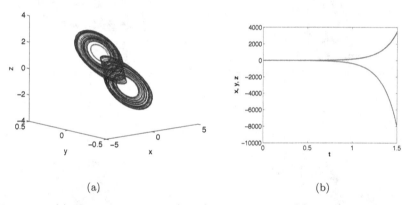

Fig. 11.9 (a) Chaotic attractor of (11.11) with $u = 0$. (b) The states of (11.11) injected $u = K_7 \bar{x}$.

$$c = \begin{pmatrix} 0 & 0 & \alpha_1 & 0 \end{pmatrix}, \quad \rho = -\alpha_1, \quad \sigma = x_1.$$

It is obvious that $\mu_1 = \min\{a, b\} \leq f'(x) \leq \mu_2 = \max\{a, b\}$.

Choosing an initial value $x_0 = (0.51, 0.41, 0.1, -0.03)^T$, when there is no control, i.e., $u = 0$, chaotic attractors of the modified Chua's circuit (11.16) for in $(x_2 - x_1), (x_1 - x_3), (x_2 - x_3)$ and $(x_3 - x_4)$ planes are shown in Fig. 11.11 with the parameters fixed at

$$\alpha_1 = 2.1428, \quad \alpha_2 = -0.2025, \quad \beta_1 = 0.09798,$$

$$\beta_2 = 0.00382, \quad a = -0.04725, \quad b = 3.15.$$

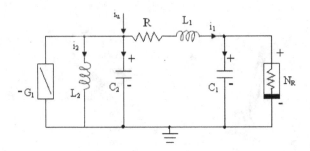

Fig. 11.10 Modified canonical Chua's circuit (11.15) injected control signal i_u.

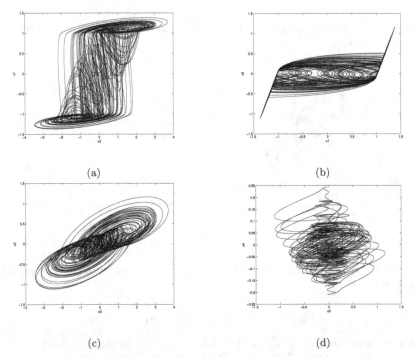

Fig. 11.11 (a) The chaotic attractor of the modified Chua's equation (11.16) with $u = 0$.
(a) $x_2 - x_1$ plane. (b) $x_1 - x_3$ plane. (c) $x_2 - x_3$ plane. (d) $x_3 - x_4$ plane.

Hyperchaotic attractors of the modified Chua circuit (11.16) in $(x_2 - x_1), (x_1 - x_3), (x_2 - x_3)$ and $(x_3 - x_4)$ planes are shown in Fig. 11.12 with parameters fixed at

$$\alpha_1 = 2.1428, \quad \alpha_2 = -0.1285, \quad \beta_1 = 0.0393,$$

11.2. Dichotomy: application to chaos control for Chua's circuit system

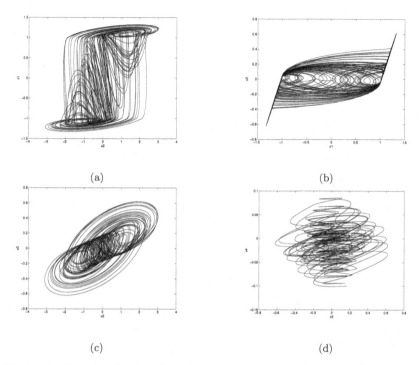

Fig. 11.12 The hyperchaotic attractor of the modified Chua's equation (11.16) with $u = 0$. (a) $x_2 - x_1$ plane. (b) $x_1 - x_3$ plane. (c) $x_2 - x_3$ plane. (d) $x_3 - x_4$ plane.

$$\beta_2 = 0.001532, \quad a = -0.0299, \quad b = 1.995,$$

and the corresponding initial value is $x_0 = (0.01, 0.01, 0.01, 0.01)^T$. Applying Theorem 11.1 to the modified Chua circuit (11.16). We can design a control matrix K such that the system (11.18) with $u = Kx$ is dichotomous. The obtained control matrices are $K_1 = 1.0e + 005(-8.3196, -0.0015, -5.7682, -0.0001)$ for chaotic attractor and $K_2 = 1.0e + 005(-2.4148, -0.0007, -1.6281, -0.0000)$ for hyperchaotic attractor, respectively. States of the system (11.16) injected control signal $u = K_1 x$ and $u = K_2 x$ are shown in Fig.11.13 and Fig. 11.14, respectively. It is obvious that the chaotic and hyperchaotic oscillations of the system disappear. In Fig. 11.13, the solution of the system (11.16) with the same initial value $x = (0.51, 0.41, 0.1, -0.03)^T$ converges to specific equilibrium point $(0.0000, 0.0000, -0.0000, -0.0342)^T$ of the system (11.16). In Fig. 11.14, the solution of the system (11.16) with the same initial value $x = (0.01, 0.01, 0.01, 0.01)^T$ converges to specific equilibrium point $(0.0000, 0.0000, -0.0000, 0.0096)^T$ of the system (11.16).

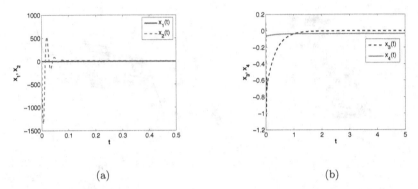

Fig. 11.13 The states of the modified Chua's equation (11.16) with $u = K_1 x$. (a) $x_1(t), x_2(t)$. (b) $x_3(t), x_4(t)$.

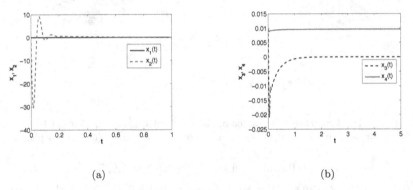

Fig. 11.14 The states of the modified Chua's equation (11.16) with $u = K_2 x$. (a) $x_1(t), x_2(t)$. (b) $x_3(t), x_4(t)$.

11.3 Kalman conjecture: application to the stabilization of Chua's circuit

As stated in Chapter 3, Kalman conjecture is true for $n = 3$. In this section we use the Kalman conjecture to discuss the stabilization problem of Chua's circuit system.

Consider an uncertain system defined by equation:

$$\dot{x}(t) = (A + \Delta A(t))x(t) \qquad (11.19)$$

where $\Delta A(t) \in \Omega$ with

$$\Omega = \left\{ \Delta A(t) = E\Delta(t)F : \Delta(t) \in R^{m \times m}, \Delta^T(t)\Delta(t) \leq I \right\}, \qquad (11.20)$$

$E \in R^{n \times m}, F \in R^{m \times n}$ are given matrices and E is full column rank.

If there exists a positive definite matrix P such that
$$(A + \Delta A(t))^T P + P(A + \Delta A(t)) < 0$$
for all $\Delta A(t) \in \Omega$, then the system (11.19) is called to be quadratically stable. Kharogonekar et al. [Khargonekar et al. (1990)] gives an equivalent condition for quadratic stability of the system (11.19), that is, the system (11.19) is quadratically stable if and only if A is a Hurwitz stable matrix and $||F(sI - A)^{-1}E||_\infty < 1$.

In the following we consider the problem of stabilization for Chua's circuit system (11.8) using Kalman conjecture. In order to control chaos we inject an external control signal u in the circuit. By variable transformations as similar as in Section 11.2, the system (11.8) injected a control signal can be written as

$$\begin{cases} \dfrac{dx}{d\tau} = k\alpha(y - x - f(x)), \\ \dfrac{dy}{d\tau} = k(x - y + z), \\ \dfrac{dz}{d\tau} = k(-\beta y - \gamma z) + u, \end{cases} \quad (11.21)$$

where u is the injected control variable, $k = \pm 1$. The nonlinear function $f(x)$ is defined as

$$f(x) = m_1 x + \frac{1}{2}(m_0 - m_1)(|x+1| - |x-1|), \quad (11.22)$$

It is obvious that $\mu_1 = \min\{m_0, m_1\} \leq f'(x) \leq \mu_2 = \max\{m_0, m_1\}$.

The stabilizing problem of Chua's circuit is stated as follows: design a linear feedback control $u = Kx_c$ such that the trivial solution of system:

$$\begin{cases} \dot{x} = k\alpha(y - x - f(x)), \\ \dot{y} = k(x - y + z), \\ \dot{z} = k(-\beta y - \gamma z) + Kx_c \end{cases} \quad (11.23)$$

is globally convergent, where $x_c = \begin{pmatrix} x & y & z \end{pmatrix}^T$. The system (11.23) is rewritten as:

$$\dot{x}_c = (A_0 + b_1 K)x_c + bf(x), \quad (11.24)$$

where

$$A_0 = \begin{pmatrix} -k\alpha & k\alpha & 0 \\ k & -k & k \\ 0 & -k\beta & -k\gamma \end{pmatrix}, \quad x_c = \begin{pmatrix} x \\ y \\ z \end{pmatrix}, \quad b = \begin{pmatrix} -k\alpha \\ 0 \\ 0 \end{pmatrix}, \quad b_1 = \begin{pmatrix} 0 \\ 0 \\ 1 \end{pmatrix},$$

$f(x)$ is defined by (11.22) and $\mu_1 = \min\{m_0, m_1\} \leq f'(x) \leq \mu_2 = \max\{m_0, m_1\}$.

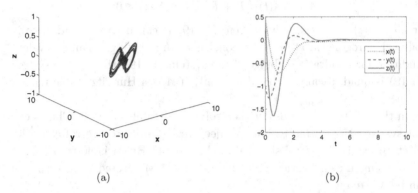

Fig. 11.15 (a) The chaotic attractor of the Chua's equation (11.21) with $u = 0$. (b) The states of Chua's circuit (11.21) injected control $u = K_1 x_c$.

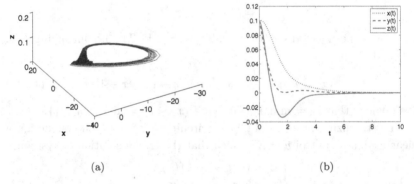

Fig. 11.16 (a) The chaotic attractor of the Chua's equation (11.21) with $u = 0$. (b) The states of Chua's circuit (11.21) injected control $u = K_2 x_c$.

Using Kalman conjecture for $n = 3$ we can obtain the following result of global convergence for the system (11.24).

Lemma 11.1. *The system (11.24) is globally convergent for given matrix K if*

(i) $A_2 + b_1 K$ is a Hurwitz stable matrix;
(ii) $\|\frac{1}{2}(\mu_2 - \mu_1) A_1 (sI - (A_2 + b_1 K))^{-1}\|_\infty < 1$;

11.3. Kalman conjecture: application to the stabilization of Chua's circuit   283

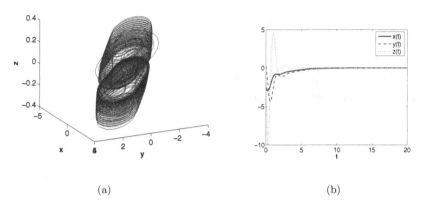

(a)   (b)

Fig. 11.17 (a) The chaotic attractor of the Chua's equation (11.21) with $u = 0$. (b) The states of Chua's circuit (11.21) injected control $u = K_3 x_c$.

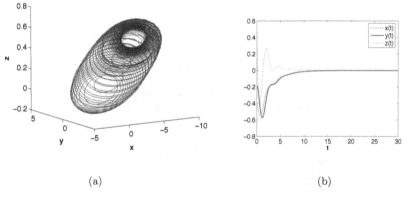

(a)   (b)

Fig. 11.18 (a) The chaotic attractor of the Chua's equation (11.21) with $u = 0$. (b) The states of Chua's circuit (11.21) injected control $u = K_4 x_c$.

where $A_2 = A_0 + \dfrac{1}{2}(\mu_1 + \mu_2)A_1$ and $A_1 = \begin{pmatrix} -k\alpha & 0 & 0 \\ 0 & 0 & 0 \\ 0 & 0 & 0 \end{pmatrix}$.

Proof. By the result of Kalman conjecture for $n = 3$, the system (11.24) is globally convergent if for any $m \in (\mu_1, \mu_2)$ the linear system

$$\dot{x}_c = A_\Delta x_c \qquad (11.25)$$

is stable, where $A_\Delta = A_0 + b_1 K + mA_1$ and $A_1 = b(1 \ \ 0 \ \ 0)$. Let

$$m = \dfrac{1}{2}[(\mu_2 - \mu_1)\Delta + \mu_2 + \mu_1].$$

Then $\mu_1 < m < \mu_2$ holds if and only if $|\Delta| < 1$ and thus

$$A_\Delta = A_0 + \frac{1}{2}(\mu_1+\mu_2)A_1 + b_1K + \frac{1}{2}(\mu_2-\mu_1)\Delta A_1 = A_2 + b_1K + \frac{1}{2}(\mu_2-\mu_1)\Delta A_1$$

where $A_2 = A_0 + \frac{1}{2}(\mu_1 + \mu_0)A_1$. By the results of quadratical stability stated in the beginning of this section, A_Δ is stable for all $|\Delta| < 1$ if and only if (i) and (ii) hold. This completes the proof. □

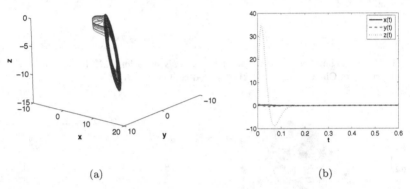

Fig. 11.19 (a) The chaotic attractor of the Chua's equation (11.21) with $u = 0$. (b) The states of Chua's circuit (11.21) injected control $u = K_5 x_c$.

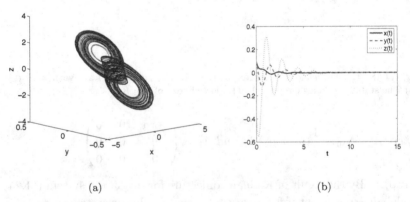

Fig. 11.20 (a) The chaotic attractor of the Chua's equation (11.21) with $u = 0$. (b) The states of Chua's circuit (11.21) injected control $u = K_6 x_c$.

Applying the bounded real lemma to the closed-loop system (11.24) and supposing $Q = P^{-1}$ and $KQ = Y$ it is easy to find that the problem of

11.3. Kalman conjecture: application to the stabilization of Chua's circuit

Table 11.3 Parameters and initial values for the gallery of attractors.

Figure	Fig. 11.15 (a)	Fig. 11.16 (a)	Fig. 11.17 (a)
α	−1.559	−1.038	−1.301
β	0.01564	0.0000969	−0.0136
γ	0.1574	0.007327	−0.02969
m_0	−0.2438	−0.09411	0.1690
m_1	−0.04251	0.0001899	−0.4767
k	−1	−1	1
initial value	$(0.5, -1, 0.5)^T$	$(0.1, 0.1, 0.1)^T$	$(-2.5, 0.2, 0.2)^T$

Figure	Fig. 11.18 (a)	Fig. 11.19 (a)	Fig. 11.20 (a)
α	−1.609	−17.38	9.0
β	0.0704	1.413	14.28
γ	0.2973	1.045	0.0
m_0	−0.1392	−1.525	−1.142
m_1	−0.2175	−0.4576	−0.7142
k	1	−1	1
initial value	$(-0.4, -0.2, 0.4)^T$	$(0.1, 0.6, -0.4)^T$	$(0.1, 0.1, -0.4)^T$

Table 11.4 Stabilizing controllers obtained by Theorem 11.2 corresponding to the Table 11.3.

Figure	corresponding chaos attractor	stabilizing controller K
Fig. 11.15 (b)	Fig. 11.15 (a)	$K_1 = (-2.7002, 5.7723, -3.0953)$
Fig. 11.16 (b)	Fig. 11.16 (a)	$K_2 = (-2.9706, 6.0595, -3.2820)$
Fig. 11.17 (b)	Fig. 11.17 (a)	$K_3 = (29.1729, -19.9361, -4.2452)$
Fig. 11.18 (b)	Fig. 11.18 (a)	$K_4 = (7.5895, -7.4170, -2.2173)$
Fig. 11.19 (b)	Fig. 11.19 (a)	$K_5 = 10^3 * (9.6143, 7.2189, -0.1170)$
Fig. 11.20 (b)	Fig. 11.20 (a)	$K_6 = (-4.9332, -9.8722, -0.0711)$

designing a matrix K satisfying conditions (i) and (ii) of Lemma 11.1 can be transformed into solving LMI problem.

Theorem 11.2. *There exists a matrix K such that the Chua's system (11.24) with injected control $u = Kx_c$ is globally convergent if there exist a positive definite matrix $Q > 0$ and a matrix Y such that*

$$\begin{pmatrix} A_2Q + QA_2^T + b_1Y + Y^Tb_1^T & I & \frac{1}{2}(\mu_2 - \mu_1)QA_1^T \\ I & -I & 0 \\ \frac{1}{2}(\mu_2 - \mu_1)A_1Q & 0 & -I \end{pmatrix} < 0 \quad (11.26)$$

holds, where $A_2 = A_0 + \frac{1}{2}(\mu_2 + \mu_1)A_1$. Furthermore, if LMI (11.26) is feasible, then $K = YQ^{-1}$.

Fig. 11.15 (a) — Fig. 11.20 (a) show the strange attractors corresponding to different parameter values given in Table 11.3.

For distinct chaotic attractors of the Chua's system shown in Fig. 11.15 (a) — Fig. 11.20 (a), by Theorem 11.2, we can design linear feedback controls such that the systems in the form of (11.21) with $u = Kx_c$ are globally convergent. That is, the chaotic attractors disappear and the state attains the trivial equilibrium $(0, 0, 0)$. Corresponding to the different parameters given in Table 11.3, the obtained control matrices $K_1 - K_6$ are shown in Table 11.4 and states of the system (11.21) with $u = Kx_c$ are shown in Fig. 11.15 (b) — Fig. 11.20 (b).

Remark 11.2. The results of Theorem 11.2 for controller design guaranteeing global convergence of Chua's circuit (11.24) with $b_1 = [1, 0, 0]^T$ and $b_1 = [0, 1, 0]^T$ are also true.

Remark 11.3. In [Chen and Dong (1993)] it is required that

$$k\alpha > 0, \gamma = 0, k\beta > 0, m_0 < 0, m_1 < 0, m_0 < m_1$$

for the stabilization of the Chua's system (11.11). The results and proof in [Chen and Dong (1993)] depend on these conditions. The methods in Section 11.2 and Section 11.3 are not limited to the above conditions exposed here. The fact can be observed from Tables 11.1 and 11.3.

11.4 An extended Chua circuit

An extended Chua circuit with two nonlinear functions was given in [Duan et al. (2005b)], see Fig. 11.21. It can be described by the differential equation (11.27).

$$\begin{cases} \dot{v}_1 = \frac{-1}{C_1 R_1} v_1 + \frac{1}{C_1 R_1} v_2 - \frac{1}{C_1} f_1(v_1), \\ \dot{v}_2 = \frac{1}{C_2 R_1} v_1 + \frac{-1}{C_2 R_1} v_2 + \frac{1}{C_2} i_3 - \frac{1}{C_2} f_2(v_2), \\ \dot{i}_3 = \frac{-1}{L_1} v_2 + \frac{-R_0}{L_1} i_3, \end{cases} \quad (11.27)$$

where f_1 and f_2 are two nonlinear functions defined as

$$f_1(v_1) = G_{12}v_1 + 0.5(G_{11} - G_{12})(|v_1 + 1| - |v_1 - 1|),$$

11.4. An extended Chua circuit

$$f_2(v_2) = G_{22}v_2 + 0.5(G_{21} - G_{22})(|v_2 + 1| - |v_2 - 1|).$$

Rewrite the above extended Chua circuit as,

$$\dot{v} = Av - Bg(y), \quad y = Cv, \tag{11.28}$$

where

$$A = \begin{pmatrix} \frac{-1}{C_1 R_1} - \frac{G_{12}}{C_1} & \frac{1}{C_1 R_1} & 0 \\ \frac{1}{C_2 R_1} & \frac{-1}{C_2 R_1} - \frac{G_{22}}{C_2} & \frac{1}{C_2} \\ 0 & \frac{-1}{L_1} & \frac{-R_0}{L_1} \end{pmatrix}, B = \begin{pmatrix} \frac{G_{11} - G_{12}}{C_1} & 0 \\ 0 & \frac{G_{21} - G_{22}}{C_2} \\ 0 & 0 \end{pmatrix},$$

$$C^T = \begin{pmatrix} 1 & 0 \\ 0 & 1 \\ 0 & 0 \end{pmatrix}, v = \begin{pmatrix} v_1 \\ v_2 \\ i_3 \end{pmatrix}, y = \begin{pmatrix} y_1 \\ y_2 \end{pmatrix}, g(y) = \begin{pmatrix} g_1(y_1) \\ g_2(y_2) \end{pmatrix},$$

$$g_1(y_1) = 0.5(|y_1 + 1| - |y_1 - 1|), g_2(y_2) = 0.5(|y_2 + 1| - |y_2 - 1|).$$

At this time, the nonlinear function g_i satisfies the sector constraint

$$0 \leq g_i(\tau)\tau \leq \tau^2, \quad i = 1, 2 \tag{11.29}$$

and $g_i(\tau)$ is differentiable for almost all $\tau \in \mathbb{R}$ and

$$0 \leq g_i'(\tau) \leq 1, \quad i = 1, 2. \tag{11.30}$$

Obviously, (11.28) can be viewed as an MIMO Lur'e system, while the canonical Chua's circuit (11.1) is an SISO Lur'e system. In [Duan et al. (2005b)], different oscillating phenomena were shown in the extended Chua circuit (11.27). By the method of Chapter 4, we can analyze the parameter domain under which system (11.27) has no bounded oscillating solutions,

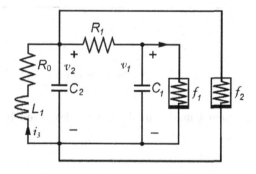

Fig. 11.21 Extended Chua circuit.

i.e., system (11.27) is dichotomous, or is absolutely stable. In these cases, the existence of chaotic attractors is impossible. For example, if we take $C_1 = 1.5, C_2 = 10, R_1 = 1, L_1 = 1/1.487, R_0 = 2, G_{11} = -1.27, G_{12} = -0.68, G_{21} = -1.27, G_{22} = -0.68$. By Theorem 4.8, we know that system (11.28) is dichotomous with the constraints (11.29) and (11.30). At this time, A is not stable, so (11.28) is not absolutely stable with these parameters. But if we take parameters $C_1 = 2, C_2 = 5, R_1 = 0.79, L_1 = 0.9, R_0 = 0, G_{11} = G_{21} = 10.7, G_{12} = G_{22} = 6$, then by Theorem 4.3 (11.28) is absolutely stable with sector condition (11.29).

Similarly to SISO Chua's circuit, if A is stable, then (11.28) is Lagrange stable.

11.5 Coupled Chua circuit

In this section, consider the following coupled Chua circuits,

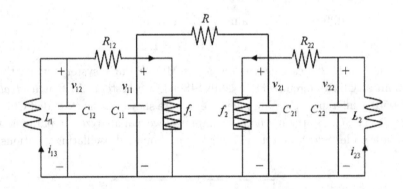

Fig. 11.22 Coupled Chua circuit.

The above coupled Chua's circuit can be described by the following equation,

$$\dot{v} = A_0 v - B_0 (f_1(v_{11}), f_2(v_{21}))^T, \qquad (11.31)$$

11.5. Coupled Chua circuit

where

$$v = \begin{pmatrix} v_1 \\ v_2 \end{pmatrix}, \ v_1 = \begin{pmatrix} v_{11} \\ v_{12} \\ i_{13} \end{pmatrix}, \ v_2 = \begin{pmatrix} v_{21} \\ v_{22} \\ i_{23} \end{pmatrix}, \ A_0 = \begin{pmatrix} A_{01} & A_{12} \\ A_{21} & A_{02} \end{pmatrix},$$

$$B_0 = \begin{pmatrix} b_{01} & 0 \\ 0 & b_{02} \end{pmatrix}, \ b_{01} = \begin{pmatrix} \frac{1}{C_{11}} \\ 0 \\ 0 \end{pmatrix}, \ b_{02} = \begin{pmatrix} \frac{1}{C_{21}} \\ 0 \\ 0 \end{pmatrix},$$

$$A_{01} = \begin{pmatrix} \frac{-1}{C_{11}R_{12}} + \frac{-1}{C_{11}R} & \frac{1}{C_{11}R_{12}} & 0 \\ \frac{1}{C_{12}R_{12}} & \frac{-1}{C_{12}R_{12}} & \frac{1}{C_{12}} \\ 0 & \frac{-1}{L_1} & 0 \end{pmatrix}, \ A_{12} = \begin{pmatrix} \frac{1}{C_{11}R} & 0 & 0 \\ 0 & 0 & 0 \\ 0 & 0 & 0 \end{pmatrix},$$

$$A_{02} = \begin{pmatrix} \frac{-1}{C_{21}R_{22}} + \frac{-1}{C_{21}R} & \frac{1}{C_{21}R_{22}} & 0 \\ \frac{1}{C_{22}R_{22}} & \frac{-1}{C_{22}R_{22}} & \frac{1}{C_{22}} \\ 0 & \frac{-1}{L_2} & 0 \end{pmatrix}, \ A_{21} = \begin{pmatrix} \frac{1}{C_{21}R} & 0 & 0 \\ 0 & 0 & 0 \\ 0 & 0 & 0 \end{pmatrix},$$

$$f_1(v_{11}) = G_{12}v_{11} + 0.5(G_{11} - G_{12})(|v_{11} + 1| - |v_{11} - 1|),$$

$$f_2(v_{21}) = G_{22}v_{21} + 0.5(G_{21} - G_{22})(|v_{21} + 1| - |v_{21} - 1|).$$

Rewrite (11.31) as the following system with Lur'e type

$$\begin{cases} \dot{v} = Av - B(g_1(y_1), g_2(y_2))^T, \\ y = Cv, \end{cases} \tag{11.32}$$

where

$$A = \begin{pmatrix} A_1 & A_{12} \\ A_{21} & A_2 \end{pmatrix}, \ B = \begin{pmatrix} b_1 & 0 \\ 0 & b_2 \end{pmatrix}, \ C = \begin{pmatrix} 1 & 0 & 0 & 0 & 0 & 0 \\ 0 & 0 & 0 & 1 & 0 & 0 \end{pmatrix}, \ y = \begin{pmatrix} y_1 \\ y_2 \end{pmatrix},$$

$$A_1 = \begin{pmatrix} \frac{-1}{C_{11}R_{12}} + \frac{-1}{C_{11}R} - \frac{G_{12}}{C_{11}} & \frac{1}{C_{11}R_{12}} & 0 \\ \frac{1}{C_{12}R_{12}} & \frac{-1}{C_{12}R_{12}} & \frac{1}{C_{12}} \\ 0 & \frac{-1}{L_1} & 0 \end{pmatrix}, \ b_1 = \begin{pmatrix} \frac{G_{11}-G_{12}}{C_{11}} \\ 0 \\ 0 \end{pmatrix},$$

$$A_2 = \begin{pmatrix} \frac{-1}{C_{21}R_{22}} + \frac{-1}{C_{21}R} - \frac{G_{22}}{C_{21}} & \frac{1}{C_{21}R_{22}} & 0 \\ \frac{1}{C_{22}R_{22}} & \frac{-1}{C_{22}R_{22}} & \frac{1}{C_{22}} \\ 0 & \frac{-1}{L_2} & 0 \end{pmatrix}, \ b_2 = \begin{pmatrix} \frac{G_{21}-G_{22}}{C_{21}} \\ 0 \\ 0 \end{pmatrix},$$

$$g_1(x) = g_2(x) = 0.5(|x + 1| - |x - 1|),$$

and A_{12}, A_{21} and v are given as in (11.31). Obviously, $g_i(x), i = 1, 2$ satisfy the sector condition (11.29) and the bounded derivative condition (11.30). Then system (11.32) can be viewed as an interconnected Lur'e system composed of two Lur'e subsystems

$$\begin{cases} \dot{v}_1 = A_1 v_1 - b_1 g_1(y_1), \\ y_1 = v_{11} \end{cases} \quad (11.33)$$

and

$$\begin{cases} \dot{v}_2 = A_2 v_2 - b_2 g_2(y_2), \\ y_2 = v_{21} \end{cases} \quad (11.34)$$

as discussed in Chapter 10.

For matrix A in (11.32), obviously we have

$$A_{12} = A'_{12} A_2, \quad A_{21} = A'_{21} A_1,$$

where

$$A'_{12} = \begin{pmatrix} \frac{1}{D_2 C_{11} R L_2 C_{22}} & 0 & \frac{1}{D_2 C_{11} R C_{21} C_{22} R_{22}} \\ 0 & 0 & 0 \\ 0 & 0 & 0 \end{pmatrix}$$

and

$$A'_{21} = \begin{pmatrix} \frac{1}{D_1 C_{21} R L_1 C_{12}} & 0 & \frac{1}{D_1 C_{21} R C_{11} C_{12} R_{12}} \\ 0 & 0 & 0 \\ 0 & 0 & 0 \end{pmatrix},$$

$D_1 = det(A_1)$, $D_2 = det(A_2)$. Then, we have

$$det(I - A'_{21} A'_{12}) = 1 - \frac{1}{D_1 D_2 C_{11} C_{21} R^2 L_1 L_2 C_{12} C_{22}}.$$

Using Theorem 10.7 in Chapter 10, we have

Theorem 11.3. *Systems (11.33), (11.34) and (11.32) cannot be absolutely stable simultaneously with sector constraint (11.29) when* $1 - \frac{1}{D_1 D_2 C_{11} C_{21} R^2 L_1 L_2 C_{12} C_{22}} < 0.$

Proof. Obviously, the conditions of Theorem 10.7 hold with $\nu_{10} = 0$ and $\nu_{20} = 0$. □

In fact, the stability of A in (11.32) is equivalent to the stability of the interconnected Chua's circuit in Fig.11.22 when $f_1(v_{11}) = G_{12} v_{11}, f_2(v_{21}) = G_{22} v_{21}$. It is possible that systems (11.33), (11.34) and

(11.32) can be absolutely stable simultaneously when $det(I - A'_{21}A'_{12}) = 1 - \frac{1}{D_1 D_2 C_{11} C_{21} R^2 L_1 L_2 C_{12} C_{22}} > 0$. For example, take

$C_{11} = 2, C_{12} = 5, R_{12} = 0.79, L_1 = 0.9, C_{21} = 5, C_{22} = 0.9,$
$R_{22} = 1.26, L_2 = 0.744, R = 10, G_{11} = 10.7, G_{12} = 6, G_{21} = 14.7, G_{22} = 35.$

At this time, $det(I - A'_{21}A'_{12}) = 1 > 0$. By Theorem 4.3 in Chapter 4, we know that systems (11.33), (11.34) and (11.32) are absolutely stable simultaneously.

On the other hand, with the parameter values chosen as

$C_{11} = 0.1, C_{12} = -1, R_{12} = -1.09, L_1 = -0.9, C_{21} = 1, C_{22} = -1.9,$
$R_{22} = -1.05, L_2 = 6.1, R = 8, G_{11} = 0.85, G_{12} = 0.9, G_{21} = 0.8, G_{22} = 0.9,$

then we know that $det(I - A'_{21}A'_{12}) = -1.0002 < 0$. By Theorem 11.3, systems (11.33), (11.34) and (11.32) cannot be absolutely stable simultaneously. Actually, systems (11.33) and (11.32) are absolutely stable for these parameter values, but (11.34) is not absolutely stable, see Fig. 11.23 for the solutions of (11.33) and (11.34) with the initial values $v_i(0) = v_2(0) = (1.2, -6.9, 5.2)^T$ and Fig. 11.24 for the solution of (11.32) with the same initial values.

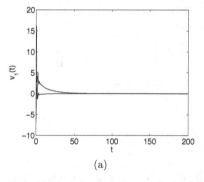 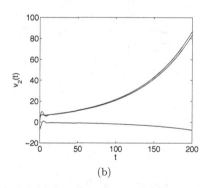

(a) (b)

Fig. 11.23 The solutions of (11.33) and (11.34). (a) $v_{11}(t), v_{12}(t), i_{13}(t)$. (b) $v_{21}(t), v_{22}(t), i_{23}(t)$.

From the above figures, we can see that system (11.34) is not absolutely stable. Its solution given in Fig. 11.23 is unbounded. But after a linear interconnection as in (11.32), (11.32) becomes absolutely stable.

The dichotomy property of coupled Chua circuits can be studied similarly. We can also study the following one-way coupled Chua circuit (Fig. 11.5).

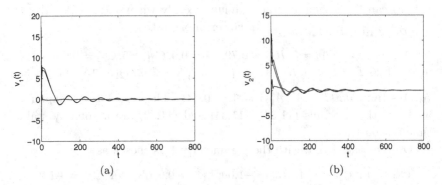

Fig. 11.24 The solution of (11.32). (a) $v_{11}(t), v_{12}(t), i_{13}(t)$. (b) $v_{21}(t), v_{22}(t), i_{23}(t)$.

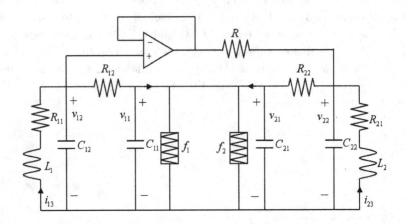

Fig. 11.25 One-way coupled Chua circuit.

A one-way coupled Chua circuit can be described by the following equation,

$$\begin{cases} \dot{v}_1 = A_1 v_1 - b_1 f_1(v_{11}), \\ \dot{v}_2 = A_2 v_2 + A_{21} v_1 - b_2 f_2(v_{21}), \end{cases} \quad (11.35)$$

where

$$A_1 = \begin{pmatrix} -\frac{1}{C_{11}R_{12}} & \frac{1}{C_{11}R_{12}} & 0 \\ \frac{1}{C_{12}R_{12}} & -\frac{1}{C_{12}R_{12}} & \frac{1}{C_{12}} \\ 0 & -\frac{1}{L_1} & -\frac{R_{11}}{L_1} \end{pmatrix}, \ b_1 = \begin{pmatrix} \frac{1}{C_{11}} \\ 0 \\ 0 \end{pmatrix}, \ b_2 = \begin{pmatrix} \frac{1}{C_{21}} \\ 0 \\ 0 \end{pmatrix},$$

$$A_2 = \begin{pmatrix} -\frac{1}{C_{21}R_{22}} & \frac{1}{C_{21}R_{22}} & 0 \\ \frac{1}{C_{22}R_{22}} & -\frac{1}{C_{22}R_{22}} - \frac{1}{C_{22}R} & \frac{1}{C_{22}} \\ 0 & -\frac{1}{L_2} & -\frac{R_{21}}{L_2} \end{pmatrix}, A_{21} = \begin{pmatrix} 0 & 0 & 0 \\ 0 & \frac{1}{C_{22}R} & 0 \\ 0 & 0 & 0 \end{pmatrix},$$

v_1, v_2, f_1 and f_2 are given as in (11.31).
Take $C_{11} = 2, C_{12} = 10, R_{11} = 1, R_{12} = 1, L_1 = 1/1.487, C_{21} = 1, C_{22} = 10, R_{21} = 0, R_{22} = 1, L_2 = 1/1.487, G_{11} = G_{21} = -1.27, G_{12} = G_{22} = -0.68$. Rewriting (11.35) as a system in Lur'e form by the above method and using Theorem 4.8 in Chapter 4, we know that the first Chua circuit in Fig. 11.25 is dichotomous, and the second Chua circuit is not dichotomous and has a chaotic solution shown as in Fig. 11.2 . After the one-way coupling as in Fig. 11.25 with a resistor $R = 1$, the coupled Chua circuits (11.35) are dichotomous by Theorem 4.8. Therefore, a chaotic Chua's circuit can be controlled by a dichotomous Chua's circuit as in Fig. 11.25 such that the resulting coupled Chua circuits are dichotomous.

11.6 Notes and references

In the past three decades, Chua's circuit has attracted a lot of attention because of its rich dynamical behavior [Chua (1994); Chua et al. (1986); Chua (1998); Chen and Dong (1998); Jorge and Chua (1999); Madan (1993); Shil'nikov (1993); Rabinder (1993)]. This simple electronic circuit is an ideal vehicle for research on chaos and bifurcation by means of both laboratory experiments and computer simulations. Except for single Chua's circuit, coupled Chua circuits were studied in [Kapitaniak et al. (1994); Imai et al. (2002)]. Chaotic Lur'e systems were studied in [Suykens et al. (1997)]. Many systems of common interest such as Chua's circuits and arrays of Chua's circuits both for unidirectional and mutual coupling can be represented in Lur'e form. Different approaches of chaos control for Chua's circuit were proposed based on fuzzy [Chen and Chen (1999); Chang (2001)], time series-based [Ogorzalek (1993)], LMIs [Tanaka et al. (1998)] and feedback control [Chen and Dong (1993)] methodologies. With the property of dichotomy, the nonexistence of chaotic or periodic solutions in single Chua's circuit was discussed in [Leonov et al. (1996)], and chaos control for Chua's circuit was studied in [Wang et al. (2006d)]. Chaos control in coupled Chua circuits was also studied under the property of dichotomy in [Duan et al. (2004b, 2005a,b)]. The hyperchaos dynamics of the modified four-order canonical Chua's electrical circuit was investigated using laboratory experiment and numerical analysis in [Thamilmaran and Lakshmanan (2004)], and chaos control was studies in [Liu et al. (2007b)].

Bibliography

Abramovitch, D. Y. (1990). Lyapunov redesign of analog phase-lock loops, *IEEE Trans. Communications*, **38**, 12, pp. 2197–2202.
Anderson, B. D. O. (1967). A system theory criterion for positive real matrices, *SIAM J. Control*, **5**, 2, pp. 171–182.
Anderson, B. D. O. and Moore, J. B. (1970). *Linear Optimal Control* (Prentice-Hall, New York).
Anderson, B. (1972). Small-gain theorem, passivity theorem and their equivalence, *Journal of the Franklin Institute-engineering and Applied Mathematics*, **293**, 2, pp. 105–115.
Anderson, B. D. O. and Vongpanitlerd, S. (1973). *Network analysis and Synthesis* (Englewood Cliffs, NJ: Prentice-Hall).
Andronov, A. A. (1955). *Collection of Works* (Moscow (in Russian)).
Ao, D., Geng, Z. Y. and Huang, L. (2008). Robust dichotomy of the Lur'e system with structured uncertainties, *International Journal of Control*, **81**, 5, pp. 778–787.
Arcak, M. and Kokotović, P. (2001). Feasibility conditions for circle criterion designs, *Systems and Control Letters*, **42**, 5, pp. 405–412.
Arcak, M., Larsen, M. and Kokotović, P. (2002). Boundedness without absolute stability in systems with stiffening nonlinearities, *European Journal of Control*, **8**, 3, pp. 243–250.
Arcak, M., and Teel, A. (2002). Input-to-state stability for a class of Lurie systems, *Automatica*, **38**, 11, pp. 1945–1949.
Arcak, M., Larsen, M. and Kokotovic, P. (2003). Circle and Popov criteria as tools for nonlinear feedback design, *Automatic*, **39**, 4, pp. 643–650.
Ascheid, G. and Meyr, H. (1982). Cycle slips in phase-locked loops: A tutorial survey, *IEEE Trans. Commun.*, **30**, 10, pp. 2228–2241.
Barabanov, N. E. (1988). On the Kalman conjecture (in Russian), *Sibirsk. Mat. Zh.*, **29**, 3, pp. 3–11.
Barnett, S. and Man, T. (1970). Comments on 'A theorem on the Lyapunov matrix equation', *IEEE Trans. Automatic Control*, **15**, 4, pp. 279–280.
Barnett, S. and Storey, C. (1967b). On the general functional matrix for a linear system, *IEEE Trans. Automatic Control*, **12**, 4, pp. 436–438.

Belevitch, V. (1970). *Classical Network Theory* (San Francisco, CA: Holden-Day).
Bernstein, D. S., Haddad, W. and Sparks, A. G. (1994). A simolified proof of the multivariable Popov criterion and an upper bound for the structured singular value with real parameter uncertainty, *In Proc. the IEEE conference on Decision and Control* (Lake Buena Vista, Florida, USA), **3** , pp. 2139–2140.
Best, R. E. (1984). *Phase-Locked Loops : Theory, Design, and Applications* (McGraw-Hill, New York).
Bialas, S. (1980). *IEEE Trans. Automatic Control*, **25**, 4, pp. 813-814.
Boccaletti, S., Latora, V., Moreno, Y., Chavez, M. and Hwang, D. U. (2006). Complex networks: structure and dynamics, *Physics Reports*, **424**, 4-5, pp. 175–308.
Boyd, S. and Desoer, C. A. (1985). Subharmonic functions and performance bounds on linear time-invariant feedback systems, *IMA J. Math. Contr. and Info.*, **2**, 2, pp. 153–170.
Boyd, S., Balakrishnan, V. and Kabamba, P. (1988). On computing the H_∞ norm of a transfer function matrix, *In Proceedings of the American Control Conference*, **88**, 1-3, pp. 396–397.
Boyd, S., Balakrishnan, V. and Kabamba, P. (1989). A bisection method for computing the H_∞ norm of a transfer matrix and related problems, *Math. Control, Signals, and Systems*, **2**, 3, pp. 207–220.
Boyd, S., Ghaoui, L. E., Feron, E. and Balakrishnan, V. (1994) *Linear Matrix Inequalities in Systems and Control Theory* (SIAM, Philadelphia).
Brockett, R. W. and Willems, J. L. (1965). Fequency-domain stability criteria-Part I, *IEEE Trans. Automatic Control*, **10**, 3, pp. 255–261.
Brockett, R. W. and Willems, J. L. (1965). Fequency-domain stability criteria-Part II, *IEEE Trans. Automatic Control*, **10**, 4, pp. 407–413.
Bruinsma, N. A. and Steinbuch, M. (1990). A fast algorithm to compute the H_∞ norm of a transfer function matrix, *Systems and Control Letters*, **14**, 4, pp. 287–293.
Chang, Y.-C. A robust tracking control for chaotic Chua's circuits via fuzzy approach, *IEEE Trans. Circuits Syst.-I*, **48**, 7, pp. 889–895.
Chen, L. and Chen, G. (1999). Fuzzy predictive control of uncertain chaotic systems using time series, *Int. J. Bifurcation and Chaos*, **9**, 4, pp. 757–768.
Chen, G. and Dong, X. (1993). Controlling chua's circutit, *J. Circuits, Systems, Computers*, **3**, 1, pp. 139-149.
Chen, G. and Dong, X. (1998). *From Chaos to Order: Methodologies, Perspectives and Applications* (World Scientific, Singapore).
Cheng, D. Z. (1988). *Geometric Theory of Nonlinear Systems* (Scientific Publishing Press, Beijing).
Chua, L. O., Komuro, M. and Matsumoto, T. (1986). The double scroll family: I and II, *IEEE Trans. Circuits Syst.-I*, **33**, 11, pp. 1072-1118.
Chua, L. O. (1994). Chua's circuit: 10 years later, *Int. J. Cir. Theor. Appl.*, **22**, 4, pp. 279–305.
Chua, L. O. (1998). *CNN: A Paradigm for Complexity* (World Scientific Series on Nonlinear Science, Series A-Vol. 31).

Corlee, M. and Shorten, R. (2009). A correct characterization of strict positive realness for MIMO systems, in *Proc. American Control Conference (ACC'09)* (St. Louis, Missouri, USA).
Curran, P. F. and Chua, L. O. (1997). Absolute stability theory and the synchronization problem, *Int. J. Bifur. Chaos*, **7**, 6, pp. 1375–1383.
Davison, E. J. (1974). The decentralized stabilization and control of a class of unknown nonlinear time-varying systems, *Automatica*, **10**, 3, pp. 309-316.
Davison, E. J. and Ozguner, U. (1983). Characterization of decentralized fixed modes for interconnected systems, *Automatica*, **19**, 2, pp. 182-196.
de Souza, C. E. and Xie, L. (1992). On the discrete-time bounded real lemma with application in the characterization of static state feedback $H\infty$ controllers, *Systems and Control Letters*, **18**, 1, pp. 61–71.
de Oliveira, M. C., Bernussou, J. and Geromel, J. C. (1999). A new discrete-time robust stability condition, *Systems and Control Letters*, **37**, 4, pp. 261–265.
de Oliveira, M. C., Geromel, J. C. and Hsu, L. (2002). A new absolute stability test for systems with state-dependent perturbations, *International Journal of Robust and Nonlinear Control*, **12**, 14, pp. 1209–1226.
de Oliveira, M. C., Bernussou, J. and Geromel, J. C. (2002). Extended H_2 and H_∞ norm characterizations and controller parameterizations for discrete-time systems, *International Journal of Control*, **75**, 9, pp. 666–679.
de Oliveira, M. C., Bernussou, J. and Geromel, J. C. (2002). Extended H_2 and H_∞ norm characterizations and controller parametrizations for discrete-time systems, *International Journal of Control*, **75**, 9, pp. 666–679.
Desoer, C. A. and Vidyasagar, M. (1975). *Feedback systems: Input-output properties* (Academic Press, San Diego).
Doyle, J. C., Glover, K., Khargonekar, P. P. and Francis, B. A. (1989). State-space solutions to standard H_2 and H_∞ control problems, *IEEE. Trans. Automatic Control*, **34**, 8, pp. 831–847.
Doyle, J. C., Packard, A. and Zhou, K. (1991). Review of LFTs, LMIs and μ, *In Proceedings of IEEE Conference on Decision and Control*, **2**, pp. 1227–1232.
Dragan, V. (1993). A small gain theorem for time-varying systems, *Applied Mathematics Letters*, **6**, 5, pp. 75–77.
Duan, Z. S., Huang, L., Wang, J. Z. and Wang, L. (2003). Harmonic control between two systems, *Acta Automatica Sinica*, **29**, 1, pp. 14–22.
Duan, Z. S., Huang, L., and Wang, L. (2004a). Multiplier design for extended strict positive realness and its applications, *Int. J. Contr.*, **77**, 17, pp. 1493–1502.
Duan, Z. S., Wang, J. Z. and Huang, L. (2004b). Multi-input and multi-output nonlinear systems: Interconnected Chua's circuit, *International Journal of Bifurcation and Chaos*, **14**, 9, pp. 3065-3081.
Duan, Z. S., Huang, L., Wang, l. and Wang, J. Z. (2004c). Some applications of small gain theorem to interconnected systems, *Systems and Control Letters*, **52**, 3-4, pp. 263–273.
Duan, Z. S., Wang, J. Z. and Huang, L. (2005a). Input and output coupled nonlinear systems, *IEEE Trans. on Circuits and Systems-I: Regular Papers*, **52**, 3, pp. 567–575.

Duan, Z. S., Wang, J. Z. and Huang, L. (2005b). Frequency-domain method for the property of dichotomy of modified Chua's equations, *International Journal of Bifurcation and Chaos*, **15**, 8, pp. 2485-2505.

Duan, Z. S., Wang, J. Z. and Huang, L. (2006a). Some special decentralized control problems in continuous-time interconnected systems, *Advances in Complex Systems*, **9**, 3, pp. 277-286.

Duan, Z. S., Zhang, J. X. and Mosca, E. (2006b). Robust H_2 and H_∞ filtering for uncertain linear systems, *Automatica*, **42**, 11, pp. 1919-1926.

Duan, Z. S., Zhang, J. X. and Mosca, E. (2006c). A Simple Design Method of Reduced-order Filters and Its Applications to Multirate Filter Bank Design, *Signal Processing*, **86**, 5, pp. 1061-1075.

Duan, Z. S., Wang, J. Z. and Huang, L. (2007a). Criteria for dichotomy and gradient-like behavior of a class of nonlinear systems with multiple equilibria, *Automatica*, **43**, 9, pp. 1583-1589.

Duan, Z. S., Wang, J. Z. and Huang, L. (2007b). Special decentralized control problems in discrete-time interconnected systems composed of two subsystems, *Systems and Control Letters*, **56**, 3, pp. 206-214.

Duan, Z. S., Wang, J. Z. and Huang, L. (2007c). A generalization of smooth Chua's equations under Lagrange stability, *International Journal of Bifurcation and Chaos*, **17**, 9, pp. 3047–3059.

Duan, Z. S., Chen, G. and Huang, L. (2007d). Complex network synchronizability: Analysis and control, *Phys. Rev. E*, **76**, pp. 056103.

Duan, Z. S., Wang, J. Z., Chen, G. R. and Huang, L. (2008). Stability analysis and decentralized control of a class of complex dynamical networks, *Automatica*, **44**, 4, pp. 1028–1035.

Fahmy, M. and Hanafy, A. (1981). Further correction to: Comments on 'On the Lyapunov matrix equation', *IEEE Trans. Automatic Control*, **26**, 2, pp. 619–619.

Feng, C. B. and Fei, S. M. (1998). *Analysis and Design of Nonlienar Control Systems* (Electrical and Industrial Publishing Press, Beijing).

Feng, C. B., Tian, Y. P. and Xin, X. (1995). *Robust Control System Design* (Southeast University Press, Nanjing).

Finsler P. (1937). Über das Vorkommen definiter und semi-definiter Formen in Scharen quadratischer Formen, *Comentarii mthematici Helvetici*, **9**, pp. 192–199.

Fradkov, A.L. and Yakubovich, V.A. (1979). The S-procedure and duality relations in nonconvex problems of quadratic programming, *Vestnic Leningrad Univ. Math.*, **6**, 1, 101-109.

Fridman, V., Tibbetts, D., Piraner, I., Guseva, N. and Kirian, D. (1997). Nonlinear chaotic and periodic vibrations in the internal combustion engine, *In Proceedings of International Conferenc on Control of Oscillations and Chaos*, **3**, pp. 403–406.

Gahinet, P. (1992). A convex parametrization of H_∞ suboptimal controllers, *In Proc. IEEE Conf. on Decision and Control*, **1**, pp. 937–942.

Gahinet, P. and Apkarian, P. (1994). A linear matrix inequality approach to H_∞ control, *International Journal of Robust and Nonlinear Control*, **4**, 4, pp.

421-448.
Gahinet, P., Nemirovski, A. and Laub, A. J. (1995) *LMI control toolbox* (The MathWorks Inc, Natick Mass).
Gahinet, P. (1996). Explicit controller formulas for LMI-based H_∞ synthesis, *Automatica*, **32**, 7, pp. 1007–1014.
Gajić, Z. (1994) *Lyapunov Matrix Equation in System Stability and Control* (Acdemic Press, San diego, Boston, New York, London, Sydney, Tokyo, Toronto).
Gao, W. B. and Huo, W. (1994). *Stability, Decentralized Control and Dynamically Hierarchical Control Foundation of Large Scale Systems* (Publishing Press of Beijing University of Aeronautics and Astronautics, Beijing).
Gardner, F. M. (1979). *Phaselock Techniques* (John Wiley & Sons, New York).
Geng, Z. Y. (2004). Stabilization of the systems with coprime factor uncertainties by convex parameterization, *Dynamics of Continuous Discrete and Impulsive Systems-Series A*, **11**, 5-6, pp. 871–884.
Green, M. and Limebeer, D. J. N. (1995). *Linear Robust Control* (Prentice Hall Englewood Cliffs, NJ).
Guzelis, C. and Chua, L. O. (1993). Stabiity analysis of generalized cellular neural networks, *Int. J. Circuit Theory Appl.*, **21**, 1, pp. 1–33.
Haddad, W. M. and Bernstein, D. S. (1993). Explicit construction of quadratic Lyapunov functions for the small gain, positivity, circle and Popov Theorems and their application to robust stability Part I: Continuous time theory, *International Journal of Robust and Nonlinear Control*, **3**, 4, pp. 313–339.
Haddad, W.M. and Kapila, V. (1997). Fixed-architecture controller synthesis for systems with input-output time-varying nonlinearities, *Int. J. Robust and Nonlinear Control*, **7**,7, pp. 675–710.
Hayes, W. D. (1953). On the equation for a damped pendulum under constant torque, *Z. Angew. Math. Phys.*, **4**, 5, pp. 398–401.
Helmersson, A. (1995). *Methods for Robust Gain Scheduling* (Link ö pings, Sweden).
Hong, Y. G. and Cheng, D. Z. (2005). *Analysis and Control of Nonlinear Systems* (Scientific Publishing Press, Beijing).
Horn. R. and Johnson, C. (1991). *Topics in Matrix Analysis* (Cambridge University Press, Cambridge).
Huang, L., Zheng, Y. P. and Zhang, D. (1964). The second method of Lyapunov and the analytical design of the optimum controller, *Acta Automatica Sinica*, **2**, pp. 203–218.
Huang, L. (1990). *Linear Algebra in Theory of Systems and Control* (Scientific Publishing Press, Beijing).
Huang, L. (2003). *Fundamental Theory on Stability and Robustness* (Scientific Publishing Press, Beijing).
Huang, A., Pivka, L. and Franz, M. (1996). Chua's equation with cubic nonlinearity, Int. J. Bifurcation and Chaos, **12**, 6, pp. 2175-2222.
Huang, C. H., Ioannou, P. A., Maroulas, J. and Safonov, M. G. (1999). Design of strictly positive real systems using constant output feedback, *IEEE Trans.*

Automatic Control, **44**, 3, pp. 569–573.
Huang, S. D. and Zhang, S. Y. (1998). Comments on ' Design of decentralized control for symmetrically interconnected systems', *Automatica*, **34**, 7, pp. 929–933.
Ikeda, M., Silyak, D. D. and Yasuda, K. (1983). Optimality of decentralized control for large scale systems, *Automatica*, **19**, 3, pp. 309–316.
Imai, T., Konishi, K., Kokame, H. and Hirata, K. (2002). An experimental suppression of spatial instability in one-way open coupled Chua's circuits, *International Journal of Bifurcation and Chaos*, **12**, 12, pp. 2897-2905.
Isidori, A. (1995). *Nonlinear Control Systems* (3rd ed. Springer, New York).
Iwasaki. T. and Skelton, R. E. (1994). All controllers for the general H_∞ control problem: LMI existence conditions and states space formulas, *Automatica*, **30**, 8, pp. 1307–1317.
Iwasaki. T. and Hara, S. (2005). Generalized KYP lemma: unified frequency domain inequalities with design applications, *IEEE Trans. Automatic Control*, **50**, 1, pp. 41–59.
Iwasaki. T., Hara, S. and Yamauchi, H. (2003). Dynamical system design from a control perpective: finite frequency positive-realness approach, *IEEE Trans. Automatic Control*, **48**, 8, pp. 1337–1354.
Iwasaki. T., Hara, S. and Fradkov, A. L. (2005). Time domain interpretations of frequency domain inequalities on (semi)finite ranges, *Systems and Control Letters*, **54**, 7, pp. 681–691.
Jia, Y. M. (2007). *Robust H_∞ Control* (Scientific Publishing Press, Beijing).
Jiang, Z. P., Teel, A. R. and Praly, L. (1994). Small gain theorem for ISS systems and applications, *Mathematics of Control, Signals, and Systems*, **7**, 2, pp. 95–120.
Jorge, L. and Chua, L. O. (1999). Hopf bifurcation and degeneracies in Chua's circuit–A perspective from a frequency domain approach, *International Journal of Bifurcation and Chaos*, **9**, 1, pp. 295-303.
Jose, A.-R., Cervantes, I., Sanchez, J.-C and Perez, S. (2005). Chua's circuit stabilization:a damping injection approach, *Int. J. Bifurcation and Chaos*, **15**, 1, pp.181–189.
Kalman, R. E. (1957). Physical and mathematical mechanisms of instability in nonlinear automatic control systems, *Trans. Amer. Soc. Mech. Eng.*, **79**, 3, pp. 553–563.
Kalman, R. E. (1963). Lyapunov functions for the problem of lur'e in automatic control, *Proc. Nat. Acad. Sci.*, **49**, pp. 201–205.
Kapila, V., Sparks, A.G. and Haizhou, P. (2001). Control of systems with actuator saturation non-linearities: an LMI approach, *Int. J. Control*, **74**, 6, pp. 586–599.
Kapitaniak, T., Chua, L. O. and Zhong, G. Q. (1994). Experimental hyperchaos in coupled Chua's circuits, *IEEE Trans. on Circuits and Systems-I*, **41**, 7, pp. 499-503.
Karanam, V. (1982). Eigenvalue bounds for algebraic Riccati and Lyapunov equations, *IEEE Trans. Automatic Control*, **27**, 2, pp. 461–463.
Kashlan, A. EL. and Geneidy, M. EL. (1996). Design of decentralized control for

symmetrically interconnected systems, *Automatica*, **32**, 3, pp. 475-476.
Kaszkurewicz, E. and Bhaya, A. (1994). On a class of global stable neural circuits, *IEEE Transactions on Circuits and Systems-I*, **41**, 2, pp. 171-174.
Khalil, H. K. (2002). *Nonlinear Systems*, 3rd edn. (Prentice Hall).
Khargonekar, P. P. and Rotea, M. A. (1988). Stabilization of uncertain systems with norm bounded uncertainty using control Lyapunov functions, *In Proc. IEEE Conf. on Decision and Control*, **1**, pp. 503-507.
Khargonekar, P. P., Petersen, I. R. and Zhou, K. (1990). Robust stabilization of uncertain linear systems: Quadratic stabilizability and H_∞ control theory, *IEEE. Trans. Automatic Control*, **35**, 3, pp. 356-361.
Komaroff, N. (1988). Simultaneous eigenvalue lower bounds for the Lyapunov matrix equation, *IEEE Trans. Automatic Control*, **33**, 1, pp. 126-128.
Komaroff, N. (1992). Upper summation and product bounds for solution eigenvalues of the Lyapunov matrix equation, *IEEE Trans. Automatic Control*, **37**, 7, pp. 1040-1042.
Krasovsky, N. N. (1952). Theorems on the stability of motions, defined by a system of two equations (in Russian), *Prikl. Matem. i Mekh.*, **16**, 5, pp. 547-554.
Kwon, W. and Pearson, A. (1977). A note on algebraic matrix Riccati equation, *IEEE Trans. Automatic Control*, **22**, 1, pp. 143-144.
Kwon, B., Youn, M. and Bien, Z. (1985). On bounds of the Riccati and Lyapunov matrix equations, *IEEE Trans. Automatic Control*, **30**, 11, pp. 1134-1135.
Kwon, B. and Youn, M. (1985). Comments on 'On some bounds in the algebraic Riccati and Lyapunov equations', *IEEE Trans. Automatic Control*, **31**, 6, pp. 591-591.
Lancaster, P. and Tismenetsky, M. (1985). *Theory of Matrices* (Academic Press, New York).
Landau, Y. D.(1979). *Adaptive control: The model reference approach* (Marcel Dekker, New York).
Leonov, G. A. (1974). Stability and oscillations of phase systems, *Sibirsk. Mat. Zh.*, **15**, 3, pp. 687-692.
Leonov, G. A. (1976). A certain class of dynamical systems with cylindrical phase space, *Siberian Math. J.*, **17**, pp. 72-90.
Leonov, G. A. (1985). Global stability of the differential-equations for phase-synchronization systems, *Differential Equations*, **21**, 2, pp. 149-156.
Leonov, G. A., Reitmann, V. and Smirnova, V. B. (1992a). *Non-local Methods for Pendulum-like Feedback Systems*, (Teubner-Texte zur Mathematik Bd. 132, B.G., Teubner Stuttgart-Leipzig).
Leonov, G. A., Burkin, I. M. and Shepeljavyi, A. L. (1992b). *Frequency Methods in Oscillation Theory* (Kluwer Academic Publishers, Dordrecht).
Leonov, G. A., Ponomarenko, D. V. and Smirnova, V. B. (1996). *Frequency-Domain Methods for Nonlinear Analysis: Theory and Applications* (World Scientific, Singapore).
Leonov, G. A., Noack, A. and Reitmann, V. (2001). Asymptotic orbital stability conditions for flows by estimates of singular values of the linearization, *Nonlinear analysis*, **44**, 8, pp. 1057-1085.

Liao, X. X. (2000). *Theory and application of stability for dynamical systems* (in Chinese) (Press of National Defense and Industy, Beijing).

Lindsey, V. C. (1972). *Synchronization Systems in Communication and Control* (Prentice-Hall, New Jersey).

Liu, Y. Z. and Chen, L. Q. (2001). *Nonlinear Vibrations* (Higher Education Press, Beijing (in Chinese)).

Liu, X., Wang, J. Z. and Huang, L. (2007a). Stabilization of a class of dynamical complex networks based on decentralized control, *Physica A*, **383**, 2, pp. 733-744.

Liu, X., Wang, J. Z. and Huang, L. (2007b). Attractors of fourth-order Chua's circuit and chaos control, *International Journal of Bifurcation and Chaos*, **17**, 8, pp. 2705–2722.

Lozano, R., Brogliato, B., Egeland, O. and Maschke, B. (2000). *Dissipative Systems Analysis and Control: Theory and Applications* (Springer-Verlag, London).

Lu, P. L., Yang, Y. and Huang, L. (2007). Robust analysis and synthesis for the nonexistence of periodic solutions in a class of nonlinear systems, *J. Optim. Theory Appl.*, **135**, 2, pp. 205–216.

Lu, P. L., Yang, Y., Li, Z. K. and Huang, L. (2008). Decentralized dynamic output feedback for global asymptotic stabilization of a class of dynamic networks, *Int. J. Control*, **81**, pp. 1054-1061.

Lu, P. L., Yang, Y. and Huang, L. (2008a). Cycles of the second kind for uncertain pendulum-like systems with several nonlinearities, *Nonlinear Analysis*, **69**, 1, pp. 35–45.

Lu, P. L., Yang, Y. and Huang, L. (2008b). Dynamic feedback control for cycle slipping in phase-controlled system, *Nonlinear Analysis*, **69**, 1, pp. 314–322.

Lur'e, A. I. and Postnikov, V. N. (1944). On the theory of stability of control systems, *Prikladnaya Matematika Mehkhanika*, **8**, 3, pp. 246–248.

Lü, J. H., Yu, X. H., Chen, G. and Cheng, D. Z. (2004). Characterizing the synchronizability of small-world dynamical networks, *IEEE Trans. Circuits Syst.-I*, **51**, 4, pp. 787-796.

Madan, R. (1993). *Chua's circuit: A paradigm for chaos* (World Scientific, Singapore).

Mareels, I. M. Y. and Hill, D. J. (1992). Monotone stability of nonlinear feedback systems, *J. Mathematical Systems, Estimation and Control*, **2**, 3, pp. 275–291.

Martensson, K. (1971). On the matrix riccati equation, *Information Sciences*, **3**, 1, pp. 17–49.

Matsumoto, T., Chua, L. O. and Kobayashi, K. (1986). Hyper chaos: Laboratory experiment and numerical confirmation, *IEEE Trans. Circuits Syst.*, **33**, 11, pp.1143–1147.

Masubuchi, I., Ohara, A. and Suda, N. (1998). LMI-based controller synthesis: a unified formulation and solution, *International Journal of Robust and Nonlinear Control*, **8**, 8, pp. 669–686.

Mitropolsky, Y. A. and Nguyen, V. D. (1997). *Applied Asymptotic Methods in Nonlinear Oscillations* (Kluwer Academic Publishers, Dordrecht).

Montemayor, J. and Womack, B. (1975). Comments on 'On the Lyapunov matrix equation', *IEEE Trans. Automatic Control*, **20**, 6, pp. 814–815.

Morary, A. (1970). Nonlinear oscillations of synchronous machines started with a pulsating rotating torque, *Trudy meshd. konf. po nelin. koleb., Kiev*, **4**.

Mori, T. (1985). On some bounds in the algebraic Riccati and Lyapunov equations, *IEEE Trans. Automatic Control*, **30**, 2, pp. 162–164.

Mori, T., Fukuma, N. and Kuwahara, M. (1986). Explicit solution and eigenvalue bounds in the Lyapunov matrix eqution, *IEEE Trans. Automatic Control*, **31**, 7, pp. 656–658.

Narendra, K. S. and Taylor, J. H. (1973). *Frequency Domain Criteria for Absolute Stability* (Academic, New York).

Nayfeh, A. H. and Mook, D. T. (1995). *Nonlinear Oscillations* (John Wiley & Sons).

Ogorzalek, M. J. (1993). Taming chaos. II. control, *IEEE Trans. Circuits Syst.-I*, **40**, 10, pp. 700–706.

Opdenacker, P. C. and Jonckheere, E. A. (1988). A contraction mapping preserving balanced reduction scheme and its infinity norm error bounds, *IEEE Trans. Automatic Control*, **35**, 2, pp. 184–189.

Park, P. (1997). A revisited Popov criterion for nonlinear Lur'e systems with sector-restrictions, *Int. J. Contr.*, **68**, 3, pp. 461–469.

Park, P. (2002). Stability criteria of sector and slope restricted Lur'e systems, *IEEE Trans. Automatic Control*, **47**, 2, pp. 308–313.

Patel, R. and Toda, M. (1978). On norm bounds for algebraic Riccati and Lyapunov equations, *IEEE Trans. Automatic Control*, **23**, 1, pp. 87–88.

Peaucelle, D., Arzelier, D., Bachelier, D. and Bermussou, J. (2000). A new robust D-stability condition for real convex polytopic uncertainty, *Systems and Control Letters*, **40**, 1, pp. 21–30.

Peng, T. (1972). A note on the Lyapunov matrix equation, *IEEE Trans. Automatic Control*, **17**, 4, pp. 565–565.

Peterson, I. R. and Hollot, C. V. (1986). A Riccati equation approach to the stabilization of uncertain linear systems, *Automatica*, **22**, 4, pp. 397–411.

Pinsler, P. (1937). Über das Vorkommen definiter und semi-definiter Formen in Scharen quadratischer Formen, *Comentarii mthematici Helvetici*, **9**, 1, pp. 188–192.

Pliss, V. A. (1964). *Nonlocal Problems of the Theory of Oscillations* (in Russian) (Nauka, Moscow-Leninggrad).

Popov, V. M. (1962). Absolute stability of nonlinear systems of automatic control, *Automation and Remote Control*, **22**, 8, pp. 857–875. Russian original in August 1961.

Popov, V. (1964). Hyperstability and optimality of aotomatic systems with several control functions, *Rev. Roum. Sci. Tech.*, **9**, 4, pp. 629–690.

Popov, V. M. (1973). *Hyperstability of Automatic Control Systems* (Springer-Verlag, New York).

Rabinder, N. M. (1993). *Chua's Circuit: A Paradigm for Chaos* (World Scientific Publishing Co . Pte. Ltd)

Rantzer, A. (1996). On the Kalman-Yakubovich-Popov lemma, *Syst. Control Let-*

ters, **28**, 1, pp. 7–10.

Rantzer, A. and Megretski, A. (1994). A convex parameterization of robustly stabilizing controllers, *IEEE Trans. Automatic Control*, **39**, 9, pp. 1082–1088.

Rasvan, V. (1998). Dynamical systems with several equilibria and natural Lyapunov functions, *Arch. Math.*, **34**, pp. 207-215.

Robel, G. (1989). On computing the infinity norm, *IEEEAC*, **34**, 8, pp. 882–884.

Rossler, O. E. (1979). An equation for hyperchaos, *Phys. Lett. A*, **71**, 2-3, pp. 155–157.

Sampei, M., Mita, T. and Nakamichi, M. (1990). An algebraic approach to H_∞ output feedback control problems, *Systems and Control Letters*, **14**, 1, pp. 13–24.

Sannemann, R. W. and Rowbotham, J. R. (1964). Unlock characteristics of the optimum type II phase-locked loop, *IEEE Trans. Aerosp. Navig. Electron.*, **11**, 1, pp. 15–24.

Scherer, C. (1990). *The Riccati Inequality and State-Space H_∞ Optimal Control* (PH. D Thesis).

Schweppe, F. C. (1973). *Uncertain Dynamic Systems* (Prentice-Hall, Englewood Cliffs, N. J.).

Serfert, G. (1952). On the existence of certain solutions of nonlinear differential equations, *Z. Angew. Math. Phys.*, **3**, 6, pp. 468–471.

Sezer, M. E. and Huseyin, O. (1978). Stabilization of linear time-invariant interconnected systems using local state feedback, *IEEE Trans. Systems, Man and Cybernetics*, **8**, 10, pp. 751-756.

Shakhgil'dyan, V. V. and Lyakhovkin, A. A. (1972). *Phase locked loops* (Radio i Svyaz', Moscow) (in Russian).

Shapiro, E. (1974). On the Lyapunov matrix equation, *IEEE Trans. Automatic Control*, **19**, 5, pp. 594–596.

Shen, T. (1996). H_∞ *Control Theory and Applications* (Tsinghua University Press).

Shi, Z. C. and Gao, W. B. (1986). Stabilization by decentralized control for large scale interconnected systems, *Large Scale Systems*, **10**, pp. 147-155.

Shil'nikov, L. P. (1993). Chua's circuits: Rigorous results and future problems, *IEEE Trans. Circuits and Systems-I*, **40**, 10, pp. 784-786.

Siljak, D. D. (1978). *Large-scale dynamic systems: stability and structure* (North-Holland, New York).

Siljak, D. D. (1991). *Decentralized Control of Complex Systems* (Academic, New York).

Siljak, D. D. and Stipanovic, D. M. (2000). Robust stabilization of nonlinear systems: the LMI approach, *Math. Prob. Eng.*, **6**, 5, pp. 461–493.

Skelton, R. E. and Iwasaki, T. (1995). Increased roles of linear algebra in control education, *IEEE Control Systems*, **15**, 4, pp. 76–90.

Slotine, J. J. E. and Li, W. (1991). *Applied Nonlinear Control* (Prentice-Hall Inc.).

Stewart, I. (2000). The Lorenz attractor exists, *Nature* **406**, pp. 948–949.

Strzhinski, V. M. (1977). *Applied Methods in The Theory of Nonlinear Oscilla-*

tions (Mir publishers, Moscow).
Sun, W. Q., Khargonekar, P. P. and Shim, D. (1994). Solution to the positive real control problem for linear time-invariant systems, *IEEE Trans. Automatic Control*, **39**, 10, pp. 2034-2046.
Suykens, J. A. K., Curran, P. F., Vandewalle, J. and Chua, L. O. (1997). Robust nonlinear H_∞ synchronization of chaotic Lur'e systems, *IEEE Trans. Circuits and Systems-I*, **44**, 10, pp. 891-904.
Suykens, J. A. K., Vandewalle, J. and De Moor, B. (1998). An absolute stability criterion for the Lur'e problem with sector and slope restricted nonlinearities, *IEEE Trans. Circuits and Systems-I*, **45**, 9, pp. 1007-1009.
Tanaka,K., Ikeda,T. and Wang, H. O. (1998). A unified approach to controlling chaos via an LMI-based fuzzy control system design, *IEEE Trans. Circuits Systems-I*, **45**, 10, pp. 1021-1040.
Tausworthe, R. (1967). Cycle slipping in phase-locked loops, *IEEE Trans. Commun. Technol.*, **15**, 3, pp. 417-421.
Thamilmaran, K. and Lakshmanan, M. (2004). Hyperchaos in a modified canonical Chua's circuit, *Int. J. Bifurcation and Chaos*, **14**, pp. 221-243.
Titchmarsh, E. C. (1952). *The Theory of Functions* (Oxford University Press).
Tricomi, F. (1933). Integrazione di unequazione differenziale presentatasi in electrotechnica, *Annali della Roma Scuola Normale Superiore de Pisa: Scienza Phys. e Mat.*, **2**, pp. 1-20.
Troch, I. (1988). Solving the discrete Lyapunov equation using the solution of the corresponding continuous Lyapunov equation and vice versa, *IEEE Trans. Automatic Control*, **33**, 10, pp. 944-946.
Tsuneda, A. (2005). A gallery of attractors from smooth Chua's equation, *Int. J. Bifurcation and Chaos*, **15**, 1, pp. 1-49.
Uhlig, F. (1979). A recurring theorem about pairs of quadratic forms and extension: A survey, *Linear Algebra and Appl.*, **25**, pp. 219-237.
Wang, S. H. (1978). An example in decentralized control systems, *IEEE Trans. Automatic Control*, **23**, 5, pp. 938-938.
Wang, S. H. and Davison, E. J. (1973). On the stabilization of decentralized control systems, *IEEE Trans. Automatic Control*, **18**, 5, pp. 473-478.
Wang, X. F. and Chen, G. (2002). Synchronization in scale-free dynamical networks: robustness and fragility, *IEEE Trans. Circuits Systems-I*, **49**, 1, pp. 54-62.
Wang, Y., Xie, L. and Souza, C. E. de. (1992). Robust control of a class of uncertain nonlinear systems, *Systems and Control Letters*, **19**, 2, pp. 139-149.
Wang, J. Z. and Huang, L. (2003). Controller order reduction with guaranteed performance via coprime factorization, *Int. J. Robust and Nonlinear Control*, **13**, 6, pp. 501-517.
Wang, J. Z., Huang, L. and Duan, Z. S. (2004). Design of controller for a class of pendulum-like system guaranteeing dichotomy, *Automatica*, **40**, 6, 1011-1016.
Wang, J. Z., Duan, Z. S. and Huang, L. (2006a). Control of a class of pendulum-like systems with Lagrange stability, *Automatica*, **42**, pp. 145-150.

Wang, J. Z., Duan, Z. S. and Huang, L. (2006b). Robust gradient-like property and control design for uncertain pendulum-like systems, *Journal of Dynamical and Control Systems*, **12**, 2, pp. 229–246.

Wang, J. Z., Duan, Z. S. and Huang, L. (2006c). Robust dichotomy for nonlinear uncertain systems, *Proceedings of the 25th Chinese Control Conference*, Ha er Bin, China.

Wang, J. Z., Duan, Z. S. and Huang, L. (2006d). Dichotomy of nonlinear systems: Application to chaos control of nonlinear electronic circuit , *Physics Letters A*, **351**, pp. 143–152.

Wang, J. Z., Duan, Z. S. and Huang, L. (2007). Dichotomy property and Lagrange stability for uncertain pendulum-like feedback systems, *Systems and Control Letters*, **56**, 2, pp. 167–172.

Wang, S., Kuo, T. and Hsu, C. (1986). Trace bounds on the solution of algebraic matrix Riccati and Lyapunov equations, *IEEE Trans. Automatic Control*, **31**, 7, pp. 654–656.

Wang, Y. H. and Zhang, S. Y. (2000). Robust control for nonlinear similar composite systems wityh uncertain parameters, *IEE Pro. Control Theory and Applications*, **147**, 1, pp. 80-86.

Wen, J. T. (1988). Time domain and frequency domain conditions for strict positive realness, *IEEE Trans. Automatic Control*, **33**, 10, pp. 988–992.

Willems, J. C. (1971). Least squares stationary optimal control and the algebraic Riccati equation, *IEEE Trans. Autom. Control*, **AC-16**, 6, pp. 621–634.

Wimmer, H. (1975). Generalizations of theorems of Lyapunov and Stein, *Linear Algebra and Its Appl.*, **10**, 2, pp. 139–146.

Wu, C. W. (2002). *Synchronization in coupled chaotic circuits and systems* (World Scientific, Singapore).

Wu, N. E. (2002). Analog phaselock loop design using Popov cirterion, in *Proceedings of the American Control Conference*, pp. 16–18.

Wu, C. W. and Chua, L. O. (1994). A unified framework for synchronization and control of dynamical systems, *International Journal of Bifurcation and Chaos*, **4**, 4, pp. 979–989.

Xu, J., Chen, G. and Shieh, L. S. (1996). Digital redesign for controlling the chaotic Chua's circuit, *IEEE Trans. Aerosp. Electron. Syst.*, **32**, 4, pp. 1488–1500.

Yakubovich, V. A. (1958). On the boundedness and global stability of solutions of certain nonlinear differential equations (in Russian), *Doklady Akad. Nauk SSSR*, **121**, 6, pp. 984–986.

Yakubovich, V. A. (1962). A solution of some matrix inequalities which appear in the automatic control theory (in Russian), *Doklady Akad. Nauk SSSR*, **143**, 6, pp. 1304–1307.

Yakubovich, V. A. (1965). Method of matrix inequalities in stability theory of nonlinear control systems, II (in Russian), *Avtomatika i Telemekhanika*, **26**, pp. 577–592.

Yakubovich, V. A. (1971). S-procedure in nonlinear control theory, *Vestnic Leningrad Univ. Math.*, **1**, pp. 62–77.

Yakubovich, V. A. (1973a). Minimization of quadratic functions under the

quadratic constraints and the necessity of a frequency condition in the quadratic criterion for absolute stability of nonlinear control systems, *Soviet Math. Doklady*, **14**, pp. 593–597.
Yakubovich, V. A. (1973b). Frequency conditions for auto-oscillations in nonlinear systems with one stationary nonlinearity, *Siberian Math. J.*, **14**, 5, pp. 768–788.
Yakubovich, V. A. (1977). S-procedure in nonlinear control theory, *Vestnic Leningrad Univ. Math.*, **4**, 73–93. In Russian ,1971.
Yakubovich, V. A. (1971). Nonconvex optimization problem: The infinite-horizon linear-quadratic control problem with quadratic constraints, *Systems and Control Letters*, **19**, 1, pp. 13–22.
Yakubovich, V. A., Leonov, G. A. and Gelig, A. K. (2004). *Stability of Stationary Sets in Control Systems with Discontinuous Nonlinearities* (World Scientific, Singapore).
Yan, X. G., Wang, J. C., Lü, X. Y. and Zhang, S. Y. (1998). Decentralized output feedbaxk robust stabilization for a class of nonlinear composite large scale systems with similarity, *IEEE Trans. Automatic Control*, **43**, 2, pp. 294-299.
Yang, G. H., Wang, J. L. and Soh, Y. C. (2001). Decentralized control of symmetric systems, *Systems and Control Letters*, **42**, 2, pp. 145-149.
Yang, G. H. and Zhang, S. Y. (1995). Structural properties of large-scale systems possessing similar structures, *Automatica*, **31**, 7, pp. 1011-1017.
Yang, Y. and Huang, L. (2003). H_∞ controller synthesis for pendulum-like systems, *Systems and Control Letters*, **50**, 4, pp. 263–276.
Yang, Y., Fu, R. and Huang, L. (2004). Robust analysis and synthesis for a class of uncertain nonlinear systems with multiple equilibria, *Systems and Control Letters*, **53**, 2, pp. 89–105.
Yang, Y., Duan, Z. S. and Huang, L. (2005). Nonexistence of periodic solutions in a class of nonlinear systems with cylindrical phase space, *International Journal of Bifurcation and Chaos*, **15**, 4, pp. 1423–1431.
Yang, Y., Duan, Z. S. and Huang, L. (2006). Design of nonlinear interconnections guaranteeing the absence of periodic solutions, *Systems and Control Letters*, **55**, 4, pp. 338–346.
Yang, Y. and Huang, L. (2007). Cycle slipping in phase synchronization systems, *Phys. Lett. A*, **362**, 2-3, pp. 183–188.
Yang, Y. Z. (2005). Controllability and observalility of a class of linear, time-invariant systems with interval plants, *Int. J. Information and Systems Sciences*, **1**, pp. 184-192.
Youla, D. C. (1961). On the factorization of rantional matrices, *IRE Trans. Information Theory*, **7**, 3, pp. 172–189.
Zames, G. (1966). On the input-output stability of time-varying nonlinear feedback systems, parts I and II, *IEEE Trans. Automatic Control*, **11**, 2, 3, pp. 228–238, 465–476.
Zhai, G., Ikeda, M. and Fujisaki, Y. (2001). Decentralized H_∞ controller Design: A matrix inequality design using a homotopy method, *Automatica*, **37**, 4, pp. 565-572.

Zhang, G., Xia, Y. and Shi, P. (2008). New bounded real lemma for discrete-time singular systems, *Automatica*, **44**, 3, pp. 886–890.

Zhou, K. and Khargonekar, P. P. (1988). An algebraic riccati equation approach to H_∞ optimization, *Systems and Control Letters*, **11**, 2, pp. 85–91.

Zhou, K. and Doyle, J. C. (1998) *Essentials of Robust Control* (Prentice Hall, Upper Saddle River, New Jersey).

Zhou, K., Doyle, J. C. and Glover, K. (1996). *Robust and Optimal Control* (Prentice Hall, Englood Cliffs, NJ).

Index

S-procedure, 15
  nonstrict inequalities, 15
  strict inequalities, 15
w-limit point, 90
w-limit set, 90

absolute stability, 75, 92
absolutely stable, 76, 88, 89
admissible controller, 37
Aizerman conjecture, 82
algebraic Lyapunov equations
  continuous-time, 7
  discrete-time, 11
algebraic Riccati equation, 30, 61
  continuous-time, 30
  discrete-time, 33
algebraic Riccati inequality, 60, 61
auto-oscillation, 207

Bakaev stabilizing, 154
Bakaev stable, 124, 126, 128, 155, 156
Banach spaces, 24
bilinear transformation, 11
bounded derivative condition, 93, 269, 290
bounded real, 34, 66
bounded real lemma, 34

Cauchy sequence, 24
Cayley transform, 66, 67
chaos attractors, 286
chaos oscillations, 270, 271

chaotic attractor, 277, 282–284, 286, 288
chaotic oscillations, 272
Chua's circuit, 169, 216, 267
  absolute stability, 267, 269
  absolutely stable, 288, 290, 291
  canonical, 267, 271
  coupled, 267, 288
  dichotomous, 279
  dichotomy, 267, 269, 270
  extended, 267, 286
  Lagrange stable, 269, 288
  modified, 273, 277, 279
  One-way coupled, 291, 292
  smooth, 237, 252, 270
  stabilization, 267, 280
circle criteria, 89
circle criterion, 75, 76, 79
control of SPR, 63
controllability, 1, 2
controller design, 270
  dichotomy, 131, 179
  gradient-like property, 137
  Lagrange stability, 139
cycle of the first kind, 208–210
cycle of the second kind, 208–210
cycle slipping, 229
cylinder system, 208
cylindrical phase space, 208

decentralized control, 237, 238, 241, 244

decentrally fixed mode, 238, 264
decentrally stabilizable, 238
detectability, 6
detectable, 6
dichotomous, 90, 107, 160, 162
dichotomy, 89, 92, 131
$dom(Ric)$, 32, 34
dynamic output feedback controller design, 214

extended strictly positive real (ESPR), 51

feedback control, 270, 281
Finsler lemma, 14
Fourier series, 220

global convergence, 286
globally convergent, 281, 282
gradient-like, 107
gradient-like property, 131

$\mathcal{H}_\infty$ control
  continuous-time, 37
  discrete-time, 42
$\mathcal{H}_\infty$ space, 24
$\mathcal{H}_\infty^-$ space, 25
Hamiltonian matrix, 62
Hurwitz stable, 6
hyperchaos, 274
hyperchaotic oscillations, 279

input and output coupling, 259
input nonlinearities
  Bakaev stabilizing, 155
  dichotomy, 159
  Lagrange stabilizing, 149
input transformation, 72
interconnected feedback, 241
interconnected Lur'e system, 237, 250, 290
interconnected pendulum-like systems, 218
interconnected structure, 240
interconnected system, 237
internally stable, 64, 65

Kalman conjecture, 82, 83, 267, 280, 282
Kalman duality principle, 3
Kalman-Yakubovič-Popov lemma, 16
Kronecker product, 7
KYP lemma, 16, 62
  continuous-time, 18
  discrete-time, 18

$\mathcal{L}_\infty$ space, 24
Lagrange stable, 107, 151, 153
large-scale system, 218, 237, 239, 240, 264
limit cycle, 207, 211
linear fractional transformation, 27, 144, 181
linear interconnected system, 238
linear matrix inequalities, 12
LMI, 63
LMIs, 12
Lur'e system, 75, 84, 269
Lyapunov inequalities
  continuous-time, 10
  discrete-time, 12

mathematical pendulum, 99, 102, 208
maximum modulus theorem, 46
multiple equilibria, 107
multiplier design, 69, 72

non-degenerate, 4
nonlinear interconnection, 219
nonlinear interconnection design, 224
norm bounded uncertainties
  dichotomy, 163
  gradient-like property, 199
normed space, 24

observability, 1, 2

pendulum-like, 102, 103, 105
pendulum-like feedback system
  Bakaev stability, 99, 124
  dichotomy, 99, 107, 112
  gradient-like property, 99, 114, 117
  Lagrange stability, 99, 118

the first canonical form, 103, 105, 112, 117
the second canonical form, 105, 107, 114
periodic oscillation, 207
periodic solution, 211
phase synchronization system, 228
phase-locked loop, 100, 197, 201, 209, 233, 234
polytope, 214
polytopic uncertainty, 213
Popov criteria, 89
Popov criterion, 80, 81
positive real (PR), 46
positive real lemma, 52, 54, 160
projection lemma, 13

quadratic cone, 124, 125
quadratically stable, 168, 201, 281

Redheffer star product, 29, 68
$\mathcal{RH}_\infty$ space, 25
$\mathcal{RH}_\infty^-$ space, 25
$\mathcal{RL}_\infty$ space, 24
robust analysis
  dichotomy, 164
robust control
  dichotomy, 168

SBR, 182
Schur complememt, 12
sector condition, 76, 269, 288, 290
simplectic matrix, 33
small gain theorem, 36, 237, 246, 247, 250, 264
SPR, 49, 52, 60, 65, 77, 136, 143–145, 160, 181, 182
stabilizability, 6
stabilizable, 6
strange attractors, 270, 286
strictly bounded real (SBR), 66
strictly positive real (SPR), 46
suboptimal $\mathcal{H}_\infty$ control, 37
synchronous single-phase motor, 100, 146

uncertain pendulum-like systems
  dichotomy, 174, 179
  gradient-like property, 187, 191
  Lagrange stability, 184
uncertainty
  H_∞, 177, 193
  additive, 175, 179, 185, 192
  multiplicative, 177, 185, 193, 196

well-posedness, 36